46th Vibration Institute Annual Training Conference 2023

AA005416

Niagara Falls, New York, USA
26-28 July 2023

ISBN: 978-1-7138-7779-0

Printed from e-media with permission by:

Curran Associates, Inc.
57 Morehouse Lane
Red Hook, NY 12571

CURRAN ASSOCIATES INC. proceedings.com

Some format issues inherent in the e-media version may also appear in this print version.

Vibration Institute
2625 Butterfield Road
Suite 128N
Oak Brook, IL 60523-3415
USA

Phone: (630) 654-2254
Fax: (630) 654-2271

information@vi-institute.org

Additional copies of this publication are available from:

Curran Associates, Inc.
57 Morehouse Lane
Red Hook, NY 12571 USA
Phone: 845-758-0400
Fax: 845-758-2633
Email: curran@proceedings.com
Web: www.proceedings.com

5 VIATC TH ANNIVERSARY

TRAINING CONFERENCE & EXPO

NIAGARA FALLS, NEW YORK | JULY 26-28, 2023

CONFERENCE PROCEEDINGS

46th Annual Training Conference & Expo

Niagara Falls, NY

www.vi-institute.org (630) 654-2254

Proceedings of the Vibration Institute Annual Training Conference
VIATC 2023
Niagara Falls, NY
July 26 – 28, 2023

Table of Contents

Vibration Institute

50 1973-2023 ANNIVERSARY

VIATC TRAINING CONFERENCE & EXPO

NIAGARA FALLS, NEW YORK JULY 26 -28, 2023

THE DIRTY DOZEN: THE TWELVE MOST DIFFICULT VIBRATION PROBLEMS I ENCOUNTERED DURING THE LAST FORTY-FIVE YEARS.

Nelson L. Baxter P.E.
President
ABM Technical Services
Nelsonbaxterabm@gmail.com

Nelson L. Baxter P.E. is president of ABM Technical Services. He has 45 years of experience in troubleshooting machinery and structural vibration problems. Work has been performed in nuclear power plants, coal power plants, hydro-electric generation, paper mills, gas turbine facilities, oil refineries and process plants. He has a B.S. in Mechanical Engineering from Rose Hulman and an M.S. in Nuclear Engineering from Purdue University. Mr. Baxter has a Category IV certification from The Vibration Institute and is a registered Professional Engineer. He has been on the Board of Directors of the Vibration Institute for over 20 years and has been an instructor of the AVA CAT IV training courses. Mr. Baxter is a recipient of the Jack Frarey award.

ABSTRACT:

During a troubleshooting career that has spanned 45 years, a number of challenges and interesting problems have been encountered. The problems over the 45-year time span ranged in difficulty from simple to very difficult. This paper looks at the twelve most difficult problems in my career that were both identified and resolved during the period from 1977 through 2022. What makes a problem difficult is rather subjective therefore some discussion is given regarding as to why a problem might have been placed on this list.

KEY WORDS: Alignment; balancing; chillers; compressors; piping; pumps; resonance; turbines

INTRODUCTION:

There are several aspects of a troubleshooting assignment that contribute to the difficulty of identifying the root cause of a problem and in coming up with a solution.

In my opinion, the following are the main issues that can contribute to the difficulty of finding a clear path that proceeds from determining what is causing a problem to its solution:

1: Technical difficulty in obtaining the proper data to identify the source. Getting good torsional vibration is an example of where it is difficult to get the needed information. Getting a good reliable tach signal is another problem that is often encountered when attempting to get useful data. Sometimes, there is poor access to the best locations where data can be obtained. A good example of this issue is vertical pumps where the pump bearings are inaccessible.

2: The complexity of the problem itself. This can be compounded if there is more than one contributing factor. For instance, a combination of acoustical and structural resonances can significantly complicate the situation.

3: The physical environment. This can cover anything from the temperature, humidity, deafening sound levels, cleanliness, noise, smoke, rain, snow to even electrical interference. A challenging physical environment can make it very difficult to get data and significantly hamper communication between those doing the field work.

4: Lack of machine information. Bearing numbers, gear tooth counts, number of vanes on a pump, number of rotor bars on a motor, pump suction and discharge pressures and electrical current readings can be vital in determining the root cause of a vibration problem. In some instances, this technical data can be difficult to find. A facility may not have any manuals or the manufacturer of the equipment may have even gone out of business.

5: Time Pressure- Some jobs are under severe time pressure. This often is the case when working on large steam turbine generators. The costs associated with when they are down can run into the hundreds of thousands or even millions of dollars per day. In certain nuclear power plant related problems, if there is a tech spec violation on a machine that is deemed necessary for the safe operation or shut down of the reactor, then there is a limited time in which the problem must be resolved, otherwise the plant must shut the plant down. The cost can quickly get into the millions. In these situations, everyone seems to have an opinion as to what is wrong and sorting through all of this unsolicited information can draw the analyst away from finding a solution.

6: Political Pressure- If there is a lot of pressure coming from above that can be very detrimental to finding the most efficient path forward. A harsh political environment or one in which management is looking for someone to blame can cause the flow of information to come to a standstill. In many instances operations or maintenance may have vital information. If the flow of that information is interrupted, then finding the source of the problem may become very difficult or even impossible to determine.

As can be seen from the above list, the degree of difficulty associated with a particular problem can vary significantly depending upon several factors.

12Th MOST DIFFICULT CASE: Nuclear Power Plant Control Rod Drive Mechanism Cooling Fan

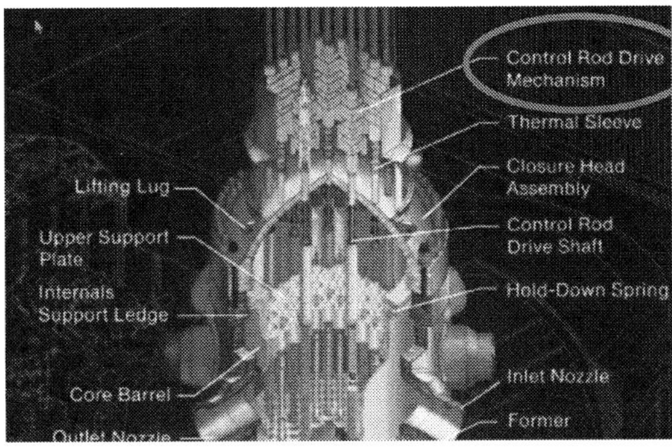

Figure 1: Nuclear Reactor with control rods shown at top

Control rods in a nuclear reactor are one of the ways the reactivity in a nuclear reactor is controlled. These rods which contain elements that absorb neutrons are lowered into the reactor core to control the fission rate. The rods are lowered by devices called control rod drives. In a Pressurized Water Reactor these drive mechanisms are located above the reactor pressure vessel head. In order to keep the drives from overheating, air is supplied to them by control rod drive mechanism cooling fans. During an outage, when the vibration on one of the fans was measured, it was three times higher than it had been in the past. The vibration was all at the fan's operating speed so it initially appeared the level of unbalance had changed on the fan. The question that arose was why did the amplitude increase. The fan was examined. The fan wheel was clean, no weights had been thrown from the fan wheel and the fan wheel did not have any cracks. To get to the root of this problem, understanding was needed in three areas. An understanding of ionizing radiation from the nuclear fission process, organic chemistry, and the vibration amplification factor.

Nuclear fusion gives off high energy gamma rays that have sufficient energy to strip off electrons, thusly called ionizing radiation. Organic chemistry comes in to the analysis because this ionizing radiation can damage the covalent bonds in rubber compounds making the rubber brittle and hard.

The amplification factor is involved in the understanding of this problem because when rubber isolators harden that changes their stiffness and thus the response of the structure to the driving force. In this case, the unbalance force had not changed, but because of the hardening of the isolators, the isolated system's natural frequency had increased, thus moving the response up the amplification curve. When the isolators were replaced, the vibration level went back to normal.

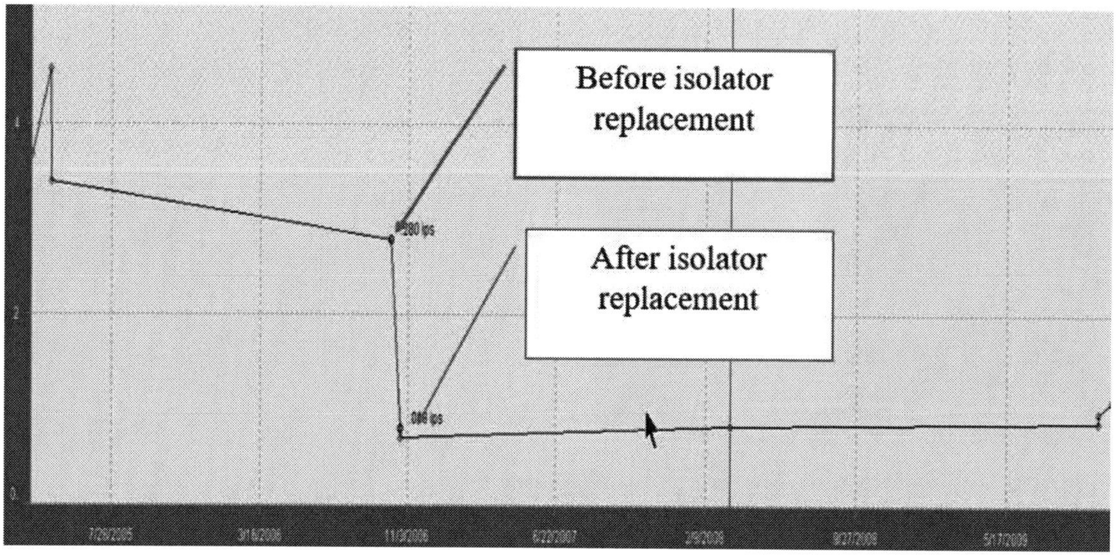

Figure 2: Vibration before and after changing out the isolators

This case was put on the list because knowledge was required in multiple areas. Those areas were understanding ionizing radiation from fission process, organic chemistry (covalent bonds in rubber compounds) and the amplification factor.

11Th MOST DIFFICULT CASE : Turbine Driven Boiler Feed Pump that was failing seals on its inboard side as well as failing couplings.

Figure 3: Boiler Feed Pump and Turbine Driver

This turbine driven pump was in a new power plant. After a short amount of operating time, the inboard pump bearing failed along with the inboard pump seals and the coupling. An alignment problem was suspected.

In order to study how the alignment moved with load, Dodd bars were installed.

DODD Bars consist of brackets with targets mounted on one machines and bars with proximity probes looking at the targets mounted on the other machine.

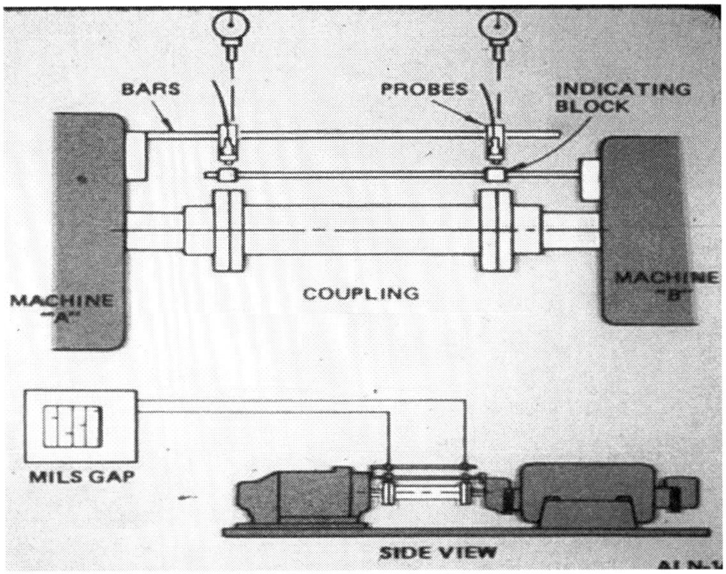

Figure 4: Diagram of Dodd bar set up

By monitoring the changes in the DC gap voltages and graphing the results it was possible to measure the amount of vertical and horizontal movement of the turbine relative to the pump.

Figure 5: Photo of Dodd bar installation.

When the turbine was coasting down, over 100 mils of movement was detected with the Dodd Bars when condenser vacuum was broken.

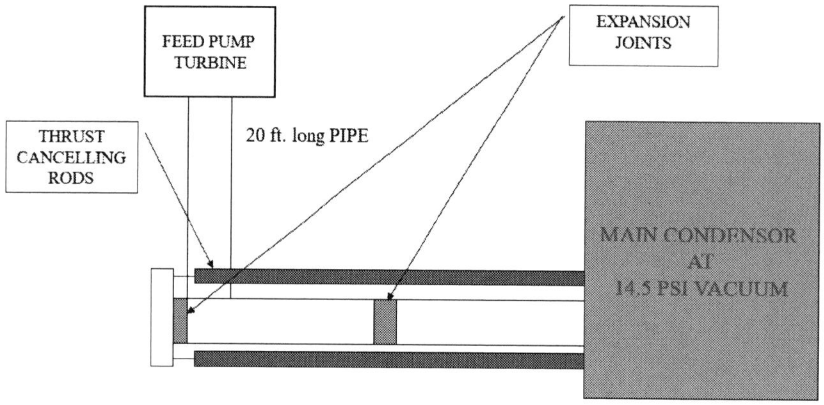

Figure 6: Diagram of feed pump and condenser piping

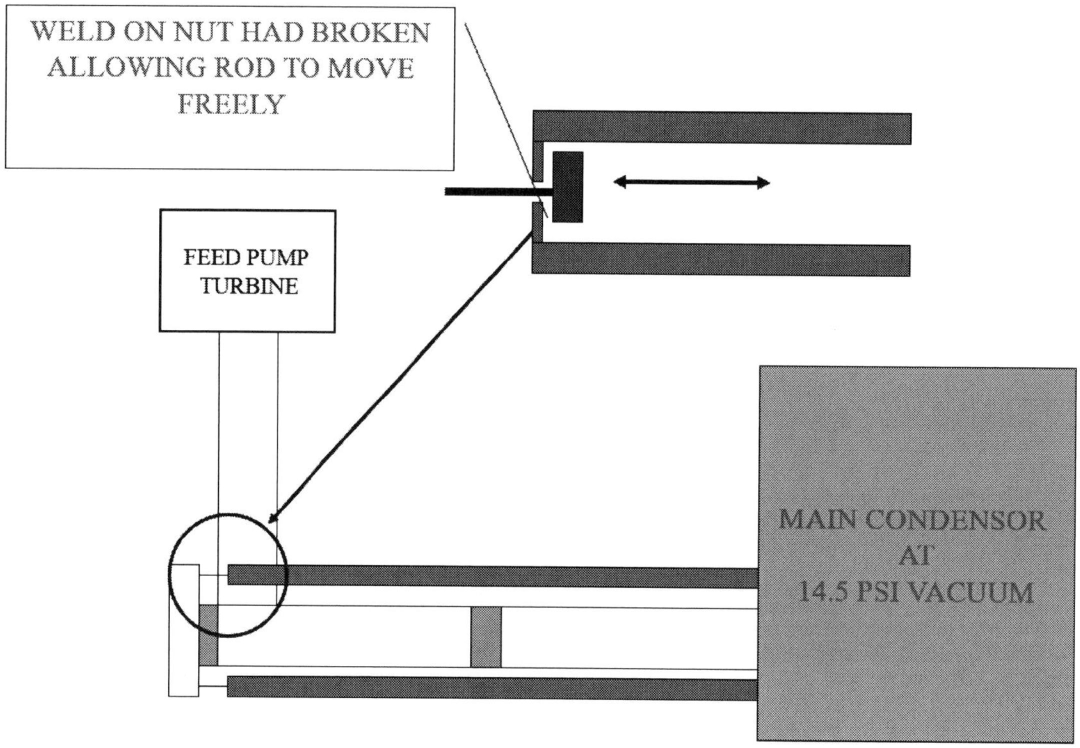

Figure 7: Problem location

When the feed pump's turbine drive condenser piping was examined, it was noticed that the expansion joints were compressed and one of them was leaking due to a crack. Flattening of the threads on one of the thrust cancelling rods showed it had been sliding back and forth in and out of the end cap. This made the thrust cancelling rods ineffective. The following force calculations show why the alignment was affected because of the non-cancelled forces.

Force Calculations:

60 " PIPE
AREA = PI*30*30=2827 square inches
2827 X 14.5 PSI=41000 lb force
41000 lb X 20 ft=820,000 ft.-lb of torque

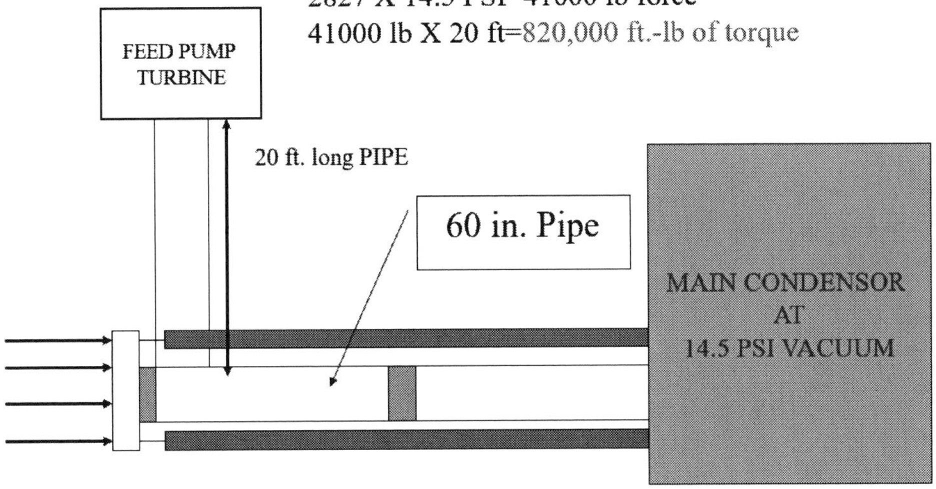

Solution:

- PUT NUT ON OUTSIDE OF END CAP

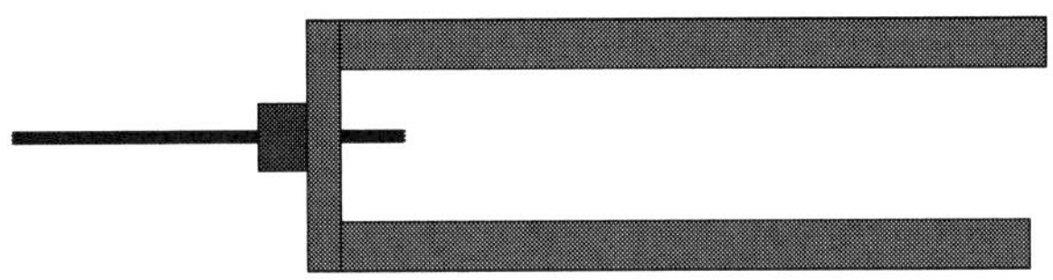

FORCE PUSHES AGAINST END CAP INSTEAD OF PULLING ON WELD

This case illustrates how a simple weld failure had a large effect. When the weld on the nut of the original design failed, that prevented the force cancelling rods from doing their job. This resulted in a torque of 820,000 lb-ft being applied to the turbine which caused a very large change in its alignment.

10th Most Difficult Case: Turbine Driven Chiller at University-Failing Turbine Inboard Bearings

Figure 8: Photo of turbine driven chiller

Problem:

The inboard turbine bearing had failed several times. The curious thing was the upper half of the bearing was always the point of failure.

Figure 9: Photo of failure of upper half of turbine inboard bearing

The alignment would vary with time. The plot below shows the movement over a two-hour period following a typical startup. The graph shows the turbine would appear to drop relative to the compressor. The movement below occurred over the first hour of operation.

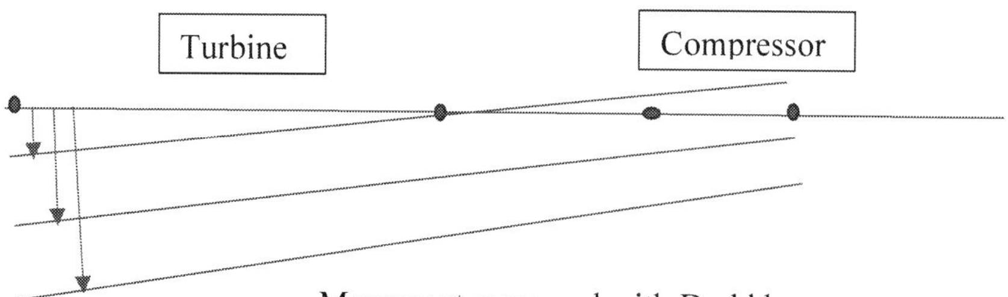

Movement measured with Dodd bars

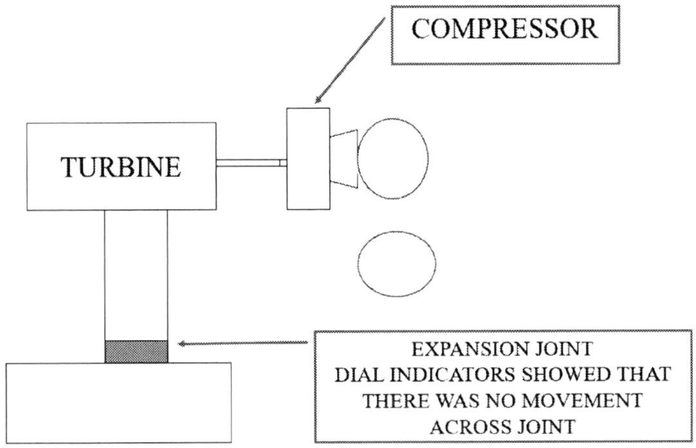

Figure 10: Machine Layout

Based upon experience with other turbines, it was thought that as vacuum was drawn the force across the expansion joint was pulling the turbine downward. However, when dial indicators were installed to verify this assumption, the readings showed that this was not the case. There was no movement of the turbine downward. This brought the focus of the investigation to the chiller. The actual problem was found by installing a laser on the chiller bearing housing and pointing it towards a target on a vertical I-Beam.

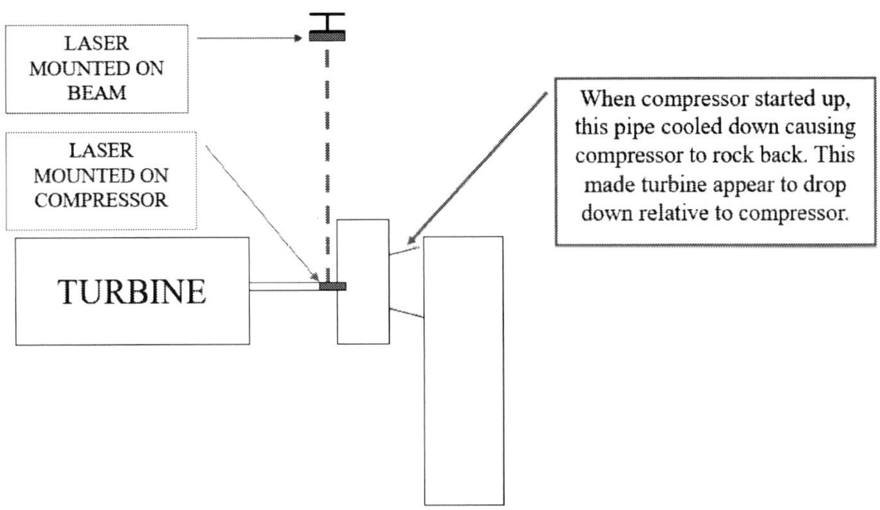

Figure 11: Laser Set up to detect compressor movement

This problem was difficult to identify because, based upon past experiences, the turbines were the machines which tended to move when vacuum was applied. This case shows it is not wise to assume anything. The solution was to install the turbine high enough to compensate for the upward movement of compressor. A tilt pad bearing was also installed on the turbine. This machine ran without any further problems for the next 30 years until it was finally retired.

9th Most Difficult Case: Vertical Pumps Breaking Shafts Multiple Times

The photo below shows the typical break occurring on vertical service water pumps at a power plant. Six shafts had broken on the these pumps.

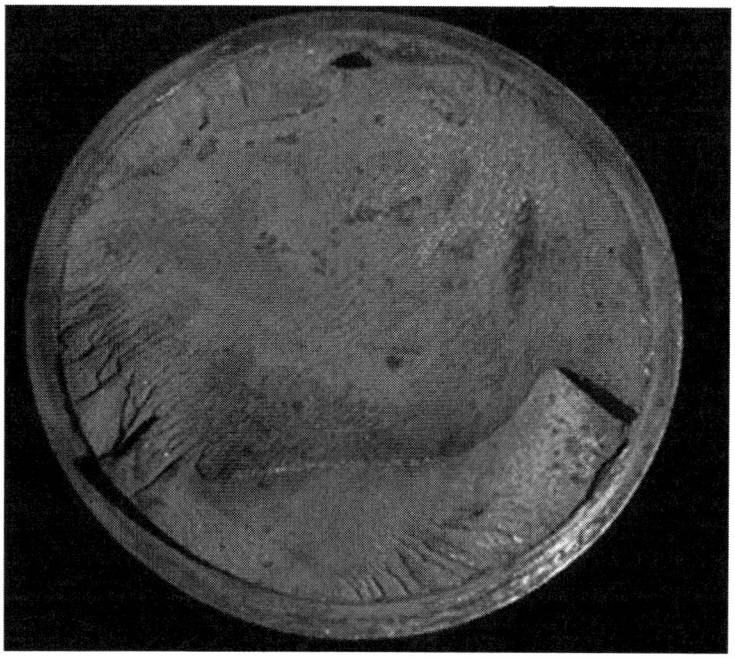

Figure 12: Photo of broken shaft

The puzzling thing was that these pumps had been in operation for many years without any problems. Over time, the pumps were overhauled and then the shaft failures started to occur.

Impeller suspended on shaft

Pump Impeller

Cross Section Diagram

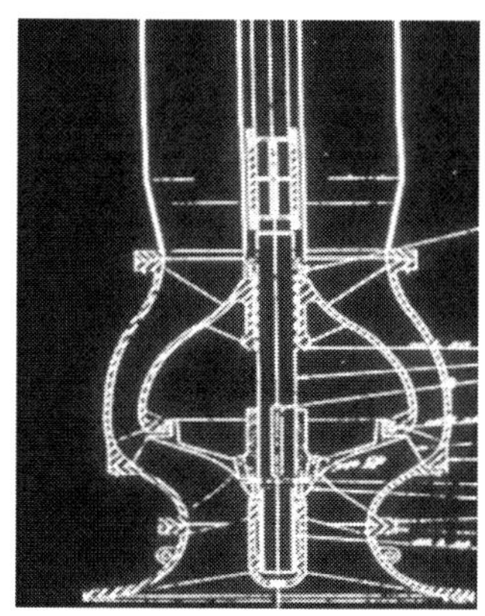

Because of the multiple failures, several tests were performed on the pumps.

Proximity probes were installed on lower bearing

Torsional Measurements were made

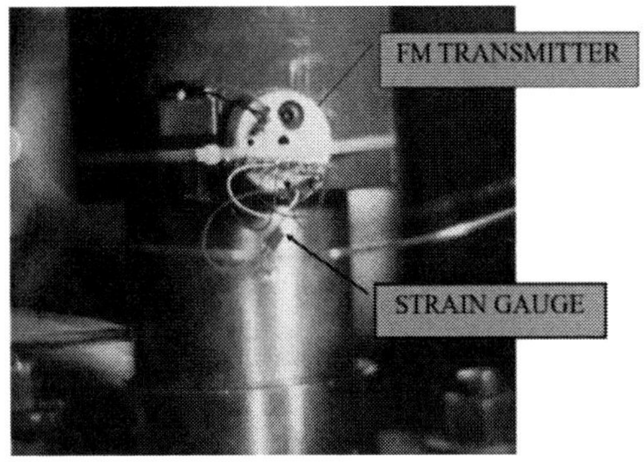

Impact Tests were performed

The impeller on its shaft was put under water for tests

Mode Shape

1st mode 375 cpm

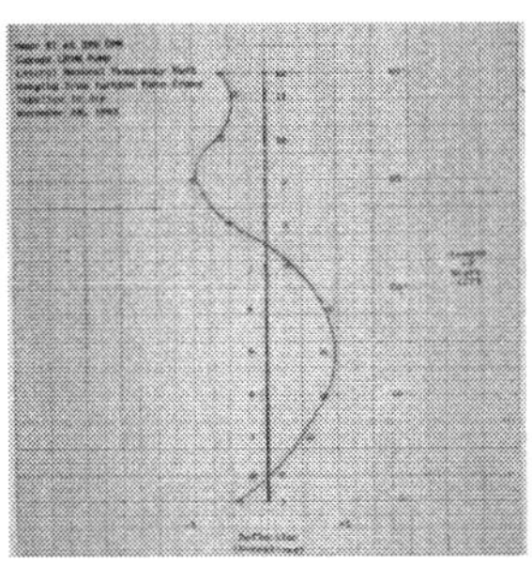

Orbit

Low Synchronous energy

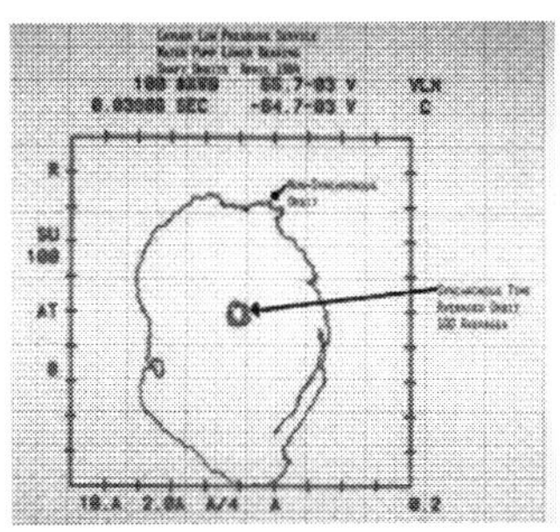

Spectra

Broad band around 1st mode

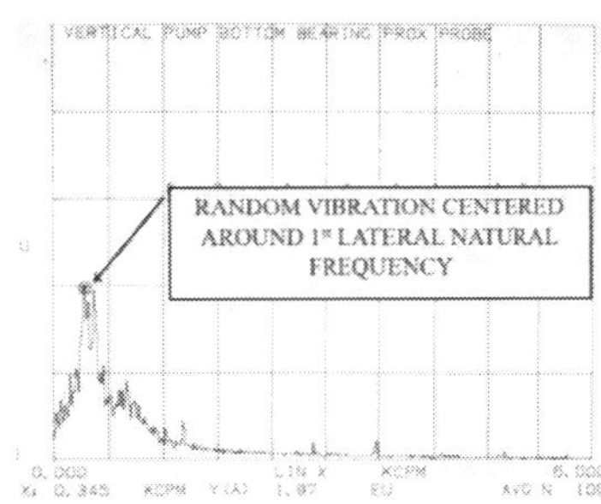

Figure 13: Test data

Test Results: When the vibration spectra were examined, there was no dominant frequency. What was found was the vibration was broad band in nature and centered around 375 cpm. This frequency matched the first bending mode of the shaft. When synchronous time averaging was applied there was almost no running speed vibration. There was also no measurable torsional vibration present. This agreed with the type of breaks that were occurring which did not show any signs of torsional fatigue.

Conclusion:

Broad band flow energy was exciting the vibration on the shaft at sub-synchronous frequencies which excited the first bending mode. This resulted in alternating bending at approximately every other revolution. Cracks then developed at the notched high stress points where the shaft couplings were installed.

Question:

Why did these pumps run for many years, then only after they were overhauled did the breaks start to occur. The pumps had run over 10 years with no failures then after the overhauls the shafts were breaking within a year of operation. This brought up the obvious question of what was changed during the overhauls?

Investigation:

Maintenance personnel were interviewed to determine if there had been any changes to the pumps when they were overhauled. Because of a pricing issue, plant personnel had elected to go with a different supplier of pump impellers. When the replacement impellers were closely examined, the angles and the thickness of the impeller vanes were found to be significantly different than the original design. This resulted in very high turbulent flow exciting the shaft's first natural frequency. When the original manufacturer impellers were put back in the pumps, the shaft failures immediately stopped.

This problem was included in the list because it took extreme effort to install the prox probes on the pump lower bearing and get the wiring through the foundation. It was also difficult to perform the natural frequency tests. The pump impellers on their long shafts had to be suspended from a crane in the turbine building. The testing also required installing a torsional measuring system. In addition to the testing issues, there was also the issue of the impellers being purchased from a pirate shop that had to be coaxed out of the maintenance personnel.

8TH MOST DIFFICULT CASE: High motor vibration on a bust duct fan at a nuclear power plant

Bust duct cooling fans are very important machines. The bust ducts house the electrical lines that carry the power (In this case 1100 Megawatts) from the generator to the main transformer. This fan consisted of a motor driver that provided power through the belts to the fan. The vibration on the motor was .345 in/sec which was above the established alarm level. The questions raised were: Was the motor having a problem or was there a structural resonance causing the high vibration levels?

Figure 14: Photo of fan room

Figure 15: Photo of motor, belt and fan bearing

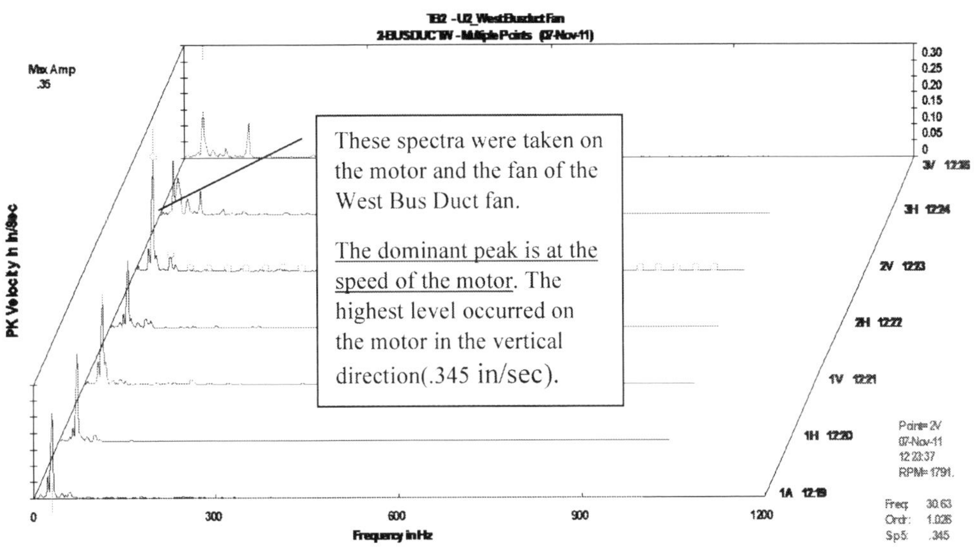

Figure 16: Spectra from motor and fan inboard bearing

The above spectral plots show the elevated vibration was all at the motor speed.

These fans are mounted in the air on metal floors that had other issues over the years. In the past, ODS videos and numerous natural frequency tests were performed on the structure. The results of these tests were that structural modifications were made on some of the fans. This history brought up on the plant personnel's' radar that there might be an underlying structural problem. It was therefore elected to perform natural frequency tests and ODS measurements.

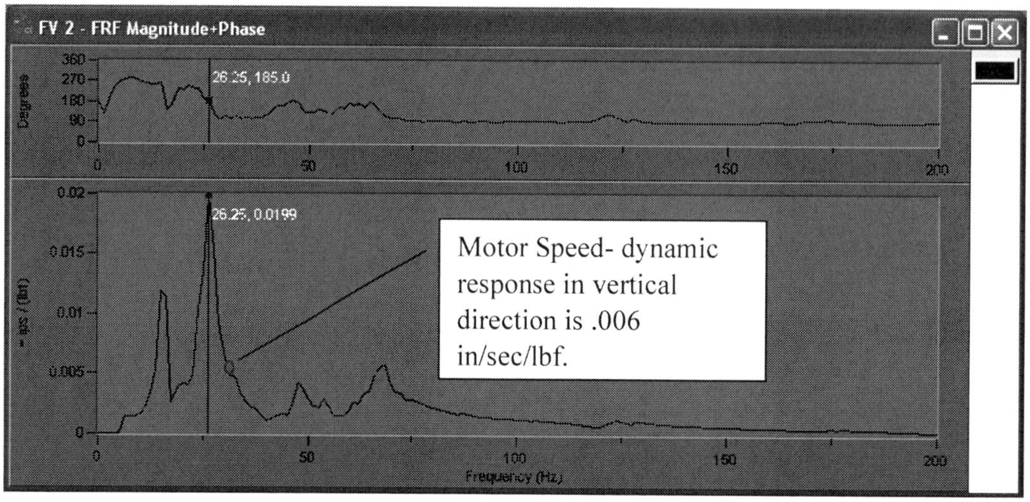

Figure 17: Motor Inboard Vertical transfer function

The above plot shows the motor speed is on the response curve and there is therefore structural amplification to any unbalance present in the motor. Similar response curves were seen on the motor outboard bearing and the metal floor.

FURTHER EXAMINATION OF THE MACHINE

It is always important to examine the structure before checking out the rotating equipment. The base bolts were checked to make sure they were tight and the support channels were inspected for the sign of any cracked welds. During this examination, it was noted that there was excessive key on both sides of the motor sheave.

Figure 18: Excess key between sheave and motor **Figure 19: Excess key outboard**

Question: Was the force from the excess key enough, when combined with the measured dynamic stiffness, enough to cause the motor vibration to be in alarm or was excess key just having a minor effect.

Comparison to ISO Standard balance grade 2.5

The motor weighed 517 lb. ISO grade 2.5 for a 517 lb motor turning 1800 RPM:

For a rigid frame, the allowable unbalance is 4.2 oz-in.

For a flexible frame the allowable unbalance is 1.7 oz-in.

This fan was mounted in the air and the vertical natural frequency on the metal base was below running speed, so the 1.7 oz-in flexible frame limit applied. There were 4 inches of excess key and the key was ¾" wide and ½" thick. The volume of the excess key was therefore 1.5 cubic inches. When taken times .283 lb/in cube the weight of the key was calculated to be .42 lb or 6.8 oz. The distance of the key from the shaft center was 1.5 inches. The unbalance was therefore 10.2 oz-in or 6 times the ISO flexible frame standard.

Calculation of response based upon FRF dynamic stiffness values

The calculated force from the 10.2 OZ-in of unbalance at 1800 RPM was 58 lbf. The dynamic stiffness from the FRF was .006 in/sec/lbf. The response to this unbalance force therefore calculates out to be .35 in/sec which is very close to the .379 in/sec measured on the motor.

Conclusions: The excess vibration was due to a combination of the excess key and to a lesser degree the motor operating on the response curve of a structural natural frequency. The solution was to install a step key. This was actually a very simple problem to solve. What made it onto the most difficult list?

1: This was a nuclear power plant. As shown in Figure 14, a ladder had to be climbed to get to the fan for testing. There were three people watching how we climbed the ladder pointing out if we did not have a constant three points of contact. This was very distracting.

2: Past problems had resulted in a lot of focus being made on structural issues when the problem could have been solved by merely installing a step key. This past history significantly lengthened the path to a very simple solution.

7th MOST DIFFICULT CASE- A GALVANIZE LINE AT A STEEL MILL

A galvanize line at a steel mill in the central part of the United States was having vibration problems. The figure below shows the components in the galvanize line. At the front of the line were the bridle rolls that pulled the plate through the galvanize tank then into the long furnace.

The entire structure which was approximately 300 feet long would vibrate excessively.

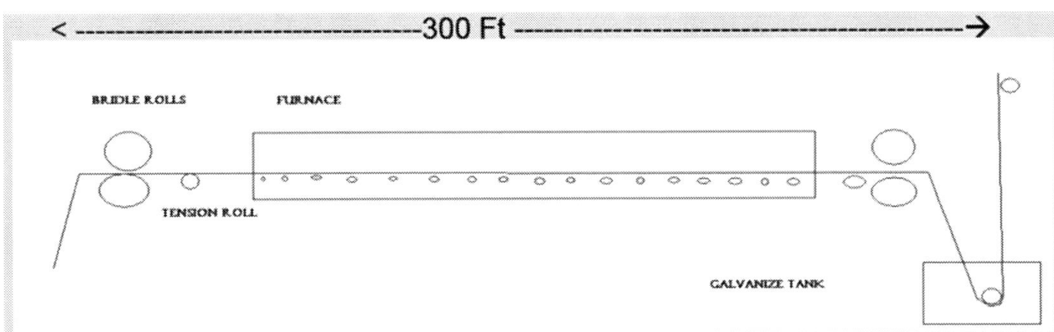

Figure 20- Drawing of galvanize line

TEST PROCEDURE:

The frequency of the vibration on the bridle rolls and the entire structure fell in the range of 6 to 7 HZ. In addition to the 6-7 HZ lateral vibration, there were also noticeable levels of oscillation observed on the drives, it was therefore elected to take torsional measurements. Torsional measurements allow a determination to be made of the magnitude of any speed oscillations present along with the frequency of those oscillations. In order to measure torsional vibration, it is necessary to acquire signals from an encoder and then demodulate the encoder pulse trains. Two techniques were used in the measurement of the torsional vibration on the galvanize line. The first was the installation of an encoder on the outboard end of the Furnace Bridle Motor 2.

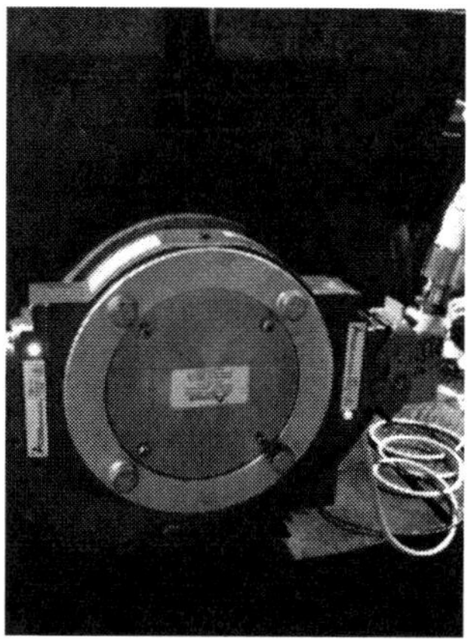

Figure 21: Encoder

Figure 22 Torsional Demodulator

The second method utilized to measure the torsional vibration involved connecting a portable encoder to a PI wheel, which could then be touched against a coupling hub or roll to measure any torsional oscillations.

Figure 23: Encoder Pi wheel demodulator

Figure 24: Encoder on roll

Test Results:

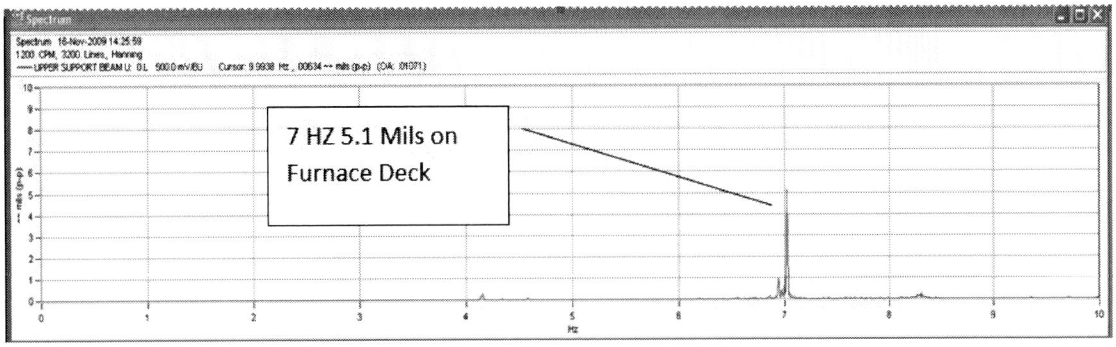

Figure 25: Vibration Spectrum from furnace deck

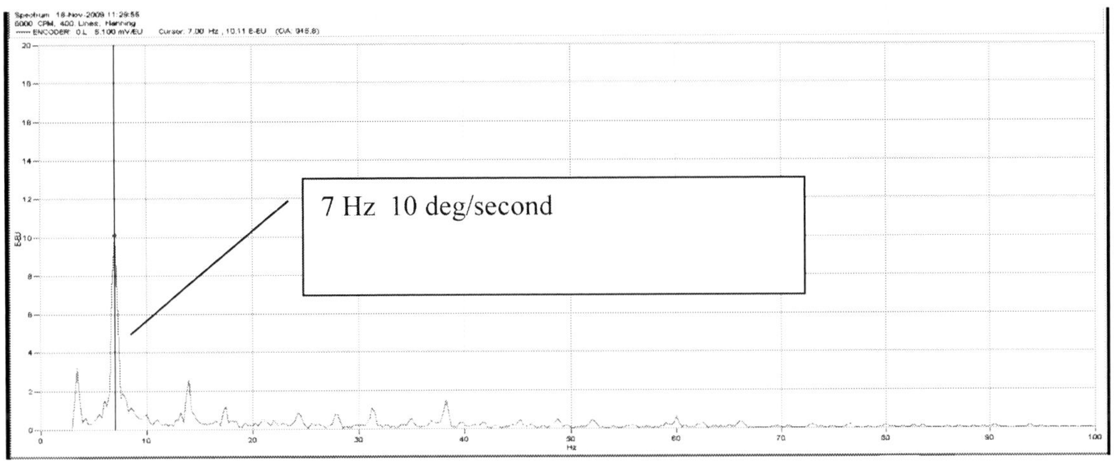

Figure 26: Torsional vibration spectrum from Bridle Roll Coupling

The frequency of the structural vibration on the furnace deck and the torsional vibration on the Bridle rolls matched exactly. The next step in the analysis was to vary the feet per minute of the steel strip. It was changed from 520 ft/minute to 450 ft per minute. The fmax was set to 50 HZ and the number off lines of resolution were set at 3200. This was done to get very accurate frequency values. The vibration frequency tracked exactly with the steel strip speed. This meant something related to the steel strip speed was causing the vibration. The next test was suggested by plant personnel. They had noticed a change in the vibration level when the tension on the steel strip was moved up and down. The proposed test was therefore to vary the amount of tension on the steel strip to determine if that was a factor in the amplitude of the torsional and lateral vibration.

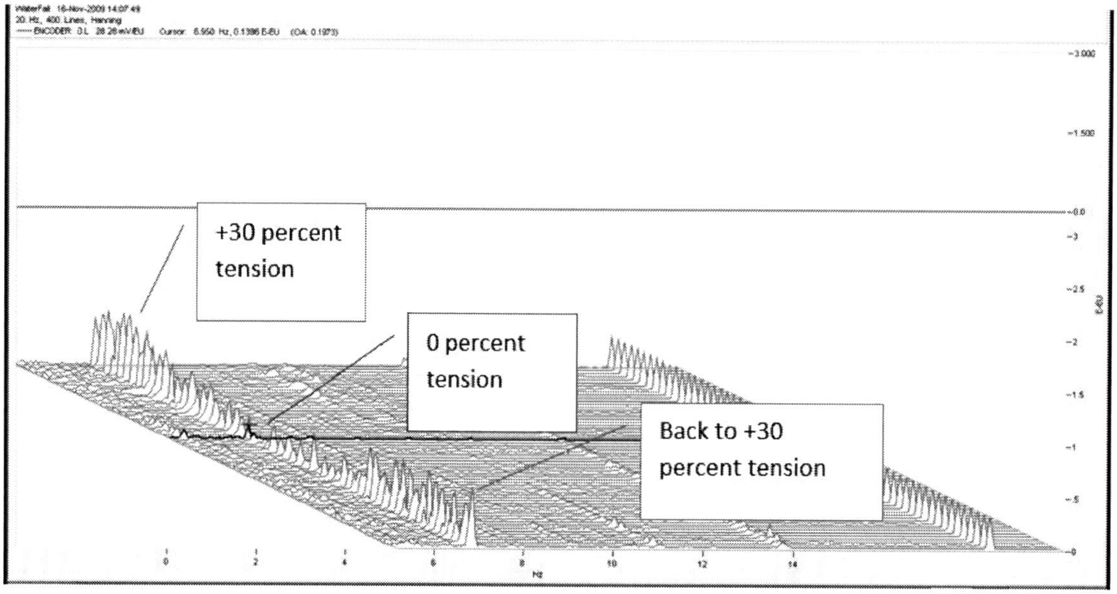

Figure 27: Effect of tension on steel strip on torsional vibration

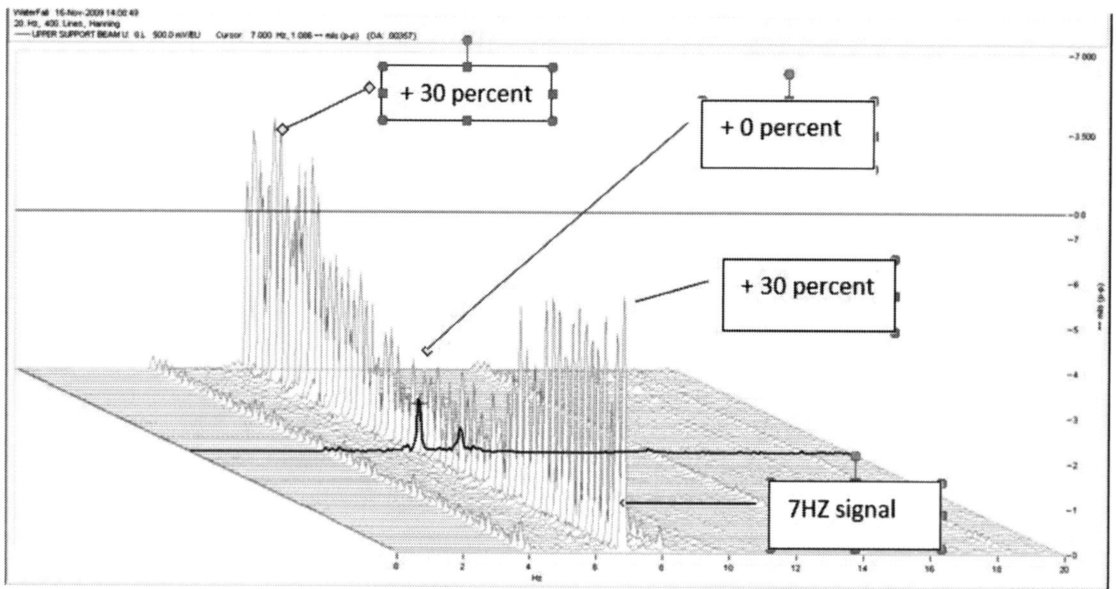

Figure 28: Effect of tension on structural vibration

<u>**Conclusions:**</u>

1: The torsional and lateral vibration were the exact same frequency.

2: The amplitude of this vibration was directly dependent on the amount of tension on the steel strip.

Question:

What was causing the 7 Hz vibration and was powerful enough to shake the entire structure?

Calculations:

7 cycles/second X 60 seconds/minute X diameter X PI= 520 feet per minute

Using the above formula, it was calculated that if a roll had a diameter of .395 ft (4.73 in) it would generate the frequency present on the structure. The same calculation was made using the vibration frequency present when the steel strip speed was 450 ft per minute and the roll diameter result came out with the exactly same diameter. Drawings were checked and it was found the furnace roll diameters were exactly 4.73 inches.

Question:

How could small furnace rolls of only 4.73" diameter cause a 300 ft long structure to vibrate?

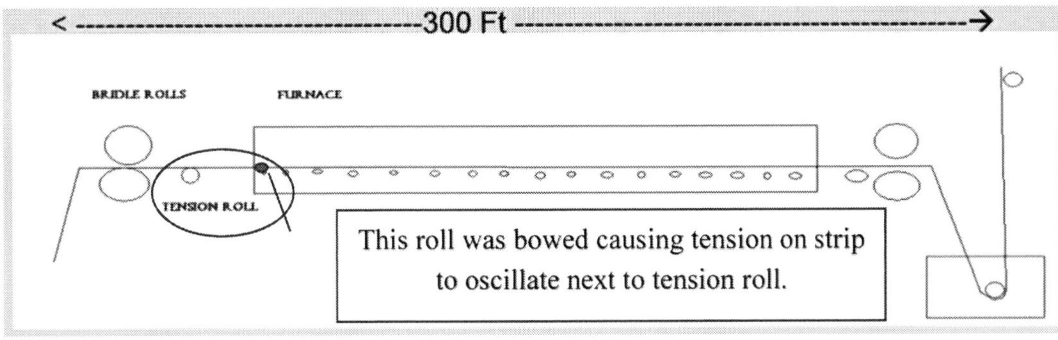

Answer:

While walking along the furnace, it was observed that the cooling line going into the last furnace roll was wobbling indicating the furnace roll had a bad bow. This was confirmed with a dial indicator. This bow was causing the tension to vary on the nearby tension control roll at the speed of the furnace roll. The control system then caused the massive bridle rolls to accelerate and de-accelerate at that same speed to compensate for the changes in tension. Two actions were taken. The first was to change out the furnace roll. The second was to tune the control system so it was less reactive.

This case was put on the most difficult list because it involved torsional vibration , lateral vibration, and the control system feedback logic. It was a case where a very small component was having a large effect on the response of a large piece of equipment. It was difficult to track down.

6th MOST DIFFICULT CASE: Large compressor that had vibration which over time would slowly build up until the compressor tripped on high vibration.

Figure 29: Photo of 40000 HP Demag Compressor

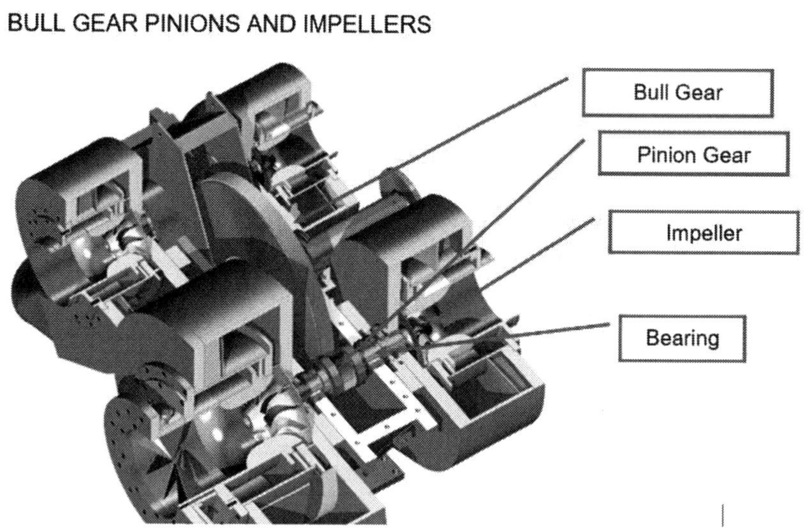

Figure 30: Layout of typical compressor.

The photo above shows the major components in a typical high speed air compressor. These compressors consist of a bull gear driven by a motor. The bull gear drives the high speed pinions that power the impellers. The bull gear is much larger than the pinion gear resulting in the speed of the pinion being much higher. In this case, the pinion speed was 17506 RPM.

Symptoms of the Problem:

The vibration on both stage 3 proximity probes would slowly increase in amplitude followed by an equally slow decrease. The period of one oscillation was approximately 7 minutes. The peak in the amplitude would get higher each cycle until the compressor finally tripped on high vibration.

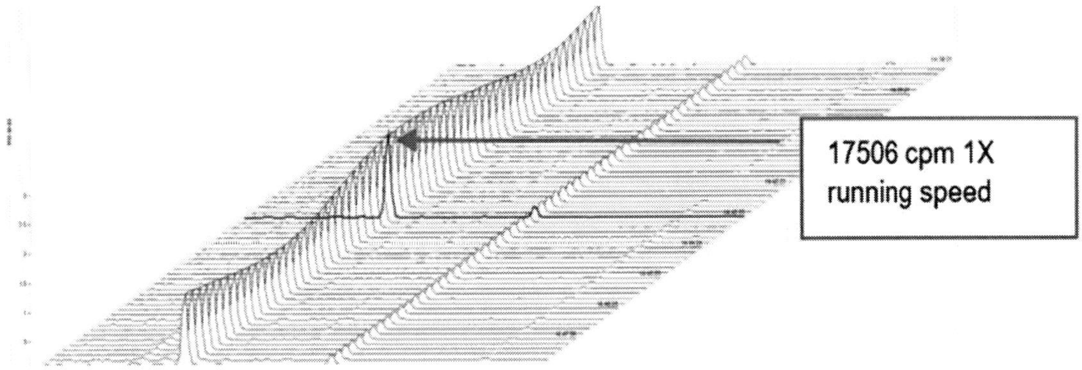

Figure 31: Waterfall plot showing vibration spectra over one cycle.

The waterfall plot above shows that all the vibration was at the speed of the 3d stage impeller.

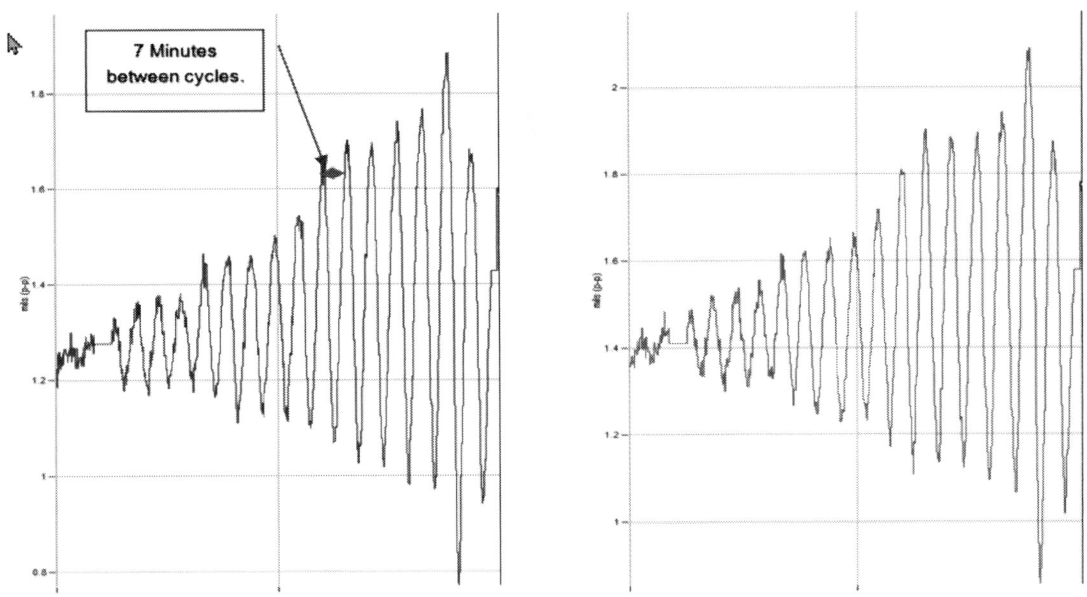

Figure 32: Two-hour time span showing trend of vibration on stage 3X and 3Y proximity probe amplitudes. Note how each cycle lasts approximately 7 minutes. Every cycle, the vibration gets progressively higher. This was disturbing to the operators in that they would see the vibration was getting worse over time.

Orbit during low part of 7 minute cycle Orbit during high point of 7 minute cycle

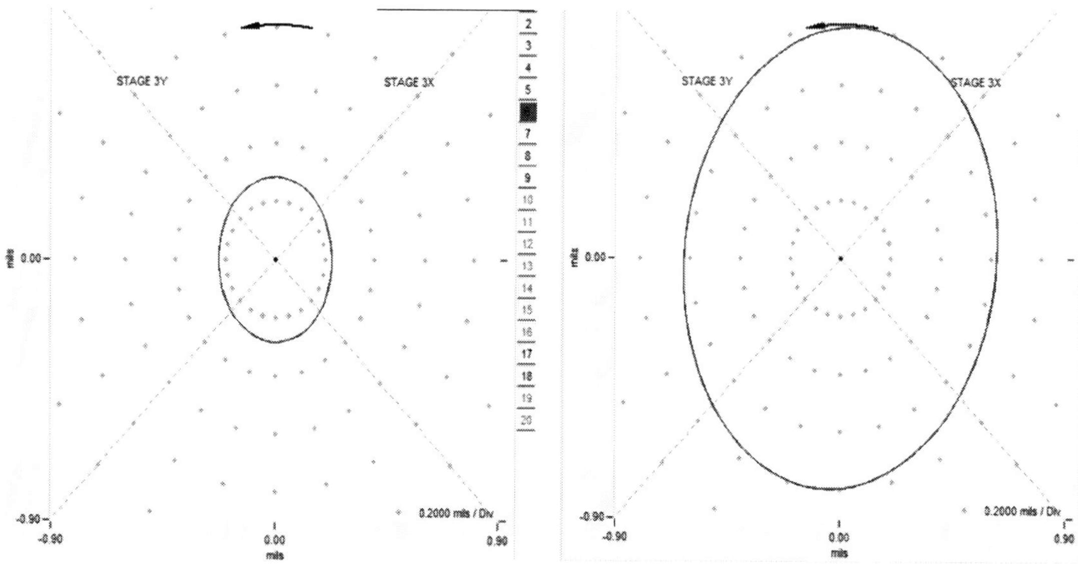

Figure 33: Stage 3 Orbit plots taken at low and high points of vibration.

This compressor was set up with a key phaser on each pinion shaft, so it was possible to read the phase of the vibration.

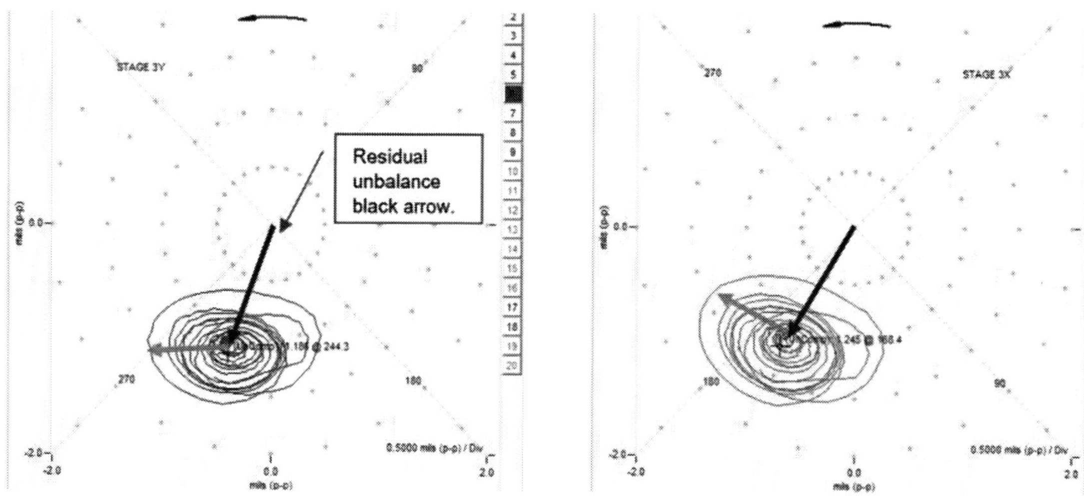

Figure 34: Polar plots of 3X and 3Y probes showing amplitude and phase during cycling of vibration amplitude.

The polar plots in the above figures show what is happening. The black arrows represent the residual imbalance on the 3d stage impeller. The green arrows show the vibration introduced by the Morton effect. The Morton effect vibration sometimes is in phase with the residual unbalance causing the amplitude to be maximum and it would sometimes be out of phase resulting in a minimum amplitude. Over time the Morton effect (green arrow) would grow in size, thus resulting in larger and large amplitude swings.

Discussion:

The Morton effect occurs as a result of uneven heating of the shaft at the high spot caused by the residual unbalance of the impeller. Because of the high-speed nature of these overhung impellers, the point of the slightly elevated shaft temperature lags the unbalance. This means the Morton effect unbalance slowly rotates around the shaft casing it to come in and out of phase with the residual unbalance. The heating also tends to slowly increase with each cycle. The solution was to change the bearing design to reduce the amount of shaft heating. This case is a good example of using various test techniques. The overall values shown on the strip chart gave insight as to the amount of time between the cycles. The orbit displays showed how much bearing clearance was being used. The polar plots which showed the cyclical relationship between the unbalance vector and the slowly rotating Morton effect vector were very important in gaining an understanding into the nature of this problem.

5th MOST DIFFICULT CASE

500-Megawatt Turbine with Hard Rub

Figure 35: Photo of Low-Pressure Rotors on Large Steam Turbine. Red circle shows a man for size comparison.

Problem:

When turbines are overhauled, new seals are installed in the casings. These seals minimize leakage between the turbine stages and through where the shaft enters the case. After an overhaul, the clearances between the rotating and stationary components are tight. If the case is distorted because of uneven heating or thermal expansion a rub can occur. When a shaft is rubbed, it will cause a hot spot on the shaft which in turn causes the shaft to expand at the point of the rub thereby resulting in the shaft bowing. This bowing can produce a high level of unbalance which in turn can aggravate the bow resulting in a positive feedback cycle. If the bow is severe enough, the system can become non-linear and exact half running speed vibration can appear. This case history shows how the presence of exact half frequency vibration appeared when an LP rotor on a 500-Megawatt turbine experienced a hard rub.

A large turbine was monitored with an advanced monitoring system following an overhaul. The following X-Y proximity probe data was recorded:

Overall amplitude plots
Time waveform plots
Spectra
Waterfall Plots
Bod'e Plots
Shaft Orbits

Test Results:

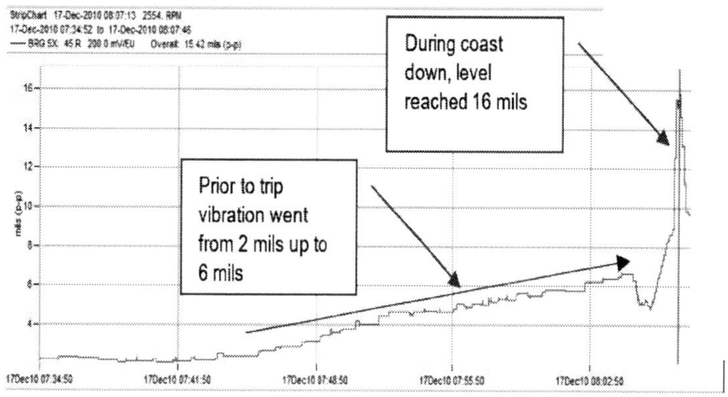

Figure 36: Trend plot showing vibration before and following vibration trip

The trend plot above shows how the vibration level increased while sitting at constant speed. Over a 12-minute period, the vibration increased from 2 mils up to 6 mils. When the turbine tripped, the amplitude reached 16 mils.

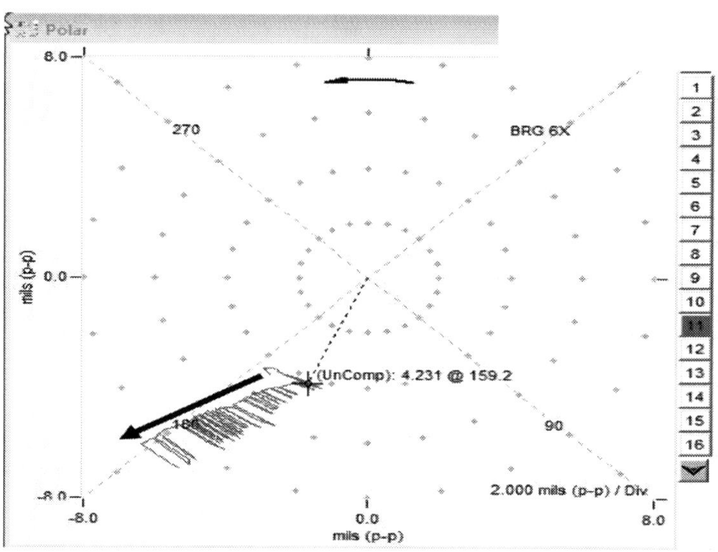

Figure 37: Polar plot showing data taken as vibration level increased.

The polar plot shows the amplitude and phase. Each point on the red line represents an amplitude and phase vector. Note how the phase is erratic. A change in phase accompanied by an increase in amplitude that occurs while the speed is constant is a strong indication a rub is occurring.

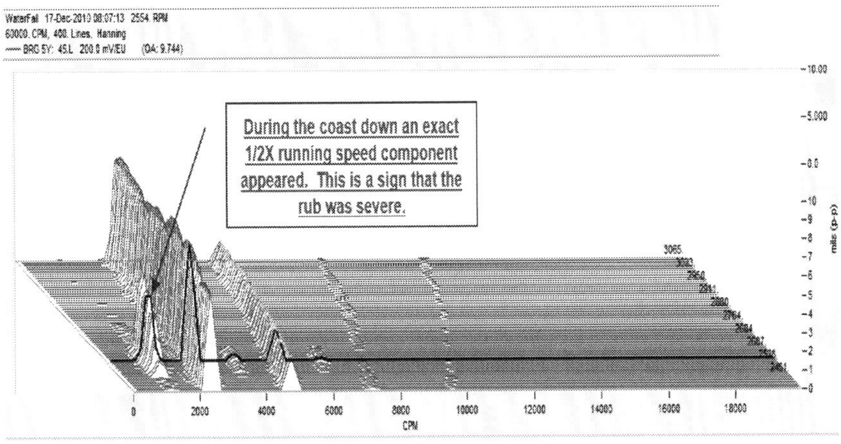

Figure 38: Map plot of proximity probe spectra as turbine coasted down from 3065 RPM to 2451 RPM.

This map plot contains the spectra taken during part of the coast down following the vibration trip. The interesting observation is that what appears to be 1/2X running speed appeared in the spectra.

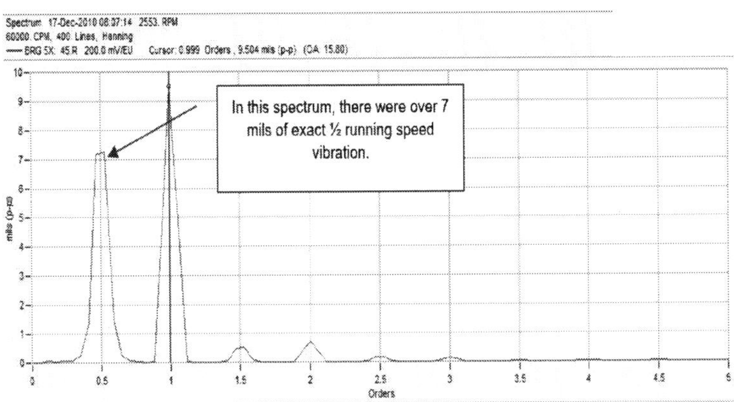

Figure 39: Single spectrum taken at 2553 RPM.

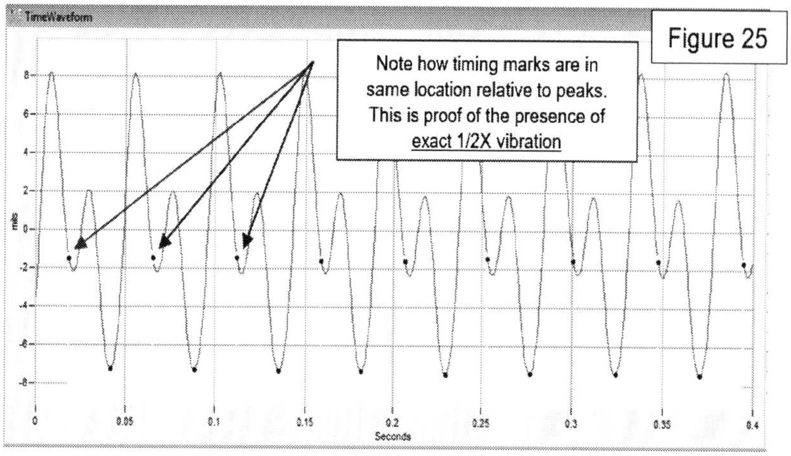

Figure 40: Time domain plot showing timing marks do not move

The question that needed to be answered was the frequency exactly half the operating speed or was it some percentage of running speed which might indicate it was oil whirl. The data in Figure 40 clearly shows the timing mark occurs every other peak in motion and that the timing mark does not move. This is proof the frequency is exactly ½ running speed.

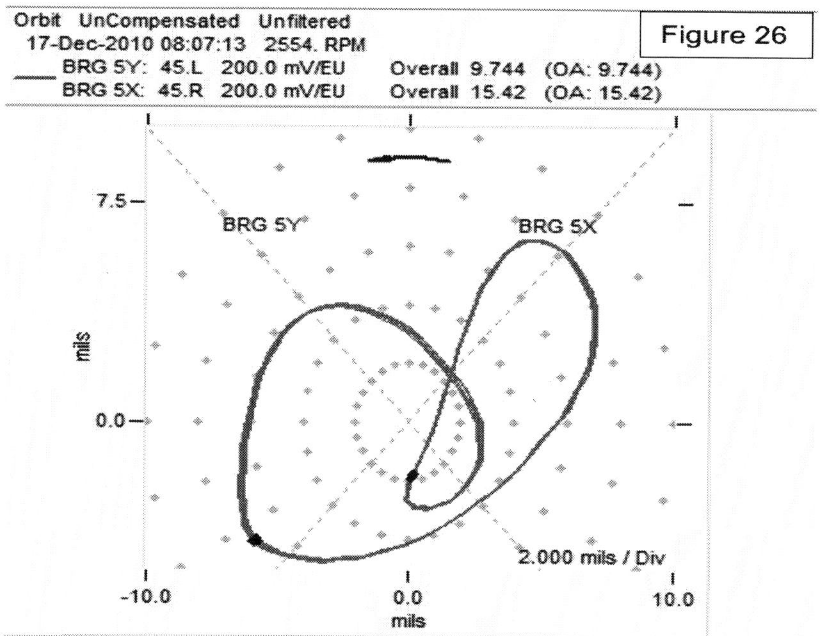

Figure 41: Orbit taken during rub.

Important Points:

1: Polar plot showed that at constant speed, both the amplitude was increasing and the phase was unsteady.
2: The map plot clearly showed the presence of what appeared to be a 1/2x vibration as the critical was traversed.
3:The time plot with the key phaser marks proved the vibration was exactly at 1/2X and was not related to something like oil whirl. This meant the rub was severe in nature.

Problem Identified:

In addition to having new seals, it was discovered the spray system used to control the temperature of the last stage blades on the LP rotor was wired wrong resulting in water being sprayed thus causing a temperature transient that resulted in thermal distortion. When the spray system was fixed, the turbine started up without further incident.

4TH MOST DIFFICULT CASE: Large Vertical Pump with resonance issues after installation of VFD

Since the introduction of reliable VFD speed control systems, it has become common practice to install them on new pump installations. With the knowledge the pumps will operate over a specified speed range, it is then possible to design the pumps and support structures to avoid the presence of natural frequencies in the projected operating speed range. This is of particular importance on vertical pumps which have a large motor mass located often far above the base on the circular pump body. Vertical pumps often have natural frequencies below their operating speed.

Sometimes, however, the change to a variable speed drive is performed on an existing vertical pump. If a design review is not performed, then there is a possibility resonance response issues will arise when the pump is put into service and the speed is lowered down to where it matches a usually 1st reed mode natural frequency.

A water company had plans to install a Variable Frequency Drive on one of its vertical pumps. While on site doing analytical work on another pump, impact tests were performed on the pump that was to have the VFD installed. Figure 42 shows that the pump, which ran at 900 rpm, had a natural frequency at 1125 cpm in the direction that was in line with its discharge pipe and a natural frequency of 750 cpm 90 degrees out from the discharge line.

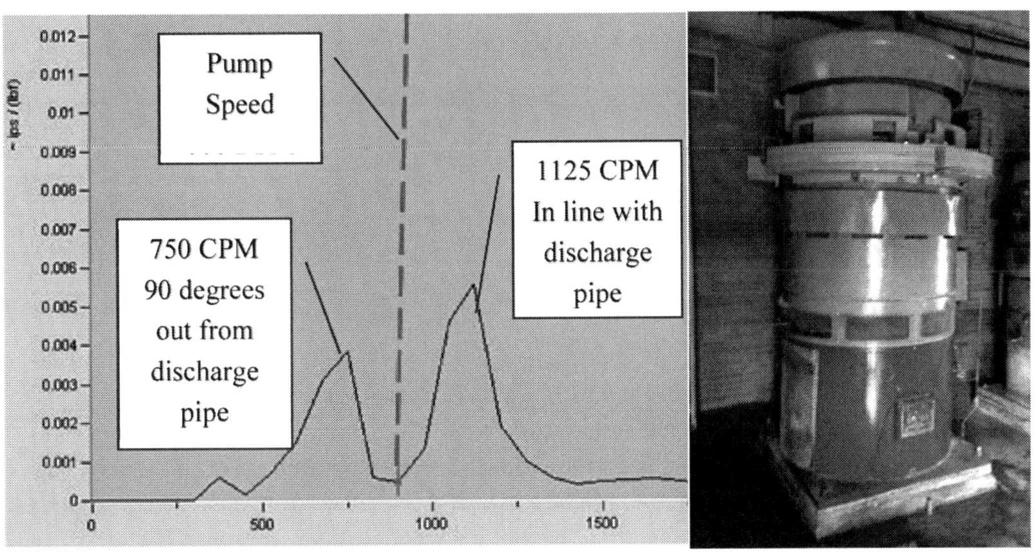

Figure 42: Vertical Pump natural frequency response

Since the operating speed was located between the two natural frequencies, there had never been a vibration problem with this constant speed pump. The issue was that the water company was planning to put a VFD on this pump. The frequency response data in Figure 42 showed that if the speed were lowered by a VFD that the vibration would significantly increase as it approached the 750 cpm natural frequency. This information was conveyed to the water company, but they said it was too late to do anything because the VFD was already ordered. They were informed they would almost certainly have a vibration problem when they lowered the pump speed.

After the installation of the VFD, the pump was tested to determine its response as the speed was lowered.

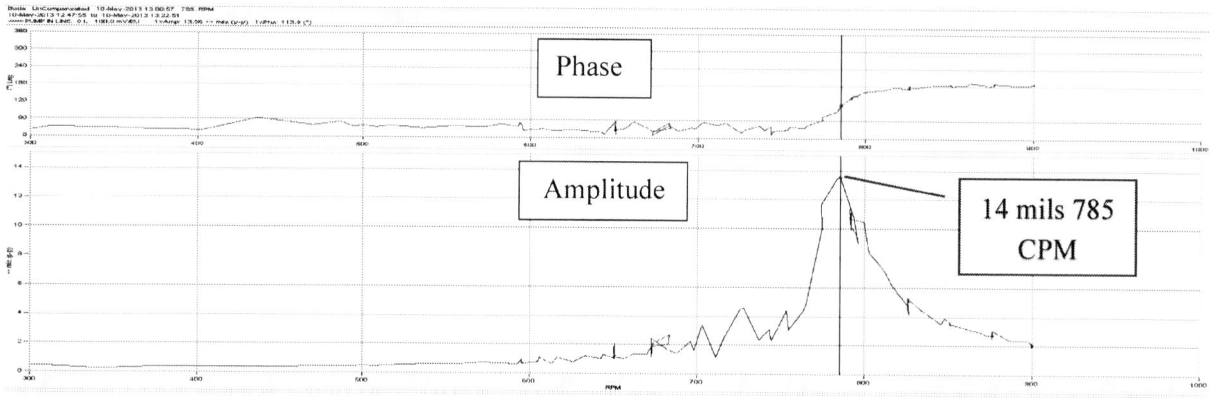

Figure 44: Vibration response curve as pump speed was lowered

As predicted by the natural frequency test, the amplitude increased significantly as the speed was reduced by the VFD. Based upon both the natural frequency test results from the impact test and from the startup-coast down data, it was determined that modifications were necessary if this pump was going to operate below its previous speed.

<u>**Solution**</u>:

The first step in determining a solution was to calculate the system damping. The following equation, which calculates the amplification factor at resonance, using the rate of phase shift was used to calculate the damping ratio.

$Q=\frac{\pi N_C \Delta\theta}{360\ X\ \Delta F}$ The Q value from the phase shift was calculated to be 23.5

$\xi = \frac{1}{2Q}$ This results in a damping ratio = .021

Two assumptions were then made. The first was that the addition of mass to the motor would not change the damping. This is a reasonable assumption because nothing in the supporting pump column was being modified. The second was the motor acted as a lumped mass. This was also felt to be a reasonable assumption because based upon several other vertical pump mode shape projects, the motor on a vertical pump generally acts like a single solid mass.

The next step in coming up with the amount of mass that would be required was to determine how much lower the natural frequency would need to be shifted in order to allow the pump to run below the Hydraulic Institute Standard acceptance level, which in this case was 6.0 mils. Using the response ratio equation and an original value of 14 mils, the calculations indicated that if the natural frequency could be shifted down 10%, then the pump at the lowest speed required by the process would run below the Hydraulic Institute level. The calculations showed the motor mass needed to be increased by 21%.

$$X/X_0 = 1/\sqrt{[1-(\frac{\omega}{\omega_n})^2]^2 + (2\xi\frac{\omega}{\omega_n})^2} \quad \text{Amplification Factor}$$

CALCULATED RESPONSE RATIOS WITH NO MASS ADDED

The graph below is the solution to the amplification factor equation using the calculated damping ratio.

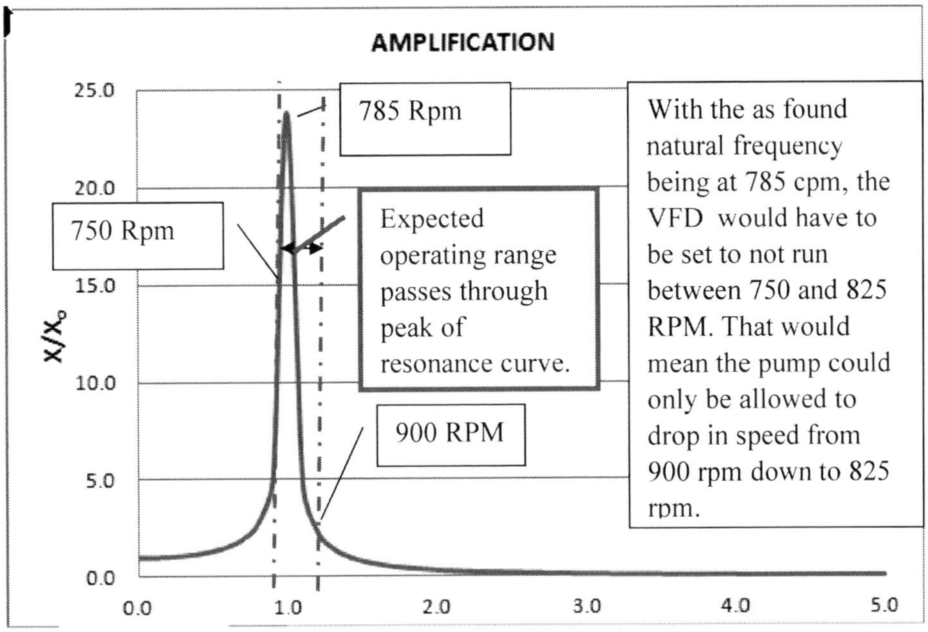

Figure 45: Response without mass added

CALCULATED RESPONSE RATIOS AFTER ADDING 21% MORE MASS

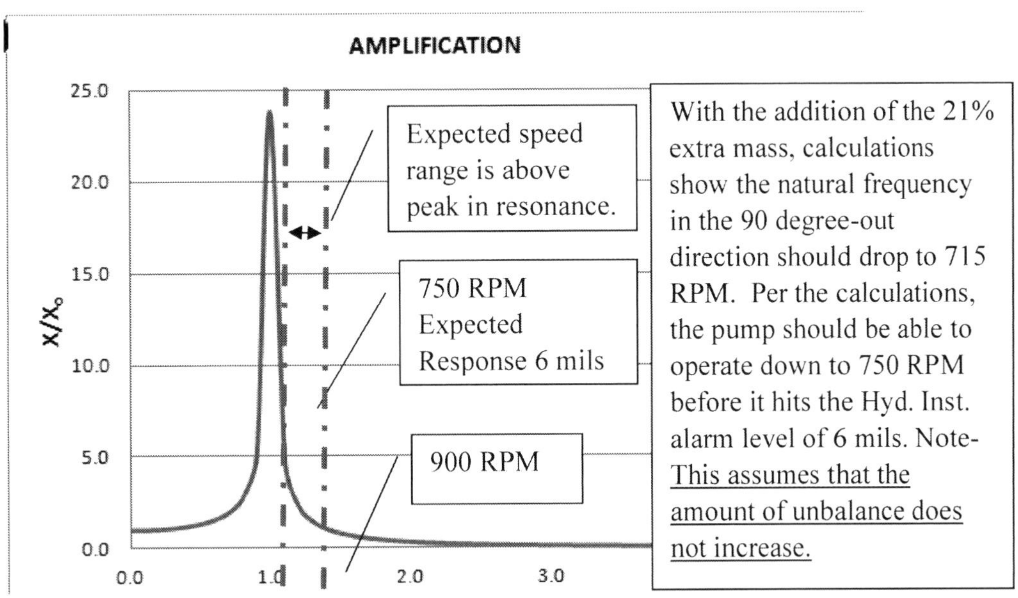

Figure 46: Response predicted after 21% mass increase (10% reduction in NF)

Special mounting brackets were designed and installed on the pump. Segmented plates were cut out with an industrial laser. The plates, which were small enough to be handled by two men, were then bolted to the brackets. A total of 3100 lbs. was added to the motor. Following the installation of the additional mass, an impact test was performed to evaluate the change in the natural frequencies. The results were better than expected. The natural frequency dropped from 785 cpm to 600 cpm.

Actual Response Curve Following 21% Mass Increase

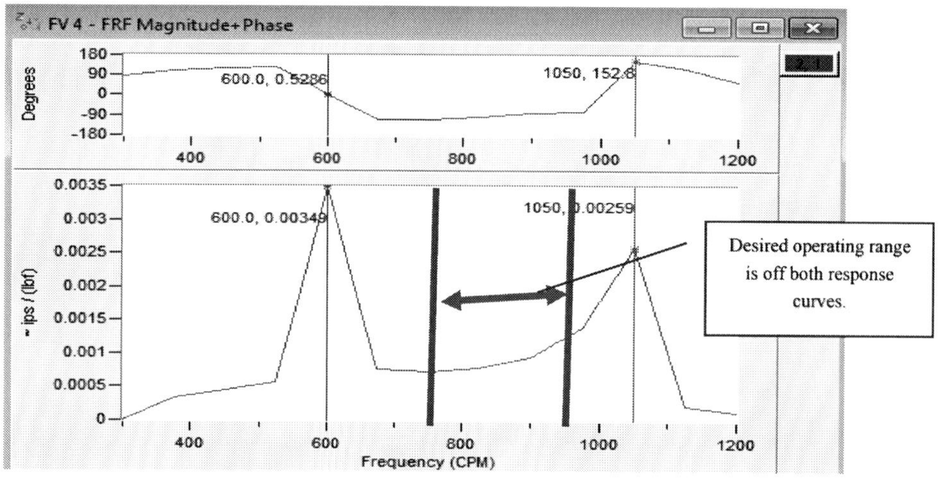

Figure 47: Natural frequencies after addition of mass

Final Results:

	Before Mass	After Mass	
In line Natural Frequency	1125	1050	
90 Degree Out Natural Frequency	785	600	
Maximum Response in 750-900 Range	14 mils	3.7 mils	74% Reduction

The solution was not ideal in that the amount of mass added had to be limited because if too much was added, then the in-line-with-the-discharge-pipe natural frequency would be drawn too near to the 900-rpm upper limit speed. It was a bit like threading a needle. Enough mass had to be added to get the 90 degree-out-direction natural frequency below the lower speed range, but not so much that it would draw the in-line-with-the-discharge-pipe natural frequency down to the higher speed value.

Vertical Pump with Mass added to reduce natural frequency

Figure 48: Photo of added mass

__Conclusions__:

It is recommended that prior to adding a VFD to a vertical pump that resonance testing be performed in the in-line-with-the-discharge-pipe direction and in the direction that is 90 degrees out from the discharge line. The very nature of vertical pumps is they have a large amount of mass (motor) mounted on a structure with large height to width ratio. This results in natural frequencies that are often near to or below running speed. Since the damping ratios of steel are low, operating near or on one of these natural frequencies can result in very high amplitudes of vibration. Since some VFDs can run above the normal line frequency, natural frequencies that are above the across-line operating speed can, in some cases, also be of a concern. This paper shows a post installation solution using some rough assumptions. The results were acceptable in that it was possible to add the right amount of weight to allow the pump to operate in the required speed range without exciting either the in-line-with-the- discharge-pipe direction or the 90 degree-out from the discharge line direction. It would, however, be advisable to consider building a computer-based model of a vertical pump prior to considering the installation of a variable speed drive. The model could then be utilized to test out various operating scenarios. A projected range of speeds that would be required for the analysis could be derived from the pump flow head curves and the system back pressures expected from the flows needed by the process. Make sure if non-VFD driven pumps are in the same header, that pump will not be driven too far back on its flow curve by the full speed pumps. Check pipe lengths to determine if there might be acoustical resonances that might get excited by variable vane pass frequencies. System hydraulics and structural natural frequency considerations must be carefully considered prior to the installation of variable speed drives, especially on vertical pumps.

3d Most Difficult Case: 500-Megawatt HP Rotor Mid Plane and Couple balance Shot

This case study involved a 500-Megawatt HP-IP Rotor with a high critical speed static unbalance and high running speed couple unbalance. During the startup of a 500-Megawatt turbine, the vibration at 1,420 RPM, when the rotor went through its critical speed reached a level of over 15 mils on the HP-IP rotor. The vibration was almost 7 mils when the machine reached 3,600 RPM. In order to gain information as to the correct balance approach, a Bently Nevada ADRE system was connected to the supervisory instrumentation. The location of the Key phaser and the zero mark on the shaft were verified. It was determined that the true phase angle was being detected by the left probe and that 90 degrees would have to be subtracted from the right-side proximity probe to get an accurate phase angle. The figures below show the resulting runout compensated Nyquist plots from the left probes at the bearing 1 & 2 locations. The plots show that when the HP-IP rotor reached the critical speed of 1,420 RPM the bearing 1 and 2 phase angles were nearly identical. This is what would be expected for a first mode response. The amplitude was 15.4 mils on bearing 1 and 13.1 mils on bearing 2.

Figure 49: 500-Megawatt HP Turbine Rotor

BEARING 1 RUNOUT COMPENSATE NYQUIST   BEARING 2 RUNOUT COMPENSATED NYQUIST

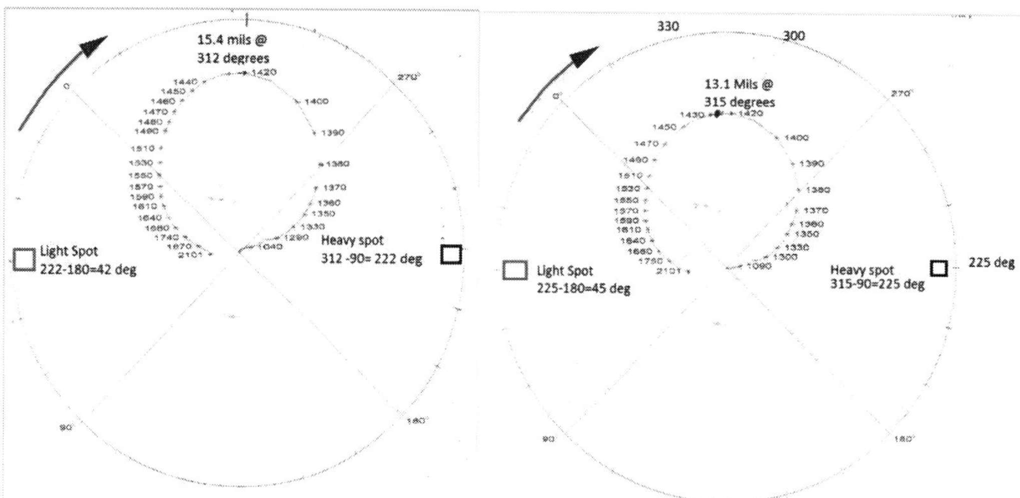

Nyquist plots showing 1st critical speed response

Figure 50: Run out compensated Nyquist plot showing 1st Critical response.

Based upon the large response at the critical speed, it was elected to try and solve the critical speed vibration issue, then attack the at-full-speed problem later. Based upon experience it was known that putting weight in the end planes would not reduce the critical speed response. The decision was therefore made to install the weights in the mid plane of the rotor. This required that both an inner cylinder and outer cylinder plug had to be removed. The temperature at the mid plane point was 1,000 degrees F and the pressure was 2,400 psi, so this decision was not one that could be made without critical considerations. In order to make sure the balance plugs did not get stuck half way in the threaded mid plane balance hole, they were set on the casing and allowed to heat up. Three balance plugs weighing 9 oz each were installed. Since the lag angle at the critical speed was 90 degrees, that number of degrees was subtracted from the critical speed phase to locate the source of the unbalance and then the correction weight was installed 180 degrees from that location. Following the installation of this mid plane balance shot, the critical speed response was below 5 mils. The at critical speed lag angle was calculated to be 104 degrees and the at critical speed sensitivity was 2.2 oz/mil.

The next step in the process was to bring the turbine to full speed. The readings on the HP-IP rotor were:

Bearing 1 Left Probe 6.3 Mils at 260 degrees

Bearing 2 Left Probe 6.9 mils at 80 degrees

Experience showed the sensitivity of couple unbalance(phase is at or close to 180 degrees difference from end to end) was 1.5 oz/mil at full speed and the couple lag angle was 68 degrees. 9.8 oz was added in plane 1 at 28 degrees and 9.8 oz was added in plane 2 at 200 degrees.

Final results:

Bearing 1 left probe 1.4 mils at 165 degrees

Bearing 2 left probe 2.4 mils at 0 degrees

This case history is a good illustration of what can happen when there are two types of unbalances in a rotor that goes through a critical and then at full speed approaches a second critical speed.

Discussion: In this case, the static (in phase) unbalance that was seen when the rotor reached its first critical speed had to be removed by a mid-plane balance shot. The author had attempted on a previous job to add a static (in phase) shot in the end planes of an identical rotor design. Thirty-six oz was added in the end planes and it had no effect on the critical speed vibration. That piece of knowledge meant a mid-plane shot was the only solution available to lower the critical speed response. Once the turbine was at the 3,600 RPM operating speed, the second mode was excited as evidenced by the 180-degree phase difference between the vibration on the two bearings. That problem was ideally suited to a couple shot (equal weights located 180 degrees from each other) in the end planes of the rotor. Use of the static couple derivation method of balancing described here has proven effective over the years at balancing large turbines.

2nd MOST DIFFICULT CASE: 500-Megawatt Turbine with Oil Whip

A 500-Megawatt steam turbine was experiencing vibration trips on its number 3 bearing.

The trips would occur if the turbine was shut down for a couple of hours, then a restart was attempted. If the turbine was started up immediately following a shut down, it would start up with no problems or if it was started up after being down for a couple of days it would also startup with not problems. It only tripped on vibration if it had been down between 2 to 8 hours.

MACHINE LAY OUT

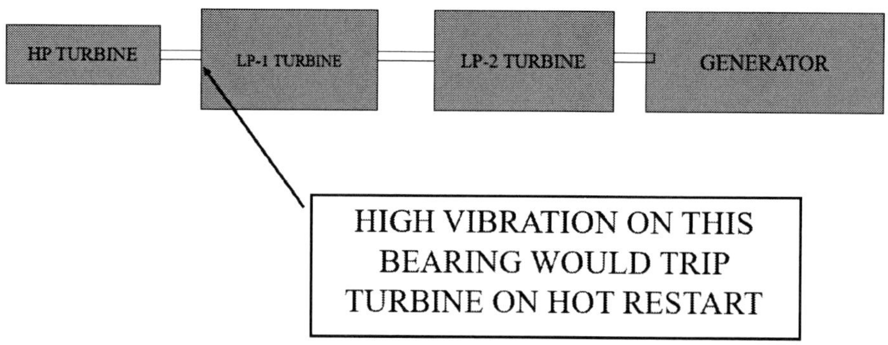

Below is how the spectrum appeared when the unit was about to trip.

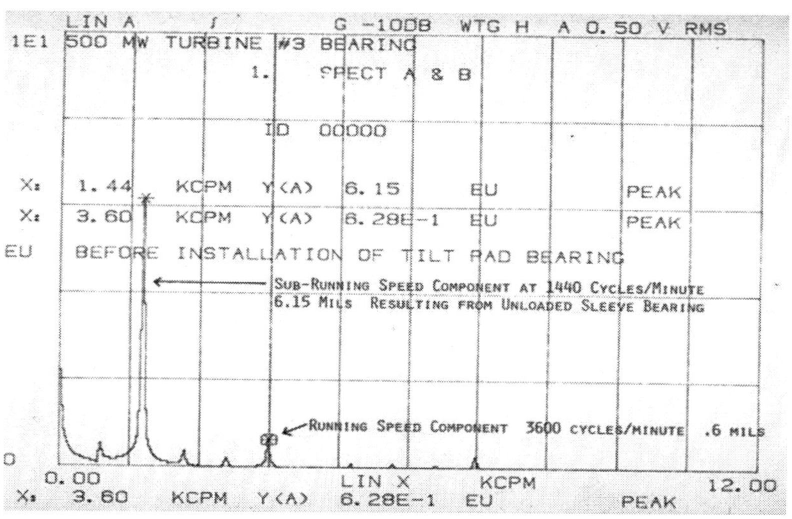

Figure 51: Spectrum showing 6.15 mils of oil whip induced 1440 cpm vibration.

Below is a map plot of what occurred during the startup.

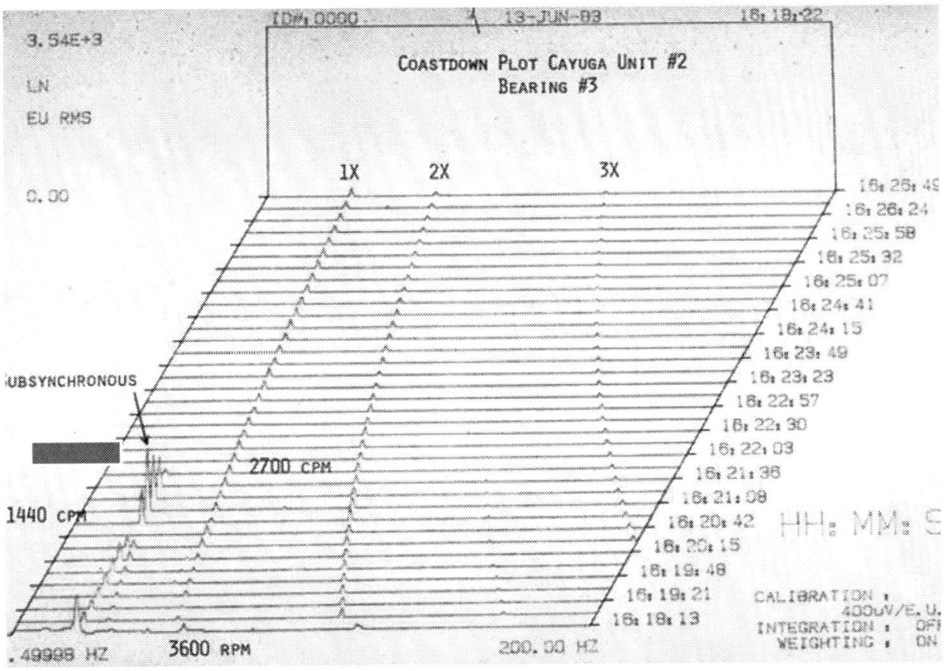

Figure 52: Map plot of development of oil whip

As the speed increased, the vibration was ok until the speed reached 2700-2800 RPM.

At that point, the 1440 cpm vibration would suddenly appear. As the speed increased, the spectral component did not increase in frequency with the speed. The frequency of the vibration matched the 1st critical speed of the High-Pressure rotor. Once twice the critical speed was reached, the 1440 cpm vibration would appear, then as the rotor speed increased, the vibration stayed locked onto the rotor's natural frequency. What was occurring was oil whip. This is a condition where the oil whirl locks onto the rotor's natural frequency.

The question was then why did this problem only occur on a hot startup and not a cold startup. The most likely source of this problem was unloading of the bearing which would therein point towards a change in the alignment. It is difficult to measure the alignment on a steam turbine, so some research was done to see if there was a way to measure vertical movement of a steam turbine's bearings. A technical paper was found where the Canadians had come up with a system that used Ethylene Glycol in a piping system to measure vertical bearing movement.

This was such a major problem that had cost the utility a large amount of money so it was decided to try and build this system to see if it could measure the vertical movement of the turbine's bearings.

ALIGNMENT TEST SYSTEM

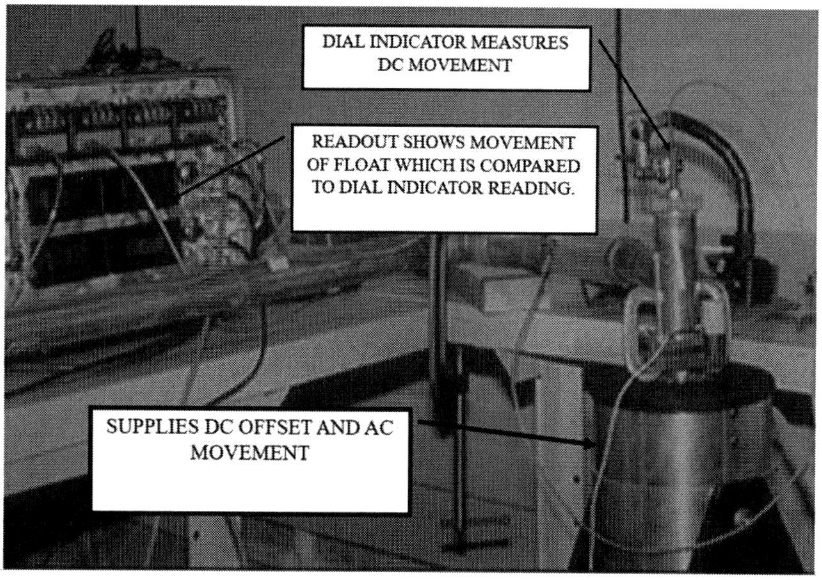

Figure 53: Lab test setup of glycol alignment system

Above is the test set up in the laboratory. In order to simulate the field situation, the copper tee containing the proximity probe and glycol was put on a shaker that was set to vibrate at 60 HZ. A DC signal was also input into the shaker to move the Tee upward at the same time it was vibrating. In order to test out the system, a dial indicator was installed to measure the actual motion of the Tee. The DC gap of the proximity probe was read at the same time. When the Tee was raised, the upward movement caused the glycol to flow out of the Tee and a float with a metal target would drop thus causing a change in the DC gap. The system proved to be quite accurate in measuring the vertical movement.

HEART OF SYSTEM

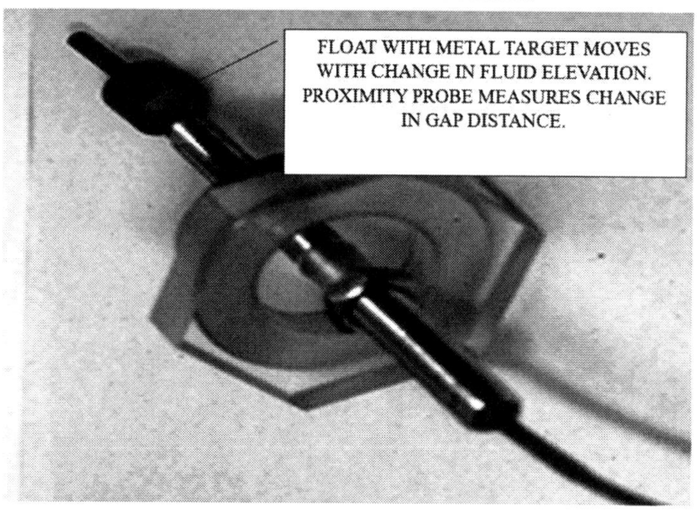

Figure 54: Proximity probe and float with metal disc

This photo shows the Tee on one of the turbine bearings. A total of four pickups were installed on the turbine. These were located on bearings 2,3,4 and 5.

Figure 55: Glycol system installed on 500 MW Turbine

One change was made to the Canadian design. That was to put a reference pickup on the foundation. The purpose of this pickup was to measure any changes in system level that might have been due to expansion of the glycol as it heated up. Any changes in the reference were therefore subtracted from any movements detected on the turbine.

Figure 56: Reference Tee

Graph showing elevation changes versus time

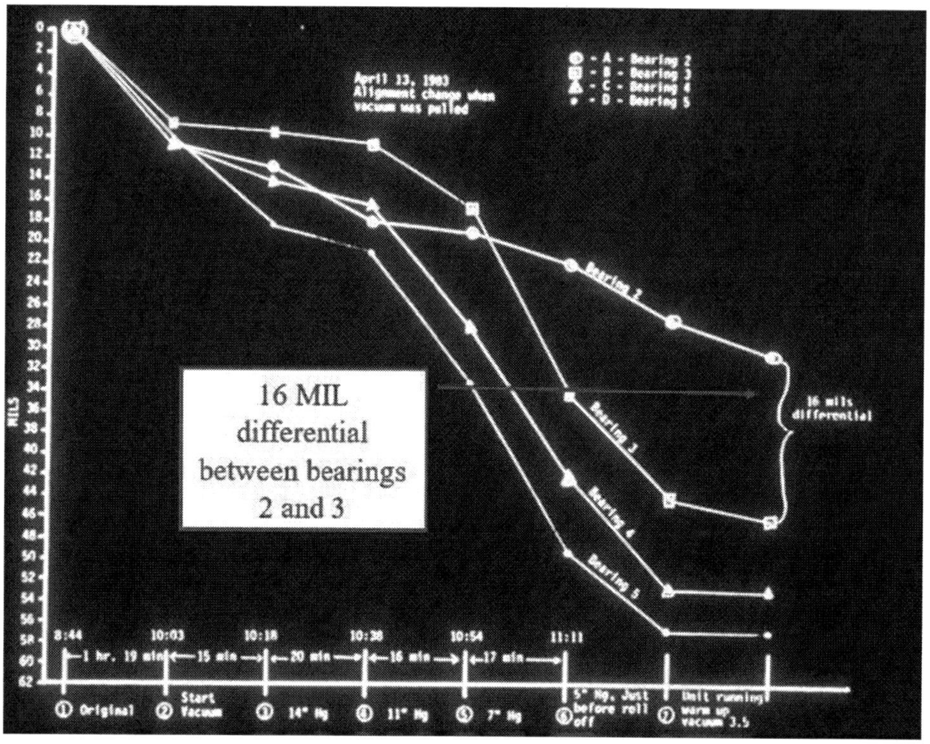

Figure 57: Graph of bearing differential movement during vacuum drawdown

Discussion:

Vacuum draw down combined with thermal differential growth unloaded bearing 2 caused it to go unstable. When both were cold, it would be stable. When both were hot, it would be stable. The problem occurred after a trip, when the thinner LP section
would cool down quicker than the thick HP section. This thermal differential when added to the vacuum draw down was so much the bearing went unstable. Bearing metal temperature readings confirmed this finding. The solution was to convert bearing 3 to a tilt pad design.

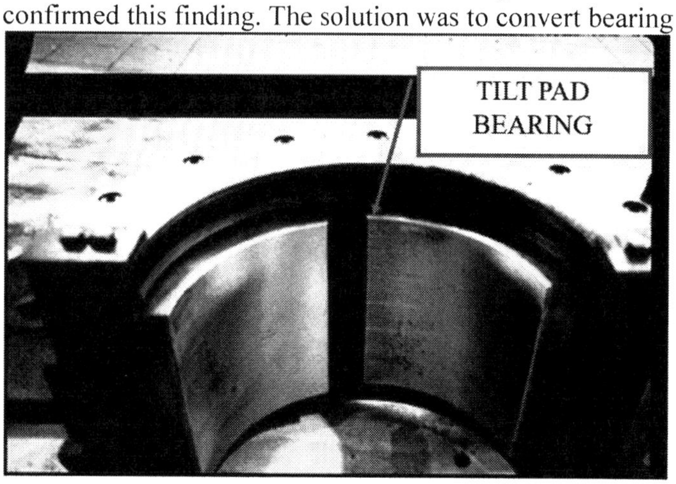

Figure 58: Tilt pad bearing installed on LP bearing 3

Test Results following conversion to a tilt pad bearing

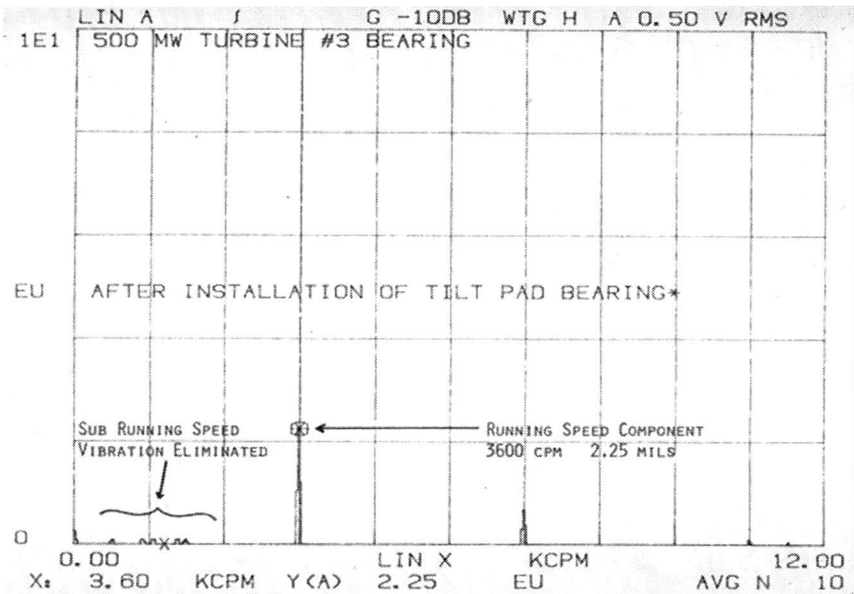

Figure 59: Spectrum after tilt pad bearing was installed

Following the installation of the tilt pad bearing, the problem was eliminated.

After this case history was published, tilt pad bearings started to be installed in the number 3 bearing location of some future turbines.

1St Most Difficult Case: High Pressure Core Injection Pump(HIPCI) at Boiling Water Nuclear Power Plant

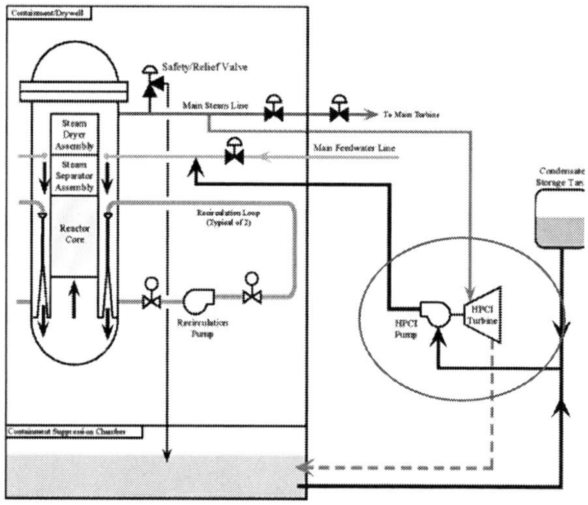

Figure 60: Diagram showing High Pressure Core Injection Pump

A HIPCI pump supplies cooling water to a BWR style nuclear reactor in the event of an emergency. It pulls water from the containment suppression chamber and pumps it into the nuclear reactor to cool the fuel rods. Even though the nuclear fission process is no longer present, the fission by-products still produce a significant amount of heat. This is therefore a very important pump that ensures the safe shutdown of a nuclear reactor. This pump is driven by a steam turbine. Previous testing in mils displacement had been the norm for this pump. When the plant started testing in velocity, this pump was found to be exhibiting very high vibration levels. The cause of the high amplitudes appeared to be at twice the speed of the pump. The vibration amplitudes had a large degree of variation.

Layout of HIPCI Pump

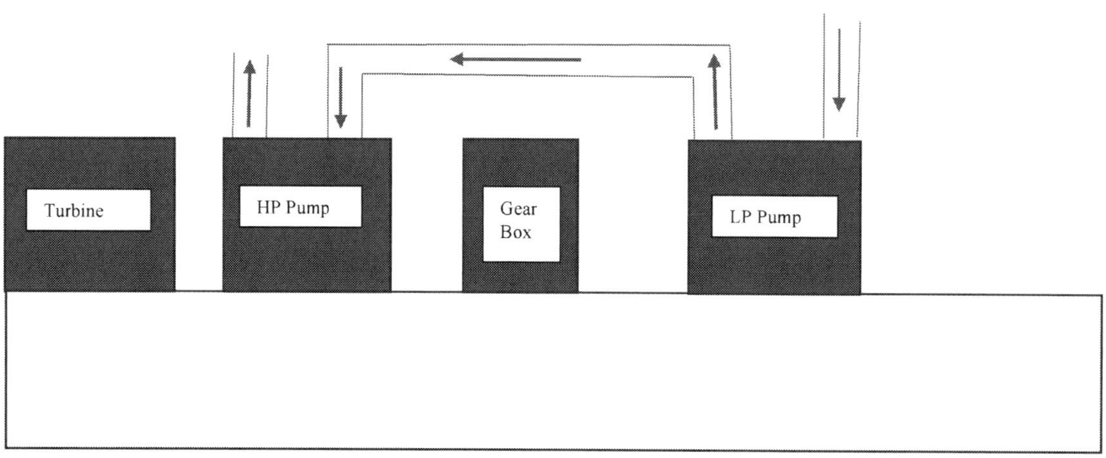

Figure 61: Layout of HIPSI Pump components

Spectrum From HP Pump inboard bearing - Both bearings had same levels

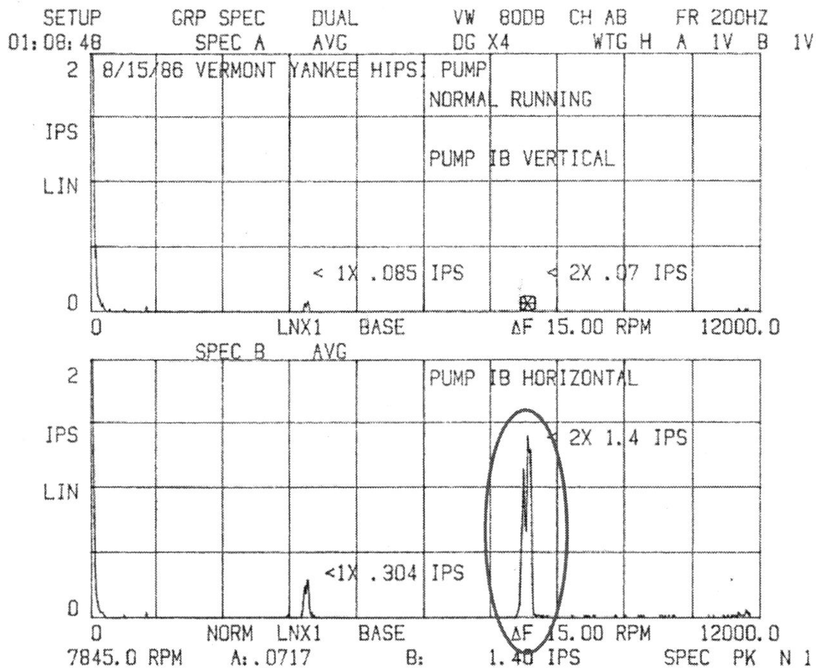

Figure 62: Vibration spectrum pump inboard bearing showing large H to V ratio

Plot below shows a change in speed of only 100 RPM cut the vibration in half.

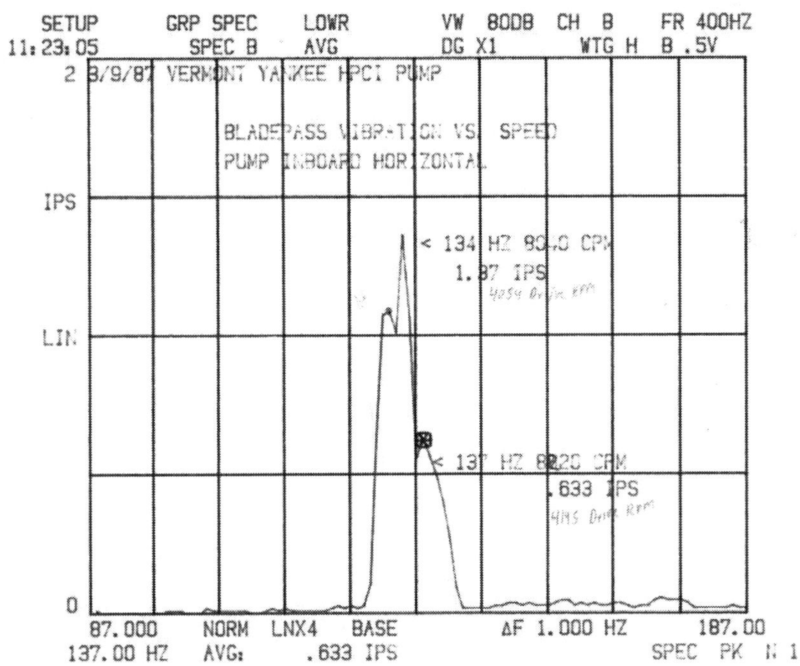

Figure 63: Peak hold plot showing speed dependency of vibration

The vibration in the horizontal direction was very high on both pump bearings with peak amplitudes of 1.2 to 1.4 in/sec. In the vertical direction, the amplitude at the apparent 2X running speed vibration was low at only .07 in/sec. The vibration was also highly speed dependent. Because of the large horizontal to vertical ratio and the significant speed dependency, resonance amplification was suspected. An impact test was performed on the HP pump case in the horizontal direction. A structural natural frequency was identified in the 7500-8000 cpm (125-133 Hz) range. Below is the mode shape of this natural frequency.

Mode Shape shows the presence of a rigid pivotal mode.

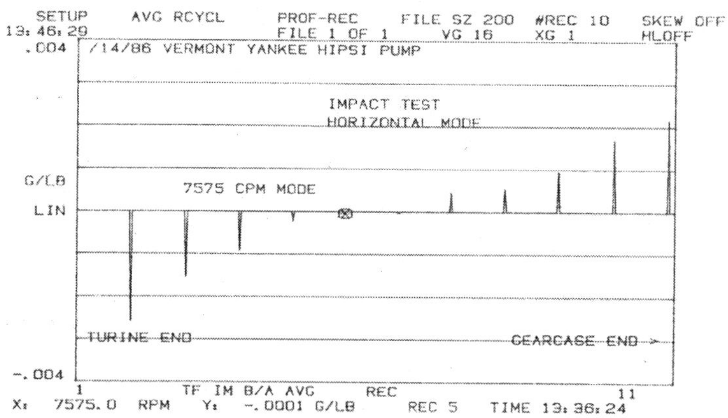

Figure 64: Mode shape of 126 Hz mode. High on ends zero in middle of pump

To confirm the presence of this mode synchronous time averaging was utilized to get a phase measurement. This measurement proved to be important in identifying the root cause of the vibration.

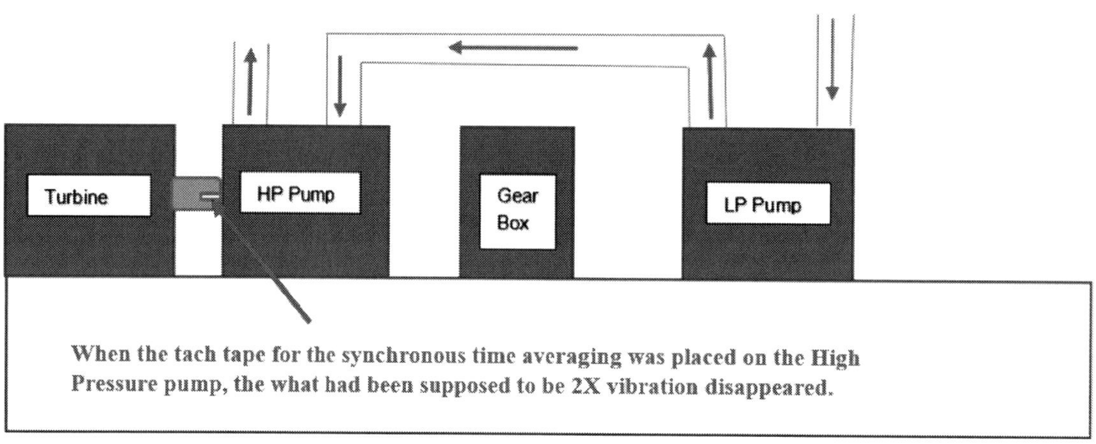

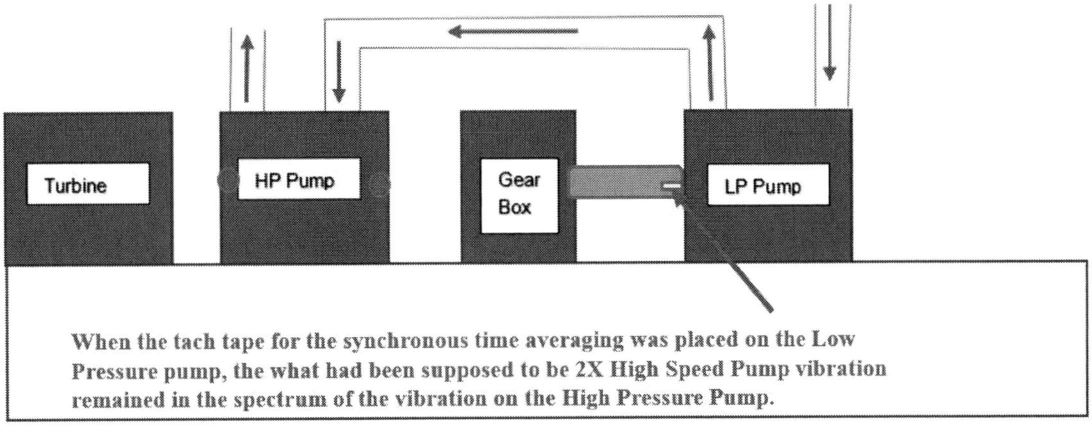

Figure 65: drawing showing placement of tach tape uses for synchronous time averaging

The reason for the high vibration was caused by several different issues.
1: The first was that the Low-Pressure Pump had a 4-vane impeller.
2: The gear box had a 2.01 reduction. When the speed reduction was combined with the number of vanes that made the vane pass vibration appear to be at almost exactly 2 times the turbine and HP pump speed. Since the turbine speed had some variation, that smeared the spectrum so it was not possible by looking at the spectrum to see the high vibration was not actually at 2X running speed.
3: As noted before, there was a horizontal structural natural frequency that fell within the 2X running speed range of the High-Pressure Turbine.
4: What made it possible for the pressure pulses in the Low-Pressure Pump to excite the High-Pressure pump was an acoustical natural frequency in the pipe connecting the two pumps. It matched the blade frequency of the Low-Pressure Pump.

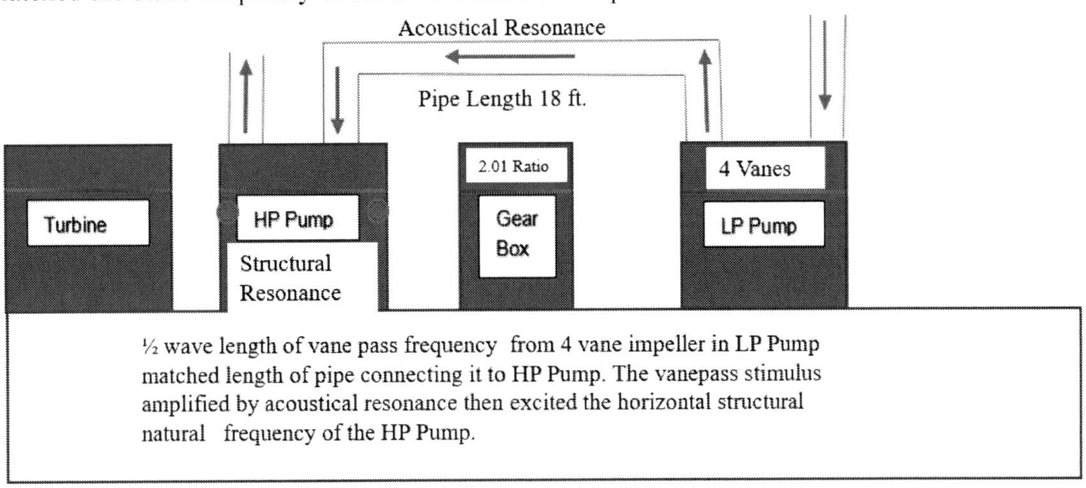

Figure 66: Location of piping resonance

<u>Solution:</u> The Low-Pressure Pump was changed from a 4-vane impeller to a 5-vane impeller. Odd numbered higher vane impellers tend to produce lower pressure pulsations. In this case the important gain was the increased vane pass frequency moved the vane pass excitation away from both the acoustical natural frequency and the structural natural frequency of the High-Pressure Pump.

Things that made this a very difficult case.

1: This pump was located at a nuclear power plant. There is always a lot of pressure when working at one of these facilities. Numerous individuals were packed in the room observing. Engineers, operations personnel, maintenance personnel, management and Nuclear Regulatory Commission personnel were all crammed in the room watching the testing.

2: The room was seven stories under the ground and since the turbine was always on hot stand by the room was very hot.

3: You only got one run in a month when operations tested the pump.

4: When alignment measurements were made, the turbine to the pump alignment was found to be out .058" so it led to a false conclusion the first trip that alignment was the source of the problem. The alignment issue was resolved and no improvement was observed. This was discouraging and resulted in a 2nd site visit.

5: The 2.01 reductio ratio of the gearbox combined with the 4 pump vanes was very close to the 2x running speed of the High-pressure pump.

6: The turbine varied in speed. That combined with item 5 made it impossible to see in the spectrum that the main frequency component was vanepass of the low-pressure pump instead of 2X running speed of the high-pressure pump.

7: There a structural resonance problem being excited by an acoustical resonance problem. This greatly complicated tracking down the source of the problem back to the low-pressure pump.

The key to identifying the problem was that the synchronous time averaging test showed the vibration was phase locked to the Low-Pressure pump instead of the High-Pressure Pump.

VIATC TRAINING CONFERENCE & EXPO

NIAGARA FALLS, NEW YORK JULY 26 -28, 2023

CASE STUDIES OF VIBRATION OF LOW-SPEED, LOW VANE-COUNT PUMPS

Robert J. Sayer, PE
Owner
Applied Structural Dynamics
Westerville, Ohio
sayerinc@bright.net

Robert Sayer, PE is the Past President of the Vibration Institute and Owner of Applied Structural Dynamics (Westerville, Ohio, USA). He has over forty-five years of experience as a mechanical and structural engineering consultant to a wide range of industries including pulp & paper, mining & metals, fossil-fuel & nuclear power generation, petrochemical, food processing, glass manufacturing, and aerospace & defense. Mr. Sayer holds engineering degrees from Ohio University (BSCE), Cleveland State University (MSIE), and Purdue University (MSCE). He is a regular contributor at technical conferences, instructor for the Vibration Institute's CAT IV training, and has published over 50 technical articles and co-authored 2 chapters in Art Crawford's "The Simplified Handbook of Vibration Analysis". He is a recipient of the Jack Frarey medal given by the Vibration Institute in recognition of outstanding contributions to the field of modal analysis.

ABSTRACT:
Sludge pumps used for the transportation of raw sewage at wastewater treatment plants, for booster applications along a wastewater distribution system, or for tailings distribution at mining plants typically are designed with a small number of vanes and operate at a relatively low speed. Vertically mounted single-stage centrifugal pumps are the most common type used for wastewater treatment applications. These pumps are usually supported on a foundation in a dry well and driven by a motor located at an elevation significantly higher than the pump (10 – 20 meters). A drive shaft is supplied between the motor and the pump and typically coupled thereto using universal joints or other flexible methods. The combination of lower speeds with lower vane count results in vane-pass pulsation frequencies that can potentially align with natural frequencies of the motor pedestal, drive shaft and/or pump foundation. For example, a 3-vane pump operating at 400 rpm will produce vane-pass pulsations at 20 Hz. Two separate cases will be used to demonstrate potential resonance issues with these types of pumps and will include the discussion of corrective action plans developed to minimize resonance issues.

KEY WORDS: Centrifugal pumps, impact testing, natural frequency, operating deflection shape, resonance, shaft vibration, shaft stick, structural dynamics, vane-pass pulsations.

CASE STUDY #1: Premature Bearing Failures for Drivetrain for a Raw Sewage Pump

Figure 1: Photo of Sludge Pump

A wastewater treatment plant installed four new raw sewage pumps. The pumping arrangement consisted of one 40 hp, 900 rpm pump (Figure 1) that operated most of the time and three 100 hp, 1,200 rpm pumps that were in standby mode for peaking purposes. All pumps discharged into a common header. Figure 2 shows the drive-train arrangement. The drive motor was located on the floor above the dry well. Double drive shafts were used to connect the motor to the pump. The motor was coupled to an upper drive shaft with a universal joint. Another universal joint coupled the upper drive shaft to the lower drive shaft. A third universal joint coupled the lower drive shaft to the input shaft of the pump. This drive train would be unstable without a guide bearing between the upper and lower shafts.

Thus, an intermediate guide bearing was located just above the universal joint between the upper and lower shafts. The primary 40 hp pump had three failures of the drive shaft bearing within the initial six months of operation. The pump was a vertically mounted, single- stage, centrifugal type with two vanes. The pump was located in a dry well adjacent to the wet well that contained incoming wastewater.

The local area serviced by the wastewater treatment plant still had some old, combined sewers that conveyed both storm and sewage to the plant. Since demand for this type of system can vary drastically, all pumps were provided with variable frequency drives. It was imperative that the primary pump be able to operate over the entire useful speed range of the VFD. Programming exclusion zones into the controls to prohibit operation at speeds where high vibration existed was not a viable option for any corrective action.

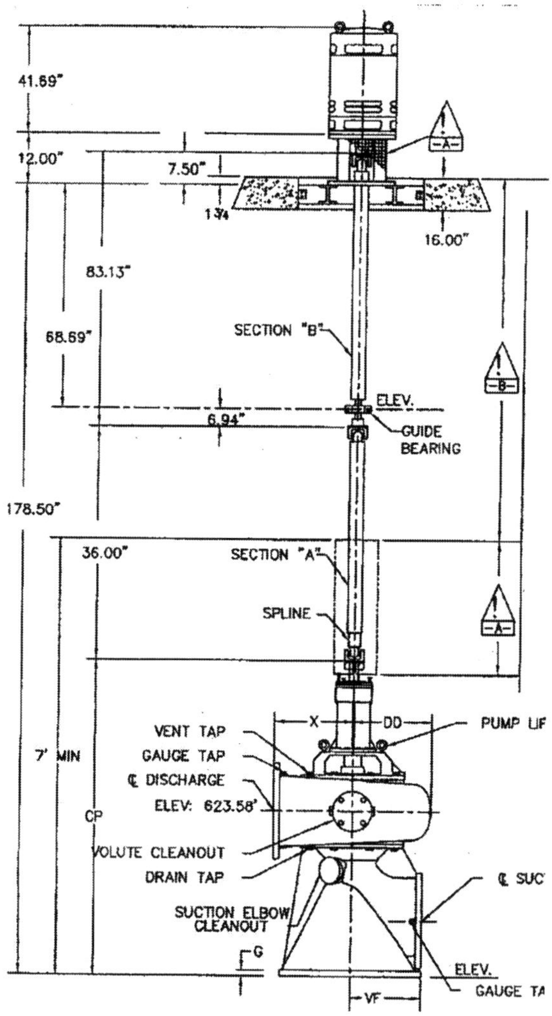

Figure 2: Pump & Motor Arrangement

Vibration data acquired from the guide bearing that had experienced several premature failures did not provide any indication of excessive vibration across the entire operating frequency range of the pump. However, at an operating speed of around 840 rpm (which occurred at a VFD setting of 93% full speed) there was a noticeable increase in vibration that reduced as the pump speed was increased to 100%. The speed sensitivity indicated possible resonance.

In cases where a speed sensitivity and possible resonance exists, but vibration levels are not excessive in the pump, the resonance is often associated with a rotating component of the system and not the stationary supports. To investigate this possibility, vibration data was acquired from one of the drive shafts, below the universal joint of the bottom shaft.

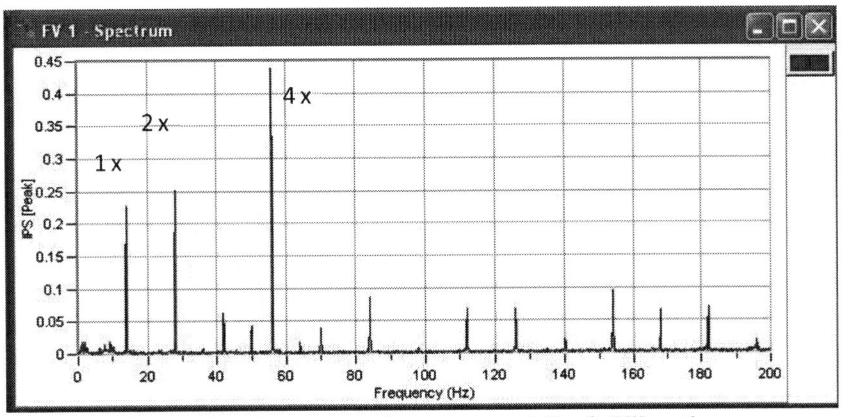

Figure 3: Frequency Spectrum of Shaft Vibration

Figure 3 is a frequency spectrum of the drive shaft vibration that was obtained using a vibration transducer attached to the end of a shaft stick. The vibration of the shaft had significant responses at operating speed (840 rpm = 14.0 Hz), 2x operating speed (28 Hz), and at 4x operating speed (56 Hz). The response at 1x operating speed could be contributed to dynamic forces produced as a result of unbalance. The response at 2x operating speed was the result of vane pass forces since the sludge pump had only two vanes (typical for these types of pumps). The response at 4x is a harmonic multiple of vane-pass, produced because the vane-pass force, although periodic, was not purely harmonic. Resonance at the 4x (56 Hz) frequency was suspected since the dynamic force produced by the harmonic of vane-pass should have been less than the primary vane-pass force, yet the vibration at the harmonic was significantly larger.

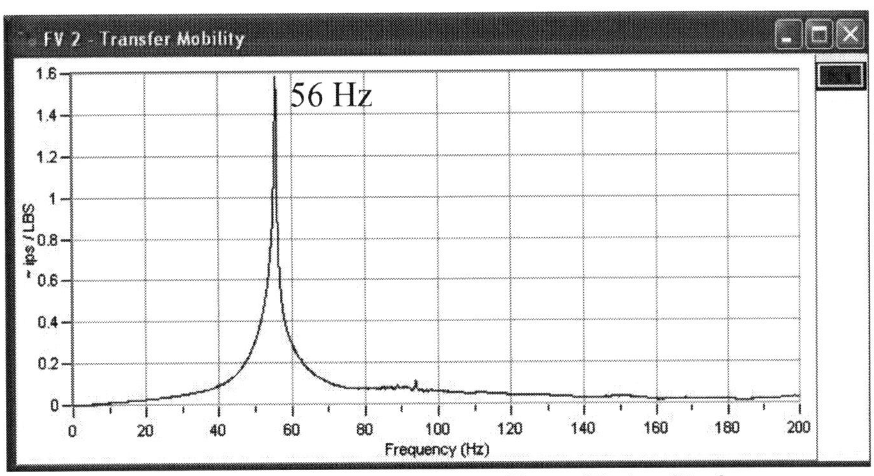

Figure 4: Natural Frequency Test of Drive Shaft

Figure 4 is the result of an impact test of the drive shaft. A lightly damped, sensitive natural frequency was clearly identified at 56 Hz. A complete experimental modal analysis (EMA) was performed to determine the mode shape associated with the natural frequency. Figure 5 is the mode shape that was determined by a curve-fit of the modal data. It is clearly the principal flexural mode of the drive shafts.

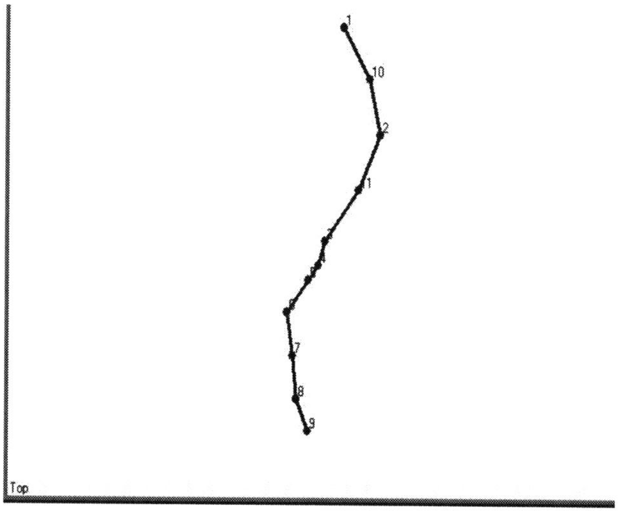

Figure 5: ODS of Drive Shaft

The flexural deflection of the upper shaft is out-of-phase with the lower shaft. This results in an s-shape pattern for the total drive train. The location of minimum deformation corresponds to that of the guide bearing. This provides insight into how the bearing could have been subjected to loadings that were not detectable by excessive vibration in the pump.

A numerical finite element analysis (FEA) of the drive train, including the structural support of the guide bearing, was performed to provide a basis upon which proposed modifications could be evaluated. Figure 6 contains the results of the FEA for the original drive train. The numerical analysis predicted the natural frequency at 56.2 Hz compared to the 56 Hz from the impact test. This indicated that the FEA provided a very good approximation of the actual conditions.

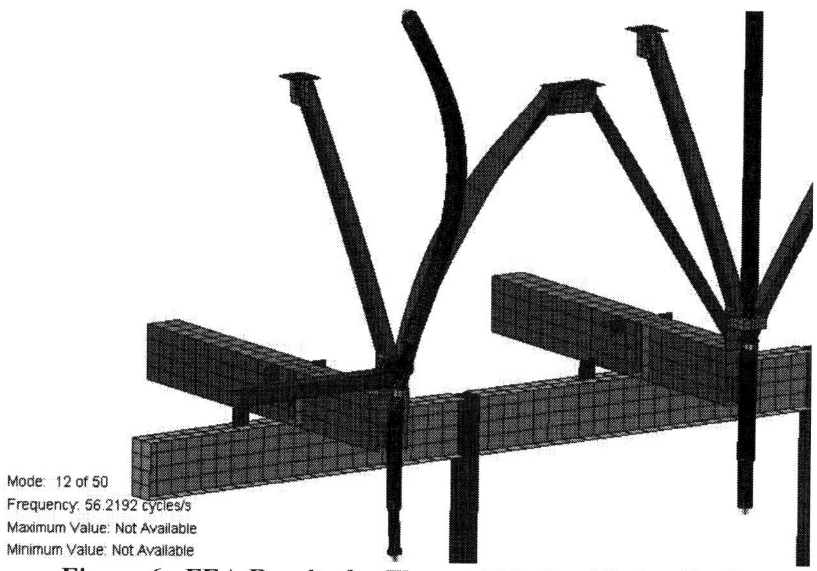

Figure 6: FEA Results for Flexural Mode of Drive Shaft

The FEA was then used to develop a final corrective action plan which included increasing the diameter of the drive shafts. This increased the stiffness with only a minor increase in mass. The structures supporting the guide bearings were also reinforced. The natural frequency of the modified system was increased to 68 Hz. This provided an acceptable separation margin between the frequency of the 4x dynamic force (60 Hz) at full speed (900 rpm) and the natural frequency.

Another structural dynamic concern for vertically mounted sludge pumps with the motors supported above on the electrical supply and control floor is the natural frequency of the motor itself. These motors are typically supported upon a pedestal that raises it above floor level. Similar to a vertical turbine pump, the motor and its support pedestal will have sway modes of natural frequency. The support pedestal is mounted on a structural floor that can also affect the natural frequencies of the sway modes.

Figure 7: Drive Motors for Pumps

Figure 7 shows the motor arrangement for the pumps in this case study. The smaller pump and motor were all that was required for normal operating conditions. The three larger pumps are for the upset condition of the PMP event. These larger pumps operate using VFD's, with a maximum motor/pump speed was 1,200 rpm (20 Hz). Vibration levels of these larger motors increased drastically as the pumps approached full speed.

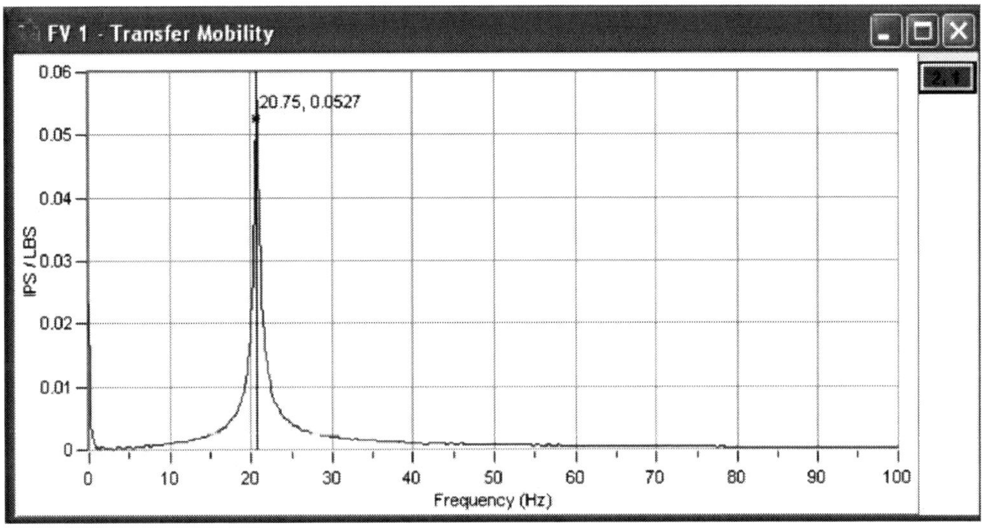

Figure 8: Natural Frequency Test of Motor on Pedestal

Resonance was suspected since the vibrations appeared to be frequency sensitive. An impact test was performed to investigate natural frequency of the motor/support assembly, the results of which are shown in Figure 8. It revealed a natural frequency at 20.7 Hz. This natural frequency was very close to the maximum operating frequency of the motor.

Since these larger units were stand-by pumps used for infrequent peaking demand, it was determined that the full-speed operation would only be required for the most extreme demand that would occur statistically once every 100 years. It was decided that the exclusion zones would be programmed into the VFD control that would prohibit operating any individual motor near the natural frequency, with manual override provided for the infrequent maximum demand period. In other words, the stand-by motors/pumps would be allowed to operate in the resonant condition, but only for short periods of very infrequent demand periods.

CASE STUDY #2: Vertical Mounted Sludge Pump (Base Plate/Foundation Flexibility)

Figure 9: Photo of Sludge Pump

Six new sludge pumps (Figure 9) were installed at a waste-water treatment plant. The pumps were vertically mounted, single-stage, three-vane, centrifugal types driven by a VFD control. The design operating speed range of the pumps was between 320 - 400 rpm. Based upon this information, the vane-pass pulsation frequency could be expected to vary between 16 Hz (at low speed) and 20 Hz (at full speed).

Two sets of bearings were located within the bearing housing connected to the top of the pump casing. Of note was the fact that the large volute pump casing was supported by four structural steel pedestals, located ninety degrees apart from each other. These pedestals were cap screwed to a base plate which was supported from two concrete piers. Two of the support pedestals were located directly above the concrete piers, while the other two pedestals were located on the base plates approximately mid-span between the concrete piers. The support provided by the foundation would be significantly different in one direction versus the other.

Each pump was driven by a 1,250 hp refurbished motor located at a floor level approximately 53 feet above the dry well floor that the pumps were located on. A double drive shaft, made of composite tube sections, was coupled to the motor shaft on the upper end and the pump shaft on the lower end. An intermediate bearing, located on a floor between the motor and pump, provided lateral constraint to the drive train. The lower part of the drive shaft can be seen coupled to the top of the pump in Figure 9.

The acceptance criterion listed in the contract specifications for these pumps was that the overall vibration level could not exceed 0.31 ips rms below 370 rpm and 0.24 ips rms above 370 rpm. Vibration data acquired from the top of the bearing housing, in the direction toward the pedestal located above the open space between concrete piers, indicated that the overall vibration exceeded specification at 340 rpm and 400 rpm, but was lower than specification at 320 rpm, 360 rpm and 380 rpm. In the other direction, the vibration was acceptable at all operating speeds.

Figure 10 contains the transfer function for an impact test of the pump. The location of the impact was at the top of the bearing assembly in a direction that would cause the main response of the pump to be resisted by the pedestals that were located between the concrete piers. A natural frequency was evident at around 21.75 Hz. This natural frequency would be sensitive to vane-pass pulsations at full speed and could be the reason for elevated vibration at full speed. But what was the reason for the excessive vibration at 340 rpm (5.67 Hz)? The vane-pass pulsation frequency would be 17 Hz, far removed from 21.75 Hz.

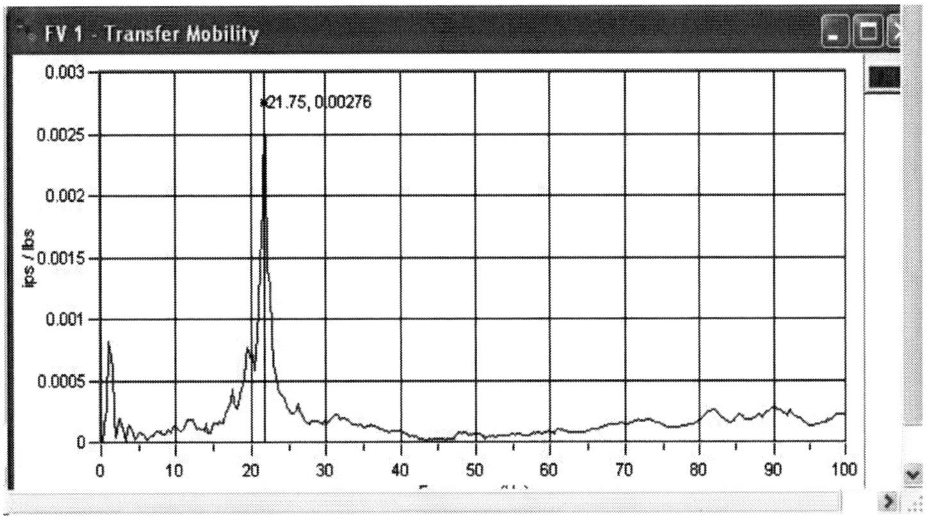

Figure 10: Natural Frequency Test of Bearing Housing

Dynamic pressure pulsation data was acquired from the discharge pipe to determine the nature of the vane-pass pressures. Frequency spectra for the pressure pulsations are included in Figure 11 through Figure 15. At lower flow rates (320 - 360 rpm), the main pulsation energy occurred at a sub-harmonic frequency (0.75x) and harmonics thereof. These sub-harmonic flow related forces subsided as the rotating speed of the pump increased (380 - 400 rpm).

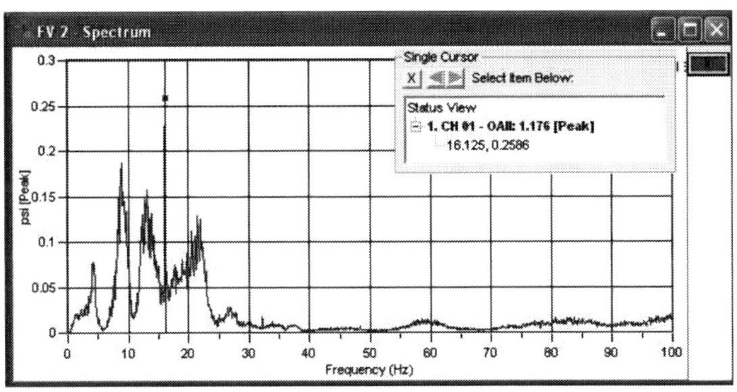

Figure 11: Pump Speed = 320 rpm (5.33 Hz) Frequency Spectrum

Pulsation Data @ 320 rpm
Overall Level = 1.18 psi
Level @ 3x = 0.25 psi
Other Broad Band Content @ 0.75x, 1.5x, and 2.25x.

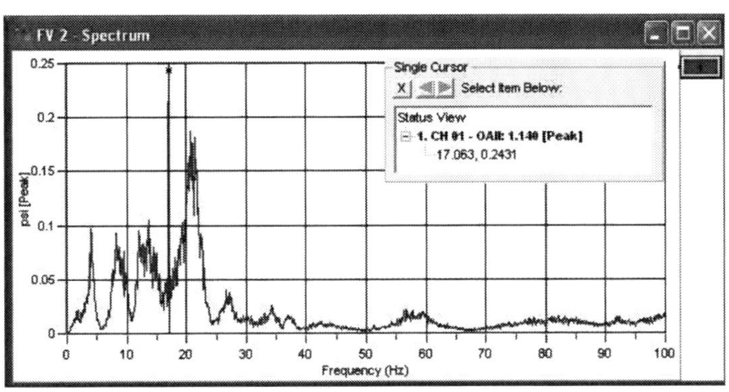

Figure 12: Pump Speed = 340 rpm (5.67 Hz) Frequency Spectrum

Pulsation Data @ 340 rpm
Overall Level = 1.14 psi
Level @ 3x = 0.24 psi
Other Broad Band Content @ 0.75x, 1.5x, 2.25x, and 3.75x.

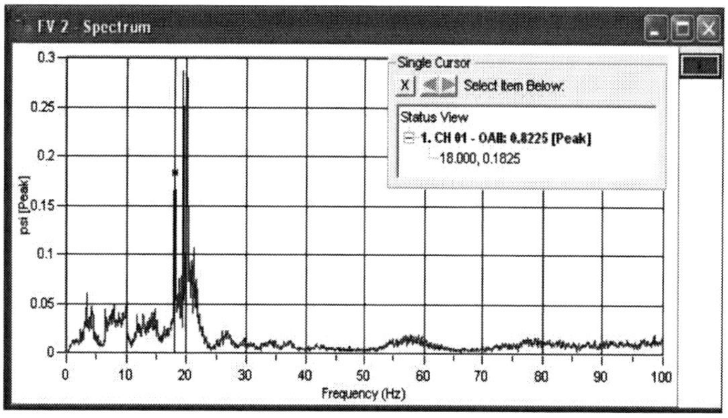

Figure 13: Pump Speed = 360 rpm (6.00 Hz) Frequency Spectrum
Pulsation Data @ 360 rpm
Overall Level = 0.82 psi
Level @ 3x = 0.19 psi
Broad Band Content still preseent but fading.

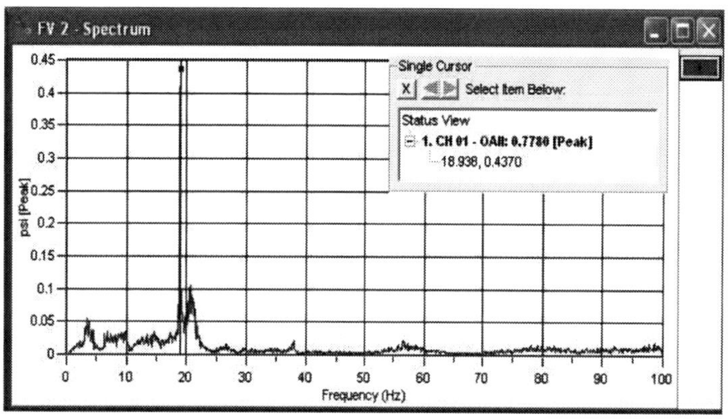

Figure 14: Pump Speed = 380 rpm (6.33 Hz) Frequency Spectrum
Pulsation Data @ 380 rpm
Overall Level = 0.78 psi
Level @ 3x = 0.44 psi
Broad Band Content Very Small

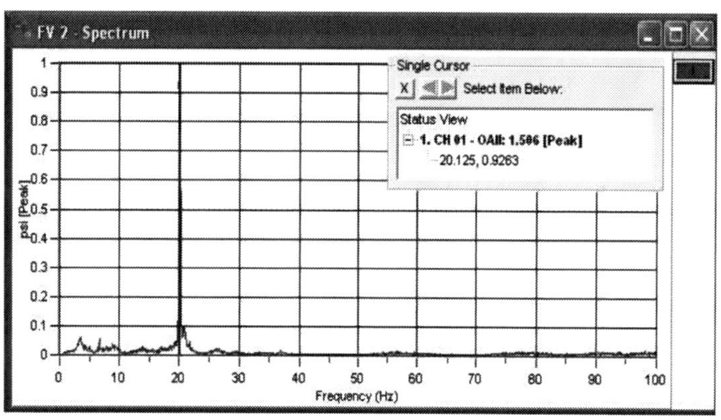

Figure 15: Pump Speed = 400 rpm (6.67 Hz) Frequency Spectrum
Pulsation Data @ 400 pm
Overall Level = 1.55 psi
Level @ 3x = 0.93 psi

At 380 rpm and 400 rpm, the pulsation spectrum was dominated by a response at vane-pass pulsation frequency. The sub-harmonic and its multiples were almost non-existent. The vane-pass pulsation frequency was very close to the natural frequency. This was probably the reason for elevated vibration levels at 400 rpm.

Figure 16 contains a frequency spectrum of bearing vibration data acquired from the top of the bearing housing in the sensitive direction with the pump operating at full speed. The overall average peak vibration level was 0.48 ips, with almost all vibration concentrated at vane-pass frequency. The nature of the vibration spectrum resembled that of the pulsation spectrum, supporting the diagnosis that the vibration was caused by vane-pass pulsations. The vibration levels clearly exceeded specification by a significant amount.

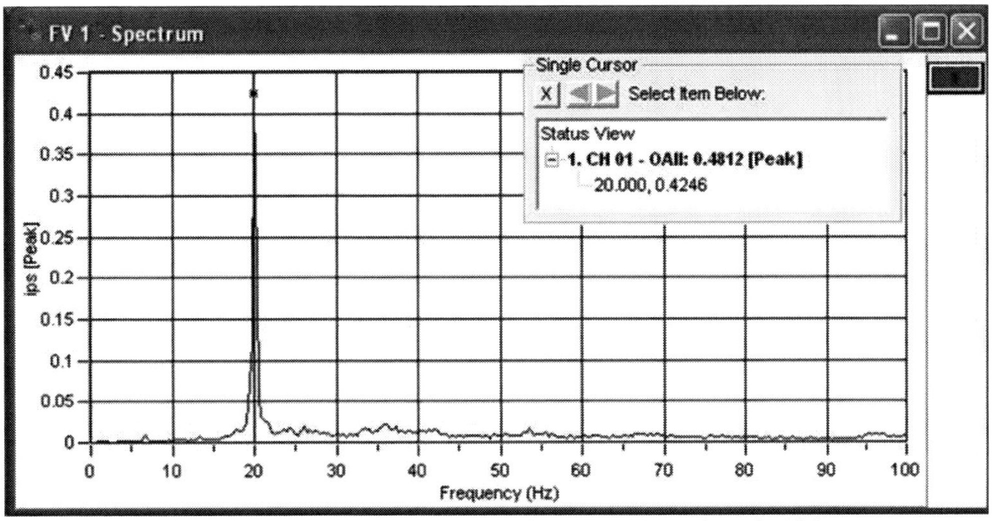

Figure 16: Frequency Spectrum of Pump Vibration @ 400 rpm

At 340 rpm, a broad band harmonic pulsation at 3.75x operating speed (21.25 Hz) was very close to the natural frequency (21.75 Hz) - see Figure 12. The pulsation at 3.75x was suspected as being a higher harmonic multiple of sub-harmonic force produced by the pump impeller at 0.75x operating speed. Pumps and fans can develop lower frequency pulsations due to rotating stall. Stall vortices are shed at a sub-harmonic frequency, typically occurring somewhere between 0.60 – 0.75x rotational speed of the pump. Stall pulsations can be detected using a dynamic pressure sensor installed in the piping which was what was done in this case.

At low flow conditions, the velocity component of the fluid at the leading edge of a vane reduces. The velocity component, due to the rotation of the impeller, remains the same. This results in an angle-of-attack that produces vortex shedding and an interruption to the flow in the passage between vanes.

In centrifugal fans and pumps, the vortex formed by the stall can setup a pattern where it is suctioned back into the rotor. The stall cell travels counter to the rotation of the impeller. This is the reason that the pulsations produced by the rotating stall condition are developed at a subharmonic frequency. Because of the interruption to flow within the impeller, it affects the build-up of vane pass pulsation pressures. Note that in the frequency spectra for pulsations acquired at 320 rpm, 340 rpm and 360 rpm, when the apparent stall condition was active, the magnitude of pulsations at vane-pass were very low. There was a distinct significant increase in vane-pass pulsation pressure after the stall condition was removed (380 – 400 rpm).

Figure 17 contains a frequency spectrum of bearing vibration data acquired from the top of the bearing housing in the sensitive direction with the pump operating at 340 rpm. The overall average peak vibration level was 0.56 ips, dominated by broad band energy around the 3.75x harmonic. There was only a small amount of vibration (0.044 ips) at vane-pass frequency. Once again, the nature of the vibration spectrum resembled that of the pulsation spectrum, supporting the diagnosis that the vibration was caused by a harmonic multiple of the rotating stall frequency. The vibration levels clearly exceeded specification by a significant amount.

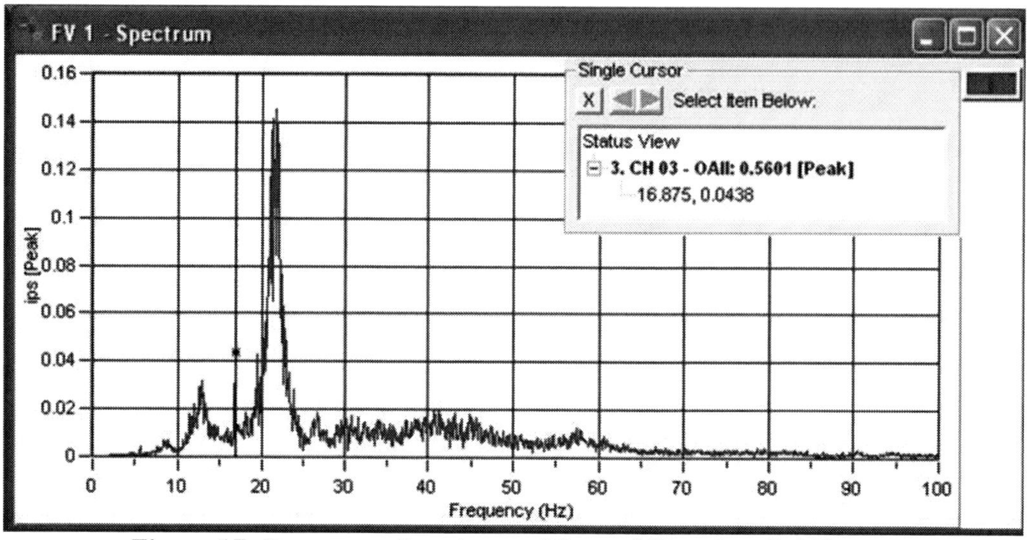

Figure 17: Frequency Spectrum of Pump Vibration @ 340 rpm

An operating deflection shape (ODS) test was performed to investigate the degree of participation of the foundation and base plate in the sensitive direction. The ODS was done at full speed. The ODS analysis provided information regarding movement during operation that was made into an animation. This animation was extremely helpful in defining areas of structural weakness. In cases of resonant excitation, the operating deflection shape animation will provide an estimate of the mode shape for the natural frequency that is being excited. Since it was determined that a natural frequency was being excited, the ODS for this case was also the mode shape. Data was acquired from eight locations on the pump bearing assembly, forty locations on the base plate (including four locations at each post that supports the volute), and eight locations on the concrete piers.

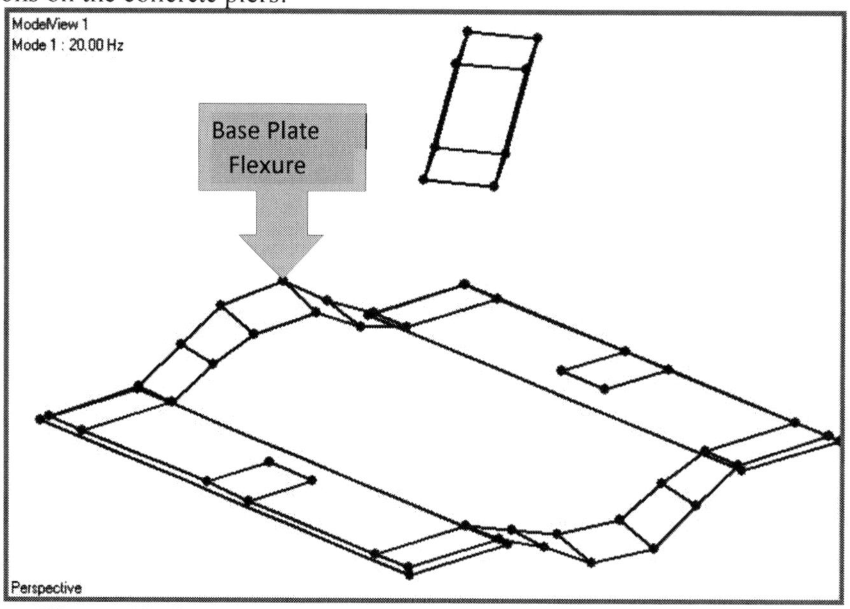

Figure 18: Snapshot of ODS Animation @ Full Speed (400 rpm)

Figure 18 is a snapshot of the ODS animation displayed to a linearly exaggerated scale. The animation clearly indicates that the vibration of the concrete foundation does not contribute to the vibration of the pump bearing assembly. A significant percentage of the bearing vibration is caused by the flexural deformation of the base plate between the foundation piers.

Figure 19: Photo of Structural Modification

Since the ODS animation approximated the mode shape for the natural frequency, anything done to increase the stiffness of the system at locations of maximum participation would increase the natural frequency and make it less sensitive to vane-pass pulsation forces. The ODS clearly identified the flexural deformation of the base plate as being the point of maximum modal participation. Based upon the ODS, a supplemental support (Figure 19) was added underneath each of the volute support pedestals. This resulted in an increase in the natural frequency from 21.75 Hz to 29 Hz. All vibration levels met the specification after the implementation of this modification.

VIATC | TRAINING CONFERENCE & EXPO

NIAGARA FALLS, NEW YORK JULY 26 -28, 2023

PRESSURE PULSATION RELATED FIXED EQUIPMENT VIBRATION

Gyorgy Szasz, PhD & Lawrence M Matta PhD, PE, CFEI
Stress Engineering Services, Inc.
13610 Westland East Blvd, Houston, TX 77041
gszasz@stress.com, lmatta@stress.com

Dr. Szasz has 26 years of experience with hands-on vibration testing, measurement and analysis. He has performed an extensive amount of field work globally in various power plants (including hydro, fossil and nuclear), refineries, pipeline stations, LNG facilities, offshore platforms, etc. He is a certified category IV vibration analyst, member of ASME, president of the VI board of directors, and a former member of the ASME Technical Committee on Sound and Vibration.

Dr. Matta's recent work has included vibration and acoustic analysis, unsteady flow phenomena, and piping and pipeline failures. He is a licensed engineer in Texas and a certified fire and explosion investigator. He specializes the areas of fluid mechanics, thermodynamics, heat transfer, and combustion. He has analyzed unsteady and pulsing flows in process and drilling equipment and vortex induced vibrations of pipelines. He has experience with pulse combustion and combustion instabilities, rockets, incineration, and commercial dryers.

ABSTRACT:
This paper presents a comprehensive overview of piping vibrations induced by internal pressure pulsations in piping systems. High amplitude pressure pulsations can induce intense vibrations in the piping system, leading to fatigue failures of the piping, supporting structure and even potential damage to the attached equipment. The paper aims to help with identification of the root causes of these piping vibrations and proposes effective mitigation strategies to reduce their intensity. Both low frequency and high frequency pulsations are visited. Some problems require analysis carried out using theoretical modeling, which can be tuned based on field measurements to accurately capture the dynamic pressure behavior in the piping system. Mitigating the vibrations may require eliminating or reducing the source of the pressure pulsations, optimizing the piping layout to minimize amplification, introducing damping, and modifying the support structure. The information provided in this paper is aimed at providing valuable insights for engineers and equipment owner operators to prevent piping failures due to pressure pulsation-induced vibrations.

KEY WORDS: pressure; pulsation; piping; pressure vessels; narrowband; broadband; vibration.

INTRODUCTION:

Dynamic pressure variations are, to some degree, always present in process piping conveying a fluid. Turbulence in the flow typically results in pressure oscillations that are random in nature. Flow through unrestricted piping generally results in low frequency broadband pressure variations, while flow through restrictions such as orifice plates and control valves can create high frequency pressure variations, particularly at higher flow rates. These high frequency flow induced pressure oscillations can be broadband or narrowband. Narrowband and periodic pressure oscillations at any frequency are also referred to as pressure pulsations and are almost always generated by attached machinery or vortex shedding. They are often the most harmful when amplified by an acoustic resonance in the piping system, and particularly so when a structural resonance is also present. They can appear in systems that have a history of good behavior after changes in process conditions (e.g., flow rate, temperature, gas composition) or piping. Dealing with these pulsation issues in a timely manner can results in:

- Improved equipment reliability via identifying and addressing problems early on.
- Reduced unplanned downtime costs and reduced maintenance costs over the long term.
- Increased safety through identifying and mitigating potential hazards in piping systems.

Mitigating pressure pulsation related vibration issues often involves eliminating the source of the excitation and/or eliminating the amplification mechanism(s). In the following sections the various sources of pressure pulsation in industry, the amplification mechanisms, and how to deal with them will be discussed.

PULSATION SOURCES:

Pressure pulsations can be generated by machinery or flow. The main source of flow induced pulsation is vortex shedding which can be due to flow around a bluff body or due to an unstable shear layer at a tee leading to a closed branch (Figure 1). Vortex shedding tends to be governed by the Strouhal number: $St = f d / U$, where f is the vortex shedding frequency, d is the characteristic dimension and U is the flow velocity [2]. The natural frequency of the leg corresponds to an acoustic wavelength of 4L. When these frequencies are close, pulsations may be generated.

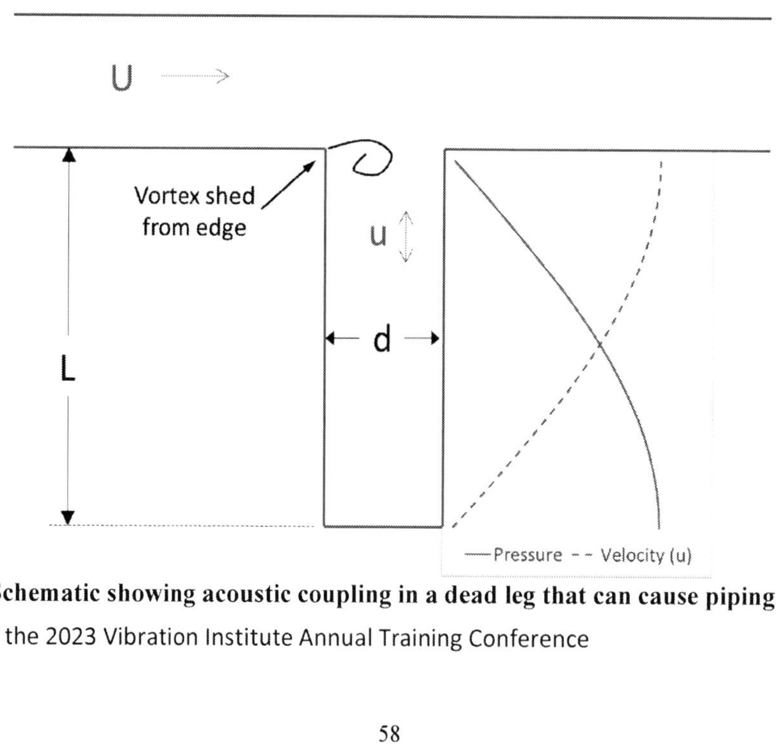

Figure 1. Schematic showing acoustic coupling in a dead leg that can cause piping vibrations.

In the case of machinery, all equipment used to move fluid operates in a periodic manner, whether through rotating impellers or reciprocating pistons, and causes fluctuations in the flow rate and pressure (pulsations) [1]. Of course, there is a great variety of pumps and compressors, and some generate steadier flow than others. The strongest pulsations are generated by low speed, low piston count positive displacement pumps and compressors, while centrifugal, axial and high-speed positive displacement machines and tend to generate lower pressure pulsations at high frequencies. A list of these machines is given below, with subsections dedicated to the most common types.

- Reciprocating Pumps and Compressors
- Centrifugal Pumps and Compressors
- Rotary Screw Compressors
- Axial Flow Compressors
- Sliding Vane Pumps and Compressors
- Internal and External Gear Pumps
- Axial Piston Pumps
- Progressing Cavity Pumps
- Peristaltic Pumps

Reciprocating Pumps and Compressors

Reciprocating machines generate pressure pulses due to their cyclical operating nature. These types of machines usually require pulsation attenuation devices such as accumulators and pulsation bottles to attenuate the pulses before discharging the flow into the pipework. Figure 2 shows two cylinders of a typical large compressor with shared pulsation bottles on both the suction and discharge sides.

Figure 3 is a schematic of a simplex pump, which has one single acting piston. As the piston moves to the left, the suction valve opens and low-pressure fluid enters the cylinder. When the direction of the piston reverses, the suction valve closes, and when the pressure of the fluid in the cylinder exceeds the pressure in the discharge line, the discharge valve opens. In the case of an incompressible fluid, there is flow in the suction line for half of the time and in the discharge line the other half of the time. This idealized flow is shown in Figure 4, and this could represent the suction or discharge side, for an incompressible fluid they will simply have opposite phase. Note that because the flow is pulsing, the peak flow rate is over 300% of the mean flow rate. Duplex, or double acting cylinders, (which use both sides of the piston) generate lower pressure pulsation than single acting cylinders for the same flow rate, but usually not low enough to avoid the need for attenuators. The ideal flow for a duplex pump is included in Figure 4 for comparison. Note that because there is flow from both directions of piston travel, peak flow rate is half of that of the simplex pump. The duplex pump still passes through a zero flow rate but it is instantaneous rather than half of the cycle. Rather than (or in addition to) using both sides of the piston, additional pistons can be attached to a common crank shaft, so that each piston's pulsations are phased so that the pump is always pushing fluid. While triplex and quintuplex pumps have much lower pulsation levels than single piston simplex pumps, they can still generate significant pressure levels, especially if an acoustic resonance occurs in the piping.

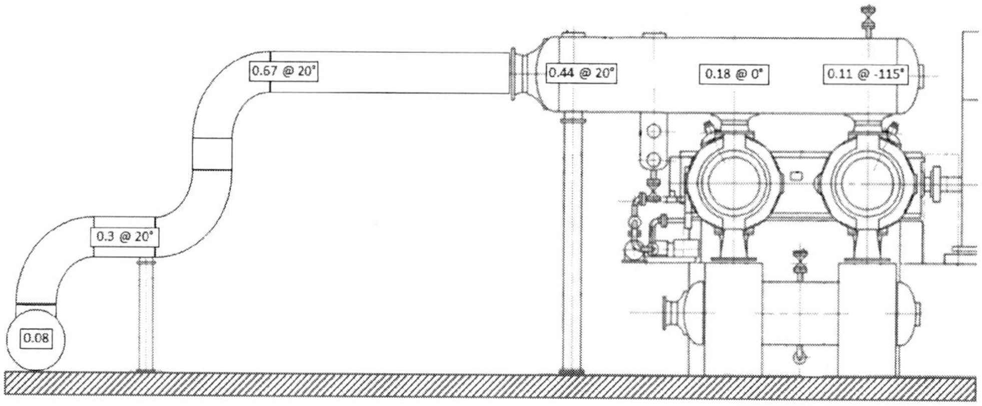

Figure 2. Reciprocating Compressor with Pulsation Bottles

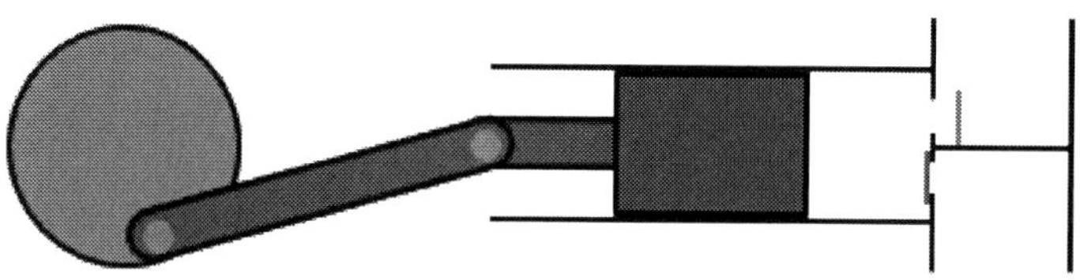

Figure 3. Simple Crank Reciprocating Compressor Schematic

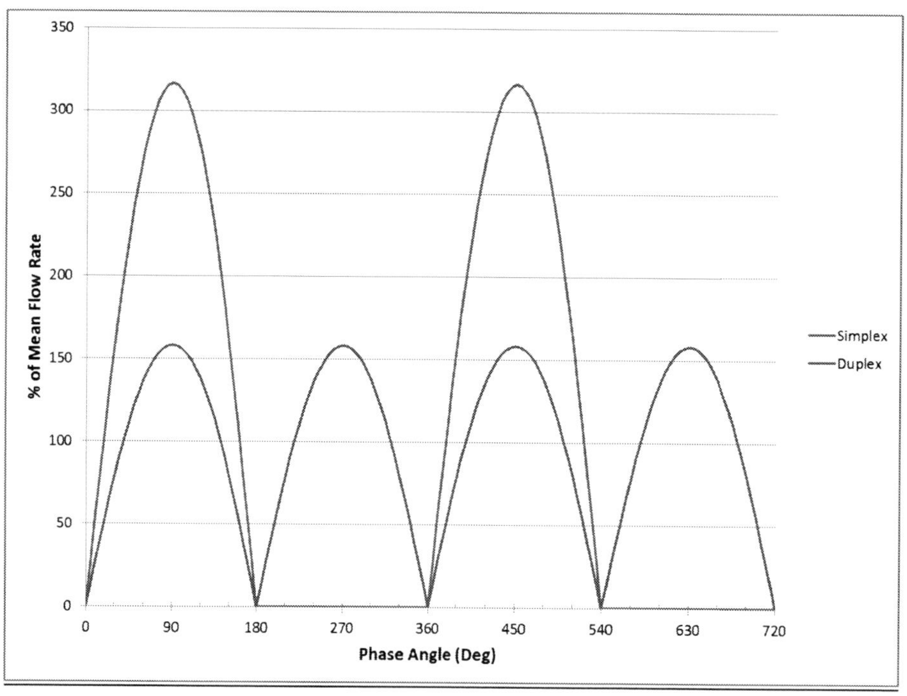

Figure 4. Reciprocating Compressor Flowrate

Centrifugal Pumps and Compressors

While very different in concept from their positive displacement relatives, dynamic pumps and compressors also operate cyclically by rotating an impellor. A major source of pulsations from centrifugal machines occurs at the vane passing frequency. Note the design of the typical centrifugal pump shown in Figure 5. The sharp edge of the casing near the outlet is referred to as the cutwater, and pulses are generated as the blades pass by this point. Higher frequencies are generated in double volute or vaned diffuser style machines, simply because the number of interaction points is increased. The amplitude of the pulses from centrifugal equipment is typically significantly lower than that of reciprocating machines, therefore attenuation devices are not normally used with them. It usually requires the presence of acoustic resonance in the piping system for the pressure pulsations to generate high vibration levels and become potentially harmful to piping integrity.

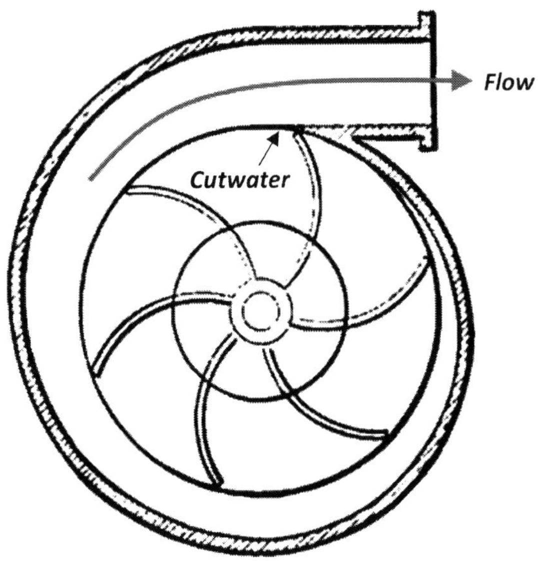

Figure 5. Centrifugal Pump Cutaway Drawing

VIBRATION FORCING BY PULSATION:

Flow interacts with the piping through pressure and friction, with pressure usually being the dominant factor in a typical operating plant environment. Unsteady or periodic flow through piping will generate corresponding pressure oscillations. Forces from small, random pressure fluctuations will often counter each other and cancel each other out, resulting in a small overall effect on vibrations. But when the pressure oscillations have dominant frequencies, the effects can be more organized and lead to high levels of forcing on the piping. Figure 6 shows an example of vibration forcing by pressure pulsations. In the schematic, one end of the pipe is connected to a reciprocating compressor, and the other end is attached to a receiver vessel. The graph shows a pressure wave shape corresponding to one particular running speed of the compressor. The amplitude of the pressure oscillation is higher at the elbow closer to the compressor than it is at the elbow closer to the receiver. Since the pressure pushes outward on the inside surfaces of the elbows, the result is that there is an alternating unbalanced force on horizontal section of pipe. Note that the net force averages to zero over time, but depending on the stiffness of the piping, there could be a vibration issue. The pressure at the elbows will create forces in the vertical direction as well, which must be considered relative to the forces at the opposite ends of the vertical sections of the piping.

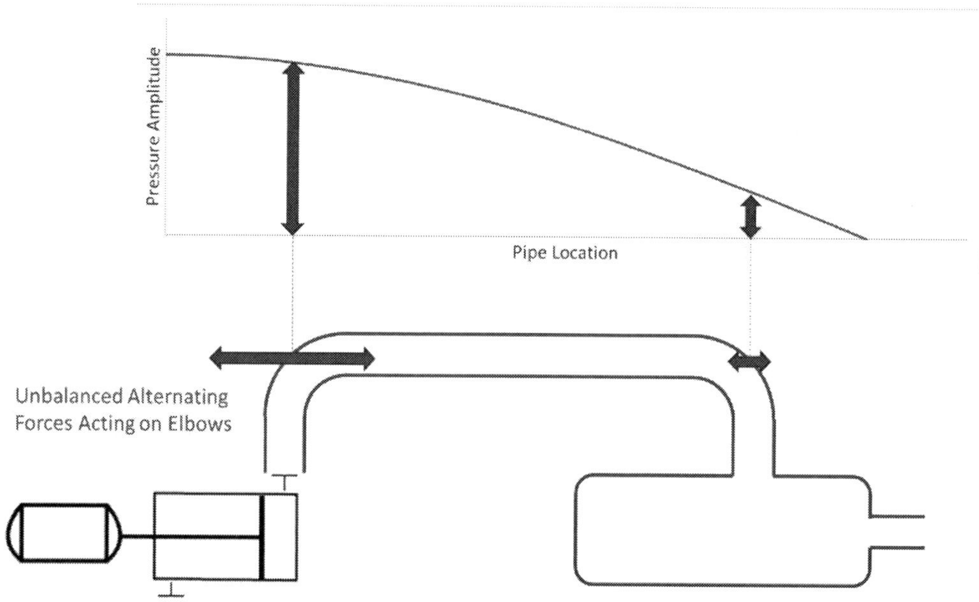

Figure 6. Schematic Showing Unbalanced Pressure Pulsations Generating Oscillating Forces

The example in Figure 6 demonstrates how pulsations generate unbalanced forces that act in the directions of sections of the piping through pressure acting on elbows, area changes, valves, etc. Keeping this in mind can help with understanding the locations where the highest vibrations may be measured and the directions in which they will be observed. Note that the response of the piping to forcing at the elbows can be complex and can result in bending mode vibrations in any or all of the pipes nearby. Also, because the energy is transmitted by the fluid pulsations, and pressure waves can travel great distances, piping vibrations can be highest a significant distance from the source of the pulsations if the pressure wave happens to align with a susceptible section of piping.

ASSESMENT APPROACH:

Pressure pulsation related vibration studies are typically initiated due to some failure event or operator observations. The first step in the assessment is the collection of appropriate vibration data. The quality of the collected data will have a significant impact on the assessment results; therefore, every effort should be made to obtain the best possible data within the constraints always present when working in the field. Before taking any data, perform a site walkdown to select vibration and dynamic pressure measurement locations. Once the locations have been identified, collect and analyze the dynamic data.

If the highest vibration amplitudes are identified a relatively long distances away from the pump/compressor, but these locations do not coincide with locations of high amplitude pressure pulsation amplitudes, then the problem is likely due to inadequate supports or mechanical resonance in the piping. In these cases, upgrading supports or installing new supports can solve the problem. Note that prior to changing or adding supports on high temperature, thermal expansion stresses must be checked.

If the vibration is greatest closest to a pump or compressor, then the problem may be excessive pulsation amplitude generated locally by the machine. The frequency content of the vibrations due to pulsations will generally be narrow band or monofrequency and will match a harmonic of the pump/compressor rotation speed. Dynamic pressure measurements acquired as close as possible to the area of the vibrations may help to confirm pressure pulsations as the cause and rule out mechanical vibrations from the fixed equipment. In the case centrifugal machines consider trimming the impeller vanes to open up the 'B-gap', switching to an odd number of vanes for double volute machines, or offsetting the impellers on multistage machines. In the case of reciprocating machines, upgrade the pulsation bottle or accumulator attached to the machine. Consider both suction and discharge sides. Sometimes an accumulator vendor can help with the selection of the proper device, in more challenging cases consult an acoustics expert.

If the highest vibration amplitudes occur relatively long distances away from the pump/compressor, and these locations coincide with locations of high-pressure pulsation amplitudes, then there is likely acoustic resonance in the connected piping. An acoustically resonant piping system is difficult to mitigate by trial and error. Solving this type of problem generally involves consulting with an acoustics expert. The expert will build a computer model of the piping using suitable software that matches the observations in the field, and then use the same model to test mitigation strategies. The most promising approach can then be implemented by the equipment owner or operator.

RESONANCE:

Resonance can become a problem in a mechanical or acoustic sense. Figure 8 shows a number of resonant mode shapes for beam bending [3], and Figure 7 shows circumferential shell modes for pipes. These figures represent some of the mechanical resonance modes that can potentially be excited by pulsations. Some longitudinal acoustic modes are shown Figure 10, and transverse modes are shown in Figure 9. Additionally, both the mechanical and acoustic mode shapes shown can combine into very complex three-dimensional modes. When the mechanical natural frequencies corresponding to a mode shape) of a pipe align (Figure 7, Figure 8) with the pulsation frequencies high vibration can result, and especially so if there are also acoustic natural frequencies (Figure 9, Figure 10) at the excitation frequencies. As discussed previously, bending modes can be forced by acoustic pressure imbalances acting at bends and restrictions in the pipe. Mechanical shell modes can be excited by acoustic transverse modes as well as by direct mechanical excitation. While complete correlation isn't required to force shell modes, it is clear from the figures that, for example, the n = 2 shell mode can be forced by an n = 2 acoustic mode.

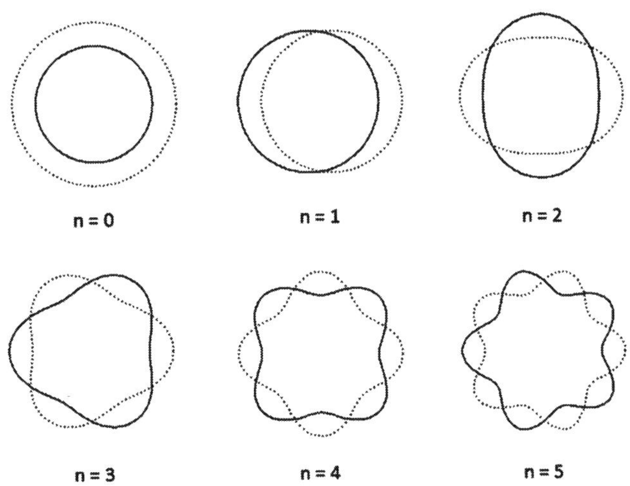

Figure 7. Circumferential shell mode patterns for a pipe.

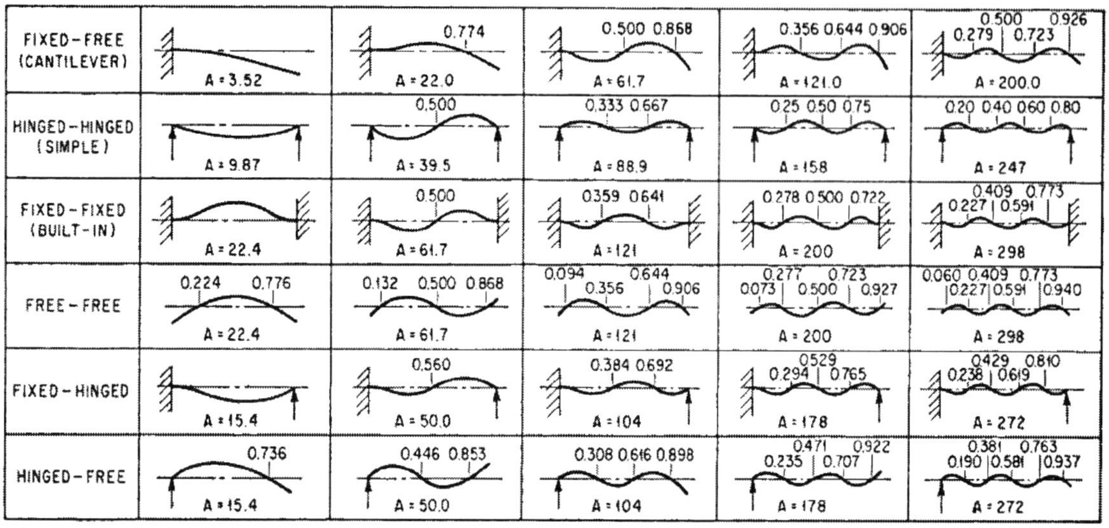

Figure 8. Beam mode vibration mode shapes from [3]

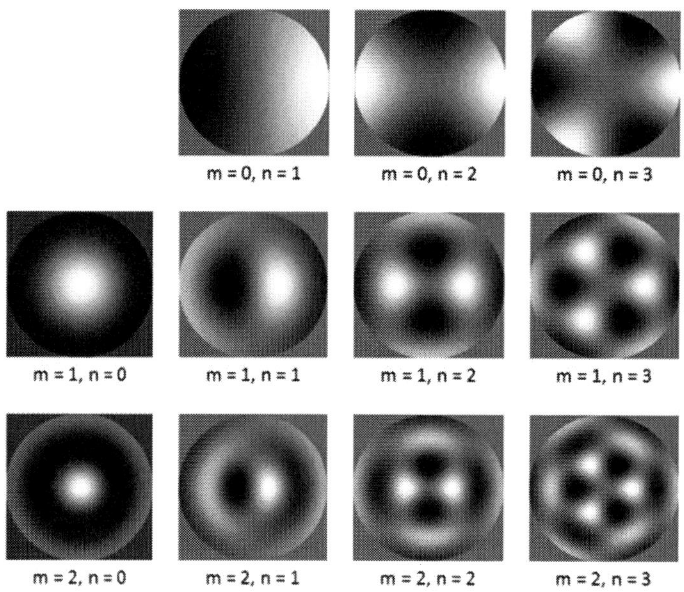

Figure 9. Cross-wall acoustic modal structure for a fluid-filled pipe with rigid walls.

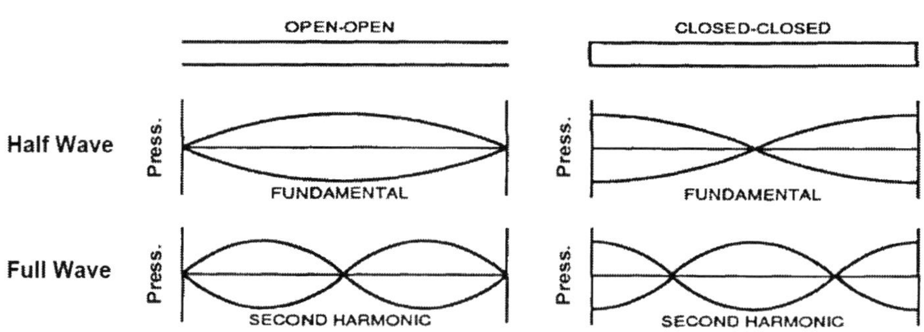

Figure 10. Longitudinal standing waves in constant diameter piping

ACOUSTIC MODELING

Simulating piping system acoustics requires solving the flow equations in the piping system. For the low frequencies typical of pulsations, the acoustic wavelengths are long compared to the diameters or the pipes, and so a one-dimensional version of the flow equations can be used. If transverse modes like those in Figure 9 are present, then a three-dimensional version of the equations must be solved instead. The acoustic oscillations are also usually considered to be isentropic, which is valid for many cases but may not be useful at, for example, very high amplitudes. Consider the following set of one-dimensional flow equations.

Conservation of mass:

$$\frac{\partial \rho}{\partial t} + \rho \frac{\partial u}{\partial x} + u \frac{\partial \rho}{\partial x} = 0$$

Conservation of momentum:

$$\rho \frac{\partial u}{\partial t} + \rho\, u \frac{\partial u}{\partial x} + \frac{\partial p}{\partial x} + \zeta\, \rho\, u^2 + 2\alpha\, \rho\, c\, u = 0$$

Isentropic relation:

$$\left| \frac{\partial p}{\partial \rho} \right|_s = c^2$$

In these equations, ρ is the density, u is the velocity in the direction of the pipe, p is the pressure, c is the isentropic speed of sound, ζ is a term representing the wall friction, α is a coefficient of viscothermal drag, s is the specific entropy, t is time, and x is distance along the pipe length.

Even this "simplified" set of equations is difficult to solve, and further assumptions are made to simplify them further. For example, assuming that there is no mean flow, the pressure oscillations are small, and the damping can be neglected leads to the classical acoustic wave equation:

$$\frac{\partial^2 p}{\partial t^2} + \rho_0\, c_0^2 \frac{\partial^2 p}{\partial x^2} = 0$$

While this version of the acoustic wave equation is very useful, neglecting the damping effects leads to erroneous solutions when investigating the effects of pulsations on real world piping systems.

Several methods have been developed to solve the one-dimensional flow equations for pulsations in pipe flow. One method is to use frequency-domain analysis of the acoustics using a transfer matrix approach. In this method, solutions of the differential equations for the acoustic behavior of piping segments are determined for harmonic forcing at a particular frequency. These piecewise solutions are matched together using transfer functions for elements like area changes or tees. The model reduces to a (potentially large) system of linear equations that are solved simultaneously. The transfer matrix method is fast, requiring a matrix to be solved for each frequency modeled. This allows for rapid simulation of multiple parameters. However, it requires some simplifications of the flow equations on damping terms like friction losses that limit its accuracy and often result in overestimates of peak amplitudes. Figure 11 and Figure 12 show examples of calculating pressure pulsation amplitude and shaking forces using commercially available software (in this case, BOSpulse by Dynaflow Research Group).

While commercially available general purpose software makes pulsation modeling considerably more convenient than having to write custom programming for each problem, there is still a significant amount of subject matter knowledge required in order to generate reasonably accurate results.

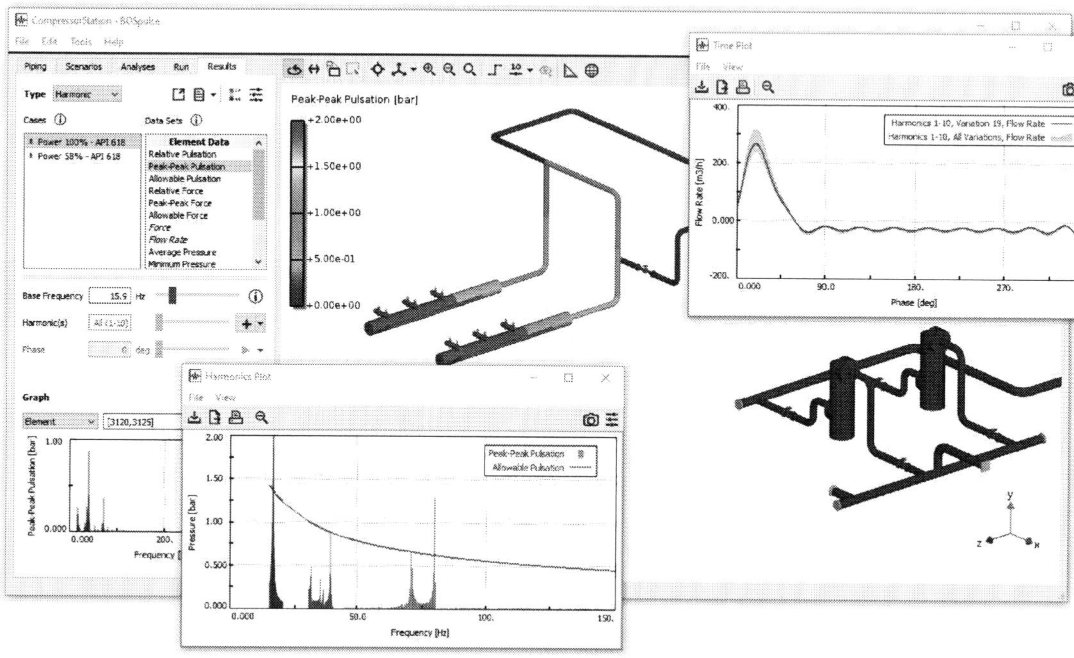

Figure 11. Pressure pulsation amplitude computation via pulsation analysis software

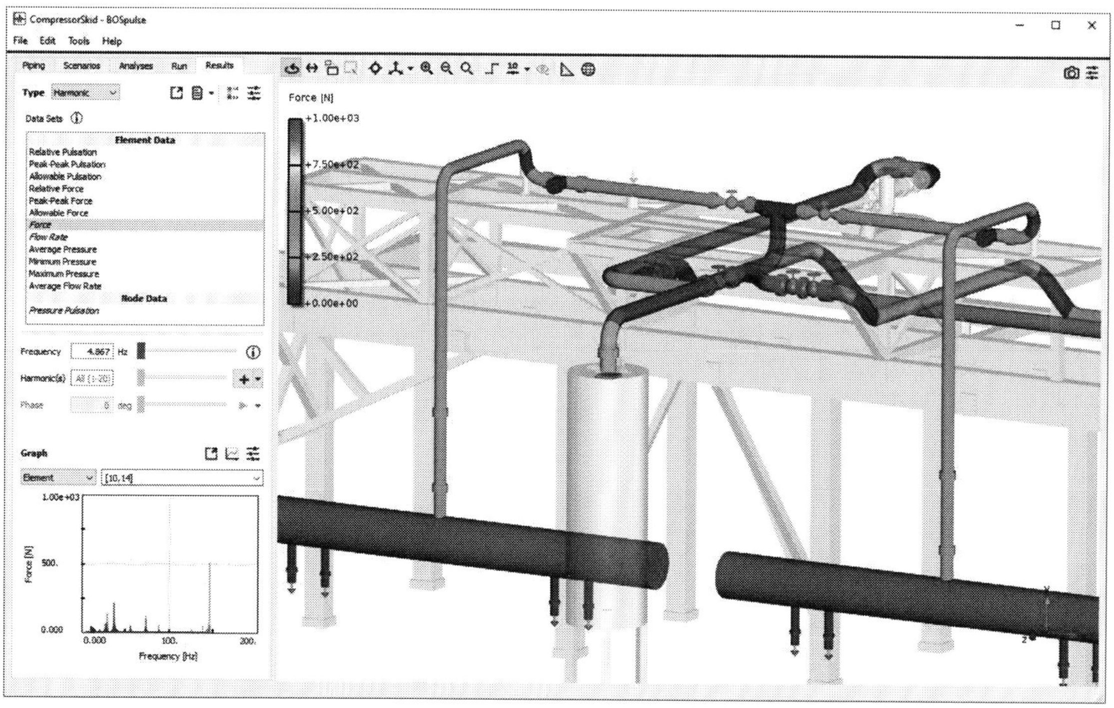

Figure 12. Shaking force determination via pulsation analysis software

An alternate method is to solve the one-dimensional flow equations in the time domain using an approach like the method of characteristics (MoC) that is commonly used for water hammer analysis. The MoC utilizes careful selection of space and time increments to reduce the differential equations of flow to algebraic equations, but the tradeoff is that it usually requires some interpolations or "adjustments" that will have some effect on accuracy. While this method is still fast compared to more general computational fluid dynamics (CFD) methods, it is significantly slower than frequency domain analysis. For time domain methods, the forcing is applied at acoustic boundaries (e.g., a compressor) and calculations are advanced in time while the disturbances propagate through the system and reach a quasi-steady state. Each change in conditions (e.g., compressor operating speed) requires a new solution (also true for frequency domain methods, but the time domain method is much slower). The benefit is that a more complete version of the fluid mechanics equations is solved, so more accurate solutions can be provided. Because it is significantly slower, a typical hybrid approach is to use the harmonic software to identify frequencies and operating conditions that may have high amplitude responses and then to run the time domain analysis for only these specific conditions.

At pulsation frequencies for which the acoustic wavelength approaches or becomes smaller than the diameter of the piping or vessel where the pulsation is occurring, the one-dimensional simplification can no longer be used. This provides a significant challenge to pulsation modelling. 3-D finite element analysis (FEA) can be used to provide mode shapes, but this approach usually assumes that mean flow and friction can be neglected. Also, applying proper boundary conditions can be a challenge. Full 3-D finite element or finite volume computational fluid dynamics (CFD) may be useful in some situations, but the number of elements required to capture 3-D flow effects in even a small piping system can be unreasonably high. In some situations, a combination of 3-D analysis where necessary and 1-D analysis where applicable may be used to make an analysis possible.

PULSATION CONTROL METHODS:

If pulsations are identified as problematic, there are various approaches that can be used, individually or in combination, to mitigate the effects. One widely used method to control pulsations from pumps and compressors is the use of orifice plates. Orifice plates damp pulsations using the friction effects of the flow through them. The plates themselves are inexpensive and can be a very effective part of a pulsation control system. Since they work by restricting the flow, however, the pulsation reduction effect comes at the cost of compressor or pump efficiency, since they add to the pressure losses in the system. The effectiveness of orifice plates at damping pulsation is affected by their location. They work best when they are positioned where the pulsation *velocity* is highest, which typically corresponds to where the pressure amplitude is low. This occurs, for example, at entrances and exits of vessels, where orifices can be effective for a wide range of frequencies. Used in other locations of piping, the effectiveness of orifice plates may be frequency dependent.

Passive silencers, including pulsation bottles, are another typical method used to control pulsations. Passive silencers work though reflections and destructive interference of the acoustic waves. The goal is to reduce the oscillating part of the flow while allowing the steady flow pass through as unhindered as possible. Accumulators in liquid systems are a type of passive silencer that work by adding a compliance to the system that change the response to pulsations.

A simple example of an acoustic passive silencer is an expansion chamber. As shown in Figure *13*, an expansion chamber is simply a section of sufficiently larger diameter piping. The graph shows the transmission loss (TL), which is a measure of how much of the incoming wave is attenuated on the outlet side of the filter. TL is defined as:

$$TL = 10\ log_{10} \left|\frac{W_i}{W_t}\right|$$

where W_i is the acoustic power of the incident wave and W_t is the power of the transmitted wave, and the units are dB. The resonance characteristics of the expansion chamber result in a filter that passes narrow bands (low TL) and attenuates at the frequencies in between. In this figure, the graph shows both the analytically derived behavior [4] and a calculation performed using pulsation modeling software for comparison. For this case, the TL is modeled nearly perfectly. This graph shows only low frequency behavior, and the behavior becomes more complicated if the pulsation frequency gets high enough for three-dimensional mode shapes to be excited.

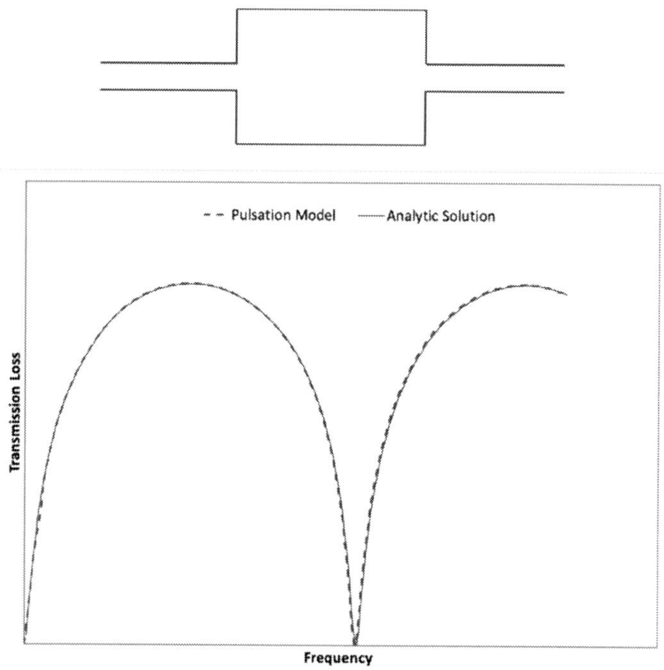

Figure 13. Expansion chamber used as a passive silencer

Most reciprocating compressor pulsation bottle designs are more complicated than a simple expansion chamber, and use a double expansion chamber, also referred to as a volume-choke-volume design. An example is shown in Figure 14. In this design, each of the silencer elements has its own resonance characteristics, and they are combined to provide low pass filtering. The pressure loss is generally higher than a simple expansion chamber, which reduces the system efficiency, but the acoustic performance is improved. The graph compares the double expansion chamber to a single expansion chamber of the same overall length. While the low-pass band is wider, the attenuation at higher frequencies is much higher for the double chamber design.

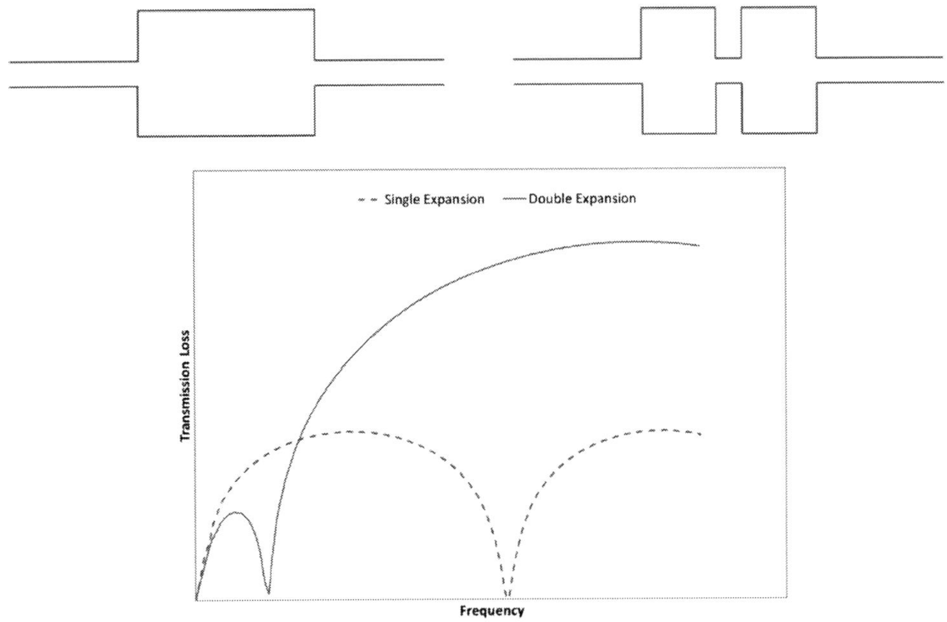

Figure 14. Double expansion chamber style pulsation bottle

When inserting filters into piping systems, it is important to note that new response frequencies can be generated. For example, Figure 15 shows model results for a pipe with a double expansion chamber. In this case, a compressor is located to the left, and the current running frequency is exciting a resonance between the pump and the filter. Note that despite the high amplitude pulsations upstream of the filter, there is little energy transmitted through the filter and into the downstream pipe. Keeping the pipe length between the compressor and pulsation bottle short keeps the resulting natural frequencies high.

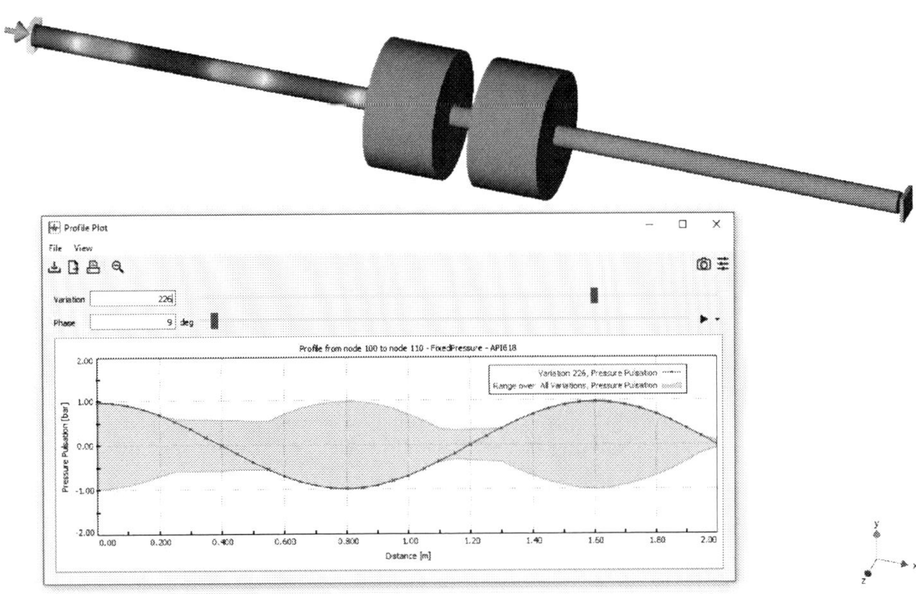

Figure 15. Model of piping with a double expansion chamber

Other types of passive silencers include Helmholtz resonators, as shown in Figure 16. The graph shows an analytical calculation of the TL [5]. Helmholtz resonators add very little drag but provide attenuation only in a narrow band near the design frequency. Multiple resonators can be used to control multiple frequencies, but the narrow band behavior usually limits their use to fixed frequency (and fixed speed of sound) devices.

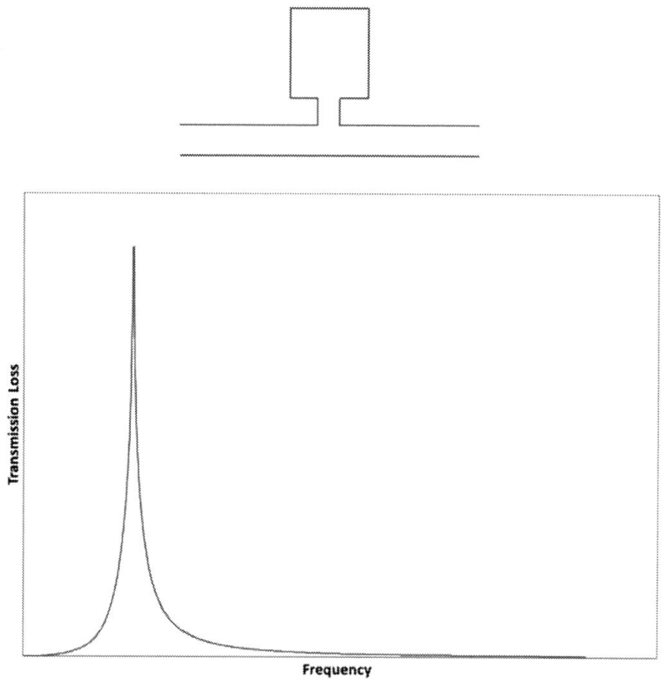

Figure 16. Helmholtz resonator schematic and TL curve

Another approach to mediating pulsation induced piping vibrations is to modify the piping system to eliminate resonances. If acoustic resonances exist, the preferred method it to change piping lengths or include diameter changes (effectively creating passive silencers) to "detune" the system away from the driving frequencies. Mechanical piping resonances can be controlled through changes in stiffness, for example by adding supports or increasing wall thicknesses.

CONCLUSIONS:

The following conclusions can be drawn from the previous sections:

- Pressure pulsation related fixed equipment vibration problems are one of the most common fixed equipment vibration problem types.
- Solving them requires an understanding of whether the problem is related to the lack of adequate pulsation control at the source, acoustic resonance, or mechanical resonance.
- Mechanical resonance can be addressed via traditional methods such as support modifications for beam mode vibrations, or pipe wall thickness increase for shell mode vibrations.
- Solving an inadequate pulsation control or acoustic resonance problem generally requires the involvement of an acoustics expert.

REFERENCES:

[1] Matta, L.M. and Szasz, G., "Vibration and Fatigue Failures at Pipeline Facilities", Proceedings of IPC 2018 (Paper No. IPC2018-78176), 12th International Pipeline Conference, September 24 - 28, 2018, Calgary, Alberta, Canada.

[2] Gulich, J.F., *Centrifugal Pumps*, 2nd Ed., Springer, 2010.
[3] Piersol, A.G., and Paez, T.L., *Harris's Shock and Vibration Handbook*, 6th Ed., McGraw Hill, 2010.
[4] Beranek, Leo L. and Ver, Istvan L., *Noise and Vibration Control Engineering, Principles and Applications*, 2nd Ed., Wiley & Sons, 2006.
[5] Barron, Randall F., *Industrial Noise Control and Acoustics*, Marcel Dekker, 2003.

VIATC | TRAINING CONFERENCE & EXPO

NIAGARA FALLS, NEW YORK JULY 26 -28, 2023

AN INDEPTH LOOK AT THE ASME OM-3 METHOD TO EVALUATE PIPING VIBRATION WITH CASE STUDIES

Nathan D. Libertowski
Staff Engineer I
E[2]G | The Equity Engineering Group, Inc.
nlibertowski@e2g.com

Michael F.P. Bifano, Ph.D., P.E.
Consulting Engineer II
E[2]G | The Equity Engineering Group, Inc.
mbifano@e2g.com

Nathan Libertowski is an engineer at The Equity Engineering Group with the Rotating Equipment, Vibration, & Dynamics (REVD) Team. He has four years of experience as an engineering consultant solving API 579 fitness-for-service (FFS) assessments, with a specialty in piping and structural vibration. Mr. Libertowski collects vibration data and performs dynamic FEA to assess vibration severity and, if required, recommends vibration mitigation techniques. He is also involved in advanced analyses such as elastic-plastic analysis of thermal transient events, crack-like flaws, creep, and fatigue. Mr. Libertowski holds engineering degrees from The Ohio State University (MSME) and Ohio Northern University (MSME). He is a member of ASME and is Category III vibration analyst certified.

Michael Bifano, Ph.D., P.E., is a consultant at E[2]G and the Team Leader of the Rotating Equipment, Vibration, & Dynamics (REVD) Team. He performs fitness-for-service (FFS) evaluations of in-service equipment using the rules of API 579-1/ASME FFS-1. He is the E[2]G technical authority on vibration measurements, diagnostics, and piping and structural vibration fatigue assessments. He is a voting member of the API 579 FFS Committee and one of the lead authors of the upcoming API 579 Part 15 methodologies on the assessment of piping vibration. Additionally, he specializes in transient thermal-mechanical fatigue analysis, computational fluid dynamics, elevated temperature creep and creep-fatigue simulations, and fracture mechanics using analytical and computational techniques. He holds a bachelor's degree in physics from Lebanon Valley College and a Ph.D. in Mechanical Engineering from Case Western Reserve University. He is also a Category IV vibration analyst.

ABSTRACT:
Methodologies for assessing piping vibration based off methods from the ASME Operation and Maintenance of Nuclear Power Plants Standard – Part 3 (ASME OM-3) are discussed. This method provides a procedure to calculate the allowable vibration limit based on parameters of the piping such as the configuration, diameter, thickness, length, unsupported valves, and the stress concentration factor (SCF). Also, modifications to the AMSE OM-3 procedure are presented. Using these methods, multiple case histories of vibrating piping are discussed.

KEY WORDS: ASME OM-3, Piping, Vibration, Piping vibration

NOMENCLATURE:

0-pk	zero-to-peak (i.e. amplitude)
A	pipe cross sectional area
A(x)	pipe cross sectional area as a function of the spatial location
ASME	American Society of Mechanical Engineers
BPVC	Boiler and Pressure Vessel Code
C_1	concentrated weight factor (per ASME OM-3)
C_2	secondary stress index (per ASME BPVC.III)
C_3	distributed weight factor (per ASME OM-3)
C_4	piping layout factor (per ASME OM-3)
C_5	frequency factor (per ASME OM-3)
CF	crest factor
D	pipe outer diameter
E	pipe elastic modulus
E(x)	pipe elastic modulus as a function of the spatial location
$S_{e,B31J}$	endurance limit (per B31J)
$S_{e,OM3}$	endurance limit (per ASME OM-3)
FEA	finite element analysis
I	pipe area moment of inertia
I(x)	pipe area moment of inertia as a function of the spatial location
ips	inches per second
K_2	local stress index (per ASME BPVC.III)
L	span length of beam
LR	long radius
M	bending moment
n	mode number
OM-3	Operation and Maintenance of Nuclear Power Plants – Part 3
psi	pounds per square inch
Q(x)	external forces on pipe as a function of spatial location
RMS	root-mean-squared
S_{weld}	material endurance limit (per BS7608)
SCF	stress concentration factor
SIF	stress intensification factor (per ASME B31.3 or B31J)
t	pipe thickness
t	time
u	vibration displacement
u(x,t)	vibration displacement as a function of spatial location and time
$u_n(x,t)$	vibration displacement as a function of spatial location and time of the nth mode shape
$V_{allow.OM3}$	allowable vibration velocity (per ASME OM-3)
$V_{allow.OM3.expanded}$	allowable vibration velocity (expanded version based on OM-3)
$v_{1,max}$	maximum vibration velocity of 1st mode shape
v_n	vibration velocity of nth mode shape
w_n	frequency of nth mode shape (rad/s)
x	spatial location along pipe
β	velocity to stress ratio (per ASME OM-3)
ρ	pipe density
ρ(x)	pipe density as a function of the spatial location
$\sigma_{1,max}$	maximum stress of 1st mode shape

INTRODUCTION:

Vibration is commonly monitored and evaluated to determine maintenance actions for rotating equipment assets; however, less emphasis is placed on condition monitoring of piping vibration in the attached piping and small-bore connections. Piping vibration can lead to failure of the pressure boundary through high-cycle fatigue cracking, leading to a loss of containment of the process fluid. Because cycling occurs at a high frequency, once a fatigue crack initiates, it can propagate through wall, causing loss of containment relatively fast. Therefore, it is often impractical and unrealistic to manage the risk of fatigue crack growth using traditional non-destructive examination (NDE) methods, since it is unlikely that a crack from vibration would be caught after it is initiated. In summation, inspection for cracks is not sufficient to determine if vibration of piping will cause issues in the future.

Similar to rotating equipment, piping vibration develops due to various unique energy sources, and a vibration analysis is often required to determine the root cause(s). Vibration remediation strategies and data acquisition techniques can differ depending on the underlying source; therefore, it is important to understand the root cause of the piping vibration.

Piping vibration fatigue assessments require knowledge of alternating stress, which can be difficult to directly measure. Instead, vibration amplitudes are commonly measured using accelerometers or similar devices and alternating stresses are calculated using computational models. These can include use of dynamic finite element analysis (FEA) by creating a dynamic model of the piping and tuning the model until it sufficiently represents the measured test data. This process can take a lot of time and resources and generally requires someone with advanced vibration analysis experience.

On the other hand, vibration can be related to stress by using screening charts or tables [1] [2]. These allow the user to simply compare the measured vibration to a table or chart to determine the severity of the vibration. The screening methods cover many different piping scenarios and, therefore, are conservative for most situations. Even with screening methods, care should be taken to ensure that the piping is relevant to the specific screening used.

If piping fails vibration screening criteria substantial resources, time, and expertise are needed to perform a complicated dynamic FEA modeling. Alternatively, fabrication and installation of revised piping supports are required, which can be costly and can lead to disruption in operations or process/equipment downtime. An alternative option is to use a hand-calculation based method, such as defined in the ASME Operation and Maintenance of Nuclear Power Plants (OM-3) [3]. In the scale of complexity, this method falls between the screening methods and full dynamic FEA by allowing for reduced conservatism compared to the screening methods without the significant resources and expertise required by dynamic FEA analysis.

The purpose of this paper is to discuss the often-overlooked OM-3 method for calculating an allowable vibration velocity. Also, by using basic beam calculations, this paper introduces some alternative ways to calculate some of the input variables to increase accuracy of the method. Lastly, two case studies are considered using this methodology.

ASME OM-3 VELOCITY LIMIT:

The ASME OM Division 2 Part 3 Section 5.1.2: Velocity Method (OM-3) [3] defines a method to calculate an allowable vibration velocity for piping spans based on parameters of the piping such as the configuration, diameter, thickness, length, unsupported mass, and the stress concentration factor (SCF). This method requires locating the exact location and direction where vibration is the highest. If the maximum measured vibration is below the allowable vibration velocity

($V_{allow.OM3}$), the vibration is deemed acceptable, assuming there are no pre-existing flaws or cracking and no other active damage mechanisms.

$V_{allow.OM3}$ is based off the free vibration of the first bending mode shape of a straight fixed-fixed beam with correction factors (C_1, C_3, C_4, C_5), resulting in Eq (1).

$$V_{allow.OM3} = \frac{C_1 C_4}{C_3 C_5} \frac{\beta(S_{el})}{C_2 K_2} \qquad (1)$$

S_{el} is the fatigue endurance limit per ASME BPVC Section III Figure I-9.1 [4] (ASME smooth bar fatigue curves) at 10^{11} cycles, while C_2 and K_2 are the secondary stress index and local stress index per ASME BPVC Section III [5], respectively. β is equal to 3.64×10^{-3} ips/psi, which is the ratio of the maximum vibration velocity to alternating stress for a straight fixed-fixed pipe in resonance at the first mode shape. In ASME OM-3, the β term is a constant; however, it is a function of parameters such as the pipe density, elastic modulus, diameter, and thickness.

There are four correction factors (C_1, C_3, C_4, C_5) that account for the fact that piping is not always a straight fixed-fixed pipe in resonance at the first mode shape.

- C_1 accounts for concentrated weights (such as valves and flanges) on the piping and is a function of the ratio of the concentrated mass to the characteristic span mass.
- C_3 accounts for distributed piping weight such as the weight of the internal fluid and insulation.
- C_4 accounts for different boundary conditions at the ends of the pipe span and for different piping configurations, such as L-bends, U-bends, and Z-bends. For a fixed-fixed beam, $C_4 = 1.0$
- C_5 accounts for if the vibration is non-resonant and is a function of the ratio of the first natural frequency (f_n) to the measured vibration frequency (f). This considers that for non-resonant vibration, the stress is proportional to the displacement, not velocity. This assumes the relative amplitude and phase of the vibration are similar to the first mode shape, even when the vibration is non-resonant.
- $C_2 K_2$ is the stress index per BPVC Section III, which is defined in BPVC Section III Division 1 Subsection NB section NB-3680 [5]. Various approaches exist in choosing the appropriate values of $C_2 K_2$. For example, it can be conservatively assumed that the largest alternating vibration stress occurs at the location with the largest stress concentration in the affected piping segment. For real-world vibration problems, this is not always the case, but this is a convenient option if insufficient data is available to identify the location with the highest alternating stress. Conservatism can be further reduced by more accurately selecting the value of $C_2 K_2$ corresponding to the location with the highest alternating stress, but this requires sufficient field test data to appropriately identify the system ODS (operating deflection shape).

The benefit of using the OM-3 method is that it allows for a relatively easy calculation of an allowable vibration limit for piping. Compared to screening methods, the OM-3 method will provide more accurate and, in most cases, larger allowable vibration limits and only requires a few inputs. Dynamic FEA will result in a more accurate and likely larger vibration limit compared to the OM-3 method; however, dynamic FEA analysis requires advanced engineering expertise, substantial data collection and recording, software, and data analysis. *In summation, the OM-3 velocity method offers a way to calculate less conservative allowable vibration limits, compared to screening limits, without having to perform dynamic FEA and advanced vibration measurement techniques.*

In-depth Review of the ASME OM-3 Velocity Limit: β (Velocity to Stress Ratio):
In OM-3, the β factor is defined as a constant and represents the ratio of the maximum vibration velocity to the maximum alternating stress for a straight fixed-fixed pipe in resonance at the first mode shape. This section provides the derivation of the β factor using Euler-Bernoulli beam theory.

The Euler-Bernoulli beam theory simplifies the beam equations by assuming the slope of the deformed beam is small and remains linearly elastic. Also, this theory assumes that cross sections of the beam will remain planer as the beam deforms, which is a valid assumption as long as the beam does not experience significant shear or torsional stresses. Lastly, this assumes that the shape of the cross section of the beam does not change, or for example, a pipe will remain circular and not ovalize. For most cases of bending mode vibration in industrial piping, these are valid assumptions. The resulting equation for dynamic motion of the beam is shown in Eq. (2). The displacement of the beam (u) is a function of position along the beam (x) and time (t).

$$\frac{\partial^2}{\partial x^2}\left(E(x)I(x)\frac{\partial^2 u}{\partial x^2}\right) = -\rho(x)A(x)\frac{\partial^2 u}{\partial t^2} + Q(x) \quad (2)$$

To simplify this equation, the beam is assumed to have a constant cross section, density, and stiffness (elastic modulus) along its length. In other words, $E(x) = E, I(x) = I, \rho(x) = \rho$, and $A(x) = A$. Also, the beam is assumed to be in "free vibration," or in other words, there are no external loads ($Q(x) = 0$). The resulting equation is shown in Eq. (3).

$$EI\frac{\partial^4 u}{\partial x^4} = -\rho A\frac{\partial^2 u}{\partial t^2} \quad (3)$$

The partial differential equation (Eq.(3)) is solved by using the separation of variables method. The solution of the displacement (u) of the beam in free vibration is a combination of all the vibration mode shapes (n), resulting in Eq. (4). In theory, there are an infinite number of possible mode shapes (n). ω_n is related to the eigenvalues of the differential equation, and α_n is a function of ω_n and the beam properties (Eq. (5)). The eigenvalues are related to the natural frequencies of the beam. $G_1, G_2, S_1, S_2, T_1, T_2$ are all unsolved constants in the solution. Constraints (or boundary conditions) must be applied to the solution to solve for these constants.

$$u(x,t) = \sum_n u_n(x,t)\text{, where}$$

$$u_n(x,t) = [G_1\cos(\alpha_n x) + G_2\cosh(\alpha_n x) + S_1\sin(\alpha_n x) + S_2\sinh(\alpha_n x)] * [T_1\cos(\omega_n t) + T_2\sin(\omega_n t)] \quad (4)$$

$$\omega_n = \alpha_n^2\sqrt{\frac{EI}{\rho A}} \quad (5)$$

This seemingly complex equation represents the displacement of the beam (u) in free vibration at its natural mode shapes assuming the beam is infinitely long and has no boundary conditions. Applying boundary conditions to the solution is required to get an exact solution and simply the equation. The OM-3 velocity method is based on the vibration of a straight beam that is fixed at both ends. In other words, at $x = 0$ and $x = L$ (L is the span length of the beam), the beam has zero displacement and zero rotation. In reality, most boundary conditions on pipes and beams will have some translational and rotational flexibility; however, assuming there is no flexibility is conservative. When the fixed-fixed boundary conditions are applied to this solution, the resulting specific solution is shown in Eq. (6) and Eq. (7).

$$u_n(x,t) = S_1[sin(\alpha_n x) - sinh(\alpha_n x) + \epsilon_n\, cos(\alpha_n x) - \epsilon_n\, cosh(\alpha_n x)] * [T_1\, cos(\omega_n t) + T_2\, sin(\omega_n t)] \qquad (6)$$

$$\epsilon_n = \frac{sin(\alpha_n L) - sinh(\alpha_n L)}{cosh(\alpha_n L) - cos(\alpha_n L)} \qquad (7)$$

In OM-3, the β term represents the ratio of the maximum vibration velocity to the maximum stress at the first natural frequency (or mode shape). The maximum vibration velocity of the first mode shape occurs at midspan ($x = L/2$) while the maximum stress will occur at the fixed boundary conditions at the ends of the pipe (x=0 or x=L). The resulting equation for β is shown in Eq. (8).

$$\beta = \frac{v_{1,max}}{\sigma_{1,max}} = \frac{\omega_1 \cdot u_1(L/2,t)}{\sigma_1(0,t)} \qquad (8)$$

The vibration velocity (v) is related to the vibration displacement (u) by ω, which represents the natural frequency of the first mode shape in units of rad/s. See Eq. (9) for details.

$$v_n = \omega_n \cdot u_n \qquad (9)$$

The equation for the maximum vibration displacement (u_1) at the first mode shape is shown in Eq. (10).

$$u_1(L/2,t) = S_1\left[sin\left(\alpha_1 \frac{L}{2}\right) - sinh\left(\alpha_1 \frac{L}{2}\right) + \epsilon_1\, cos\left(\alpha_1 \frac{L}{2}\right) - \epsilon_1\, cosh\left(\alpha_1 \frac{L}{2}\right)\right] * [T_1\, cos(\omega_1 t) + T_2\, sin(\omega_1 t)] \qquad (10)$$

The equation for stress in Euler-Bernoulli beams is a function of the bending moment (M), the diameter (D), and the area moment of inertia (I) (see Eq. (11)). The bending moment in a Euler-Bernoulli beam is a function of the elastic modulus (E), the area moment of inertia (I), and the second derivative of the displacement with respect to x (see Eq (12)).

$$\sigma_1(0,t) = M_1(0,t)\frac{D}{2I} \qquad (11)$$

$$M_1(0,t) = EI\frac{\partial^2 u_1(0,t)}{\partial x^2} \qquad (12)$$

Combining Eq. (11) and Eq. (12) and substituting the second derivative of the displacement with respect to x results in Eq. (13).

$$\sigma_1(0,t) = \frac{-4}{\sqrt{1+\left(\frac{D-2t}{D}\right)^2}} \sqrt{\rho E}\omega_1\epsilon_1 S_1 * [T_1 \cos(\omega_1 t) + T_2 \sin(\omega_1 t)] \quad (13)$$

The equation for β in Eq. (8) can be solved by substituting Eq. (10) and Eq. (13), which results in Eq. (14). The important thing to note is that the time-dependent variables and the frequency (ω) cancel out.

$$\beta = \frac{-1}{4\sqrt{\rho E}} \cdot \sqrt{1+\left(\frac{D-2t}{D}\right)^2} \cdot \frac{\sin\left(\alpha_1 \frac{L}{2}\right) - \sinh\left(\alpha_1 \frac{L}{2}\right) + \epsilon_1 \cos\left(\alpha_1 \frac{L}{2}\right) - \epsilon_1 \cosh\left(\alpha_1 \frac{L}{2}\right)}{\epsilon_1} \quad (14)$$

By numerically solving the boundary condition equations, it can be determined that for the first mode shape, $\alpha_1 L = 4.730$. Substituting the value of $\alpha_1 L = 4.730$ into Eq. (14) results in Eq. (15), which gives an equation to calculate β.

$$\beta = \frac{0.397}{\sqrt{\rho E}} \cdot \sqrt{1+\left(\frac{D-2t}{D}\right)^2} \quad (15)$$

In OM-3, β is defined as a single value of $3.64\text{x}10^{-3}$ ips/psi; however, through this derivation, it is shown that β is more accurately a function of the pipe density, elastic modulus, diameter, and thickness (Eq. (15)), and can be further extended to more accurately include alternative boundary condition assumptions if needed, however, that is beyond the scope of this work. Overall, Eq. (15) allows for a more accurate prediction of the allowable vibration velocity than assuming a single value.

It is important to note that the equation derived for β is only applicable to specific scenarios. The derivation is also valuable since it defines the assumptions and, therefore, limitations of method. Listed below is a summary of the assumptions made to solve for β (Eq. (15)).

- Euler-Bernoulli beams: The slope of the deformed beam is small, the cross sections of the beam will remain planer as the beam deforms, and the shape of the cross section of the beam does not change as the beam deforms.
- Along the beam length, there is a constant cross section, density, and stiffness (elastic modulus).
- The beam is straight (i.e., no bends or branches).
- The vibration of the beam is resonant and is vibration at its first natural mode shape.
- The beam has a finite length, L, and has fixed boundary conditions (no displacement or rotation) on each side of the beam.

Using these assumptions, most real-world piping would not be applicable for this method; however, the use of the correction factors in OM-3 (C_1, C_3, C_4, C_5) allows the method to be used for some types of piping layouts that violate the assumptions used to derive β.

- If the pipe has flanges or valves, the pipe no longer has a constant cross section and stiffness and, therefore, violates the assumptions used to derive β; however, the correction factor C_1 accounts for the effect of concentrated weights and, therefore, can capture the effects of extra weight from flanges and valves. Care should be taken though when using the C_1 correction factor because it only accounts for the extra weight from flanges and valves, but it does not account for the extra stiffness that flanges and valves have compared to the nominal pipe. If the length of the flanges and values is short compared to the span length, the effects of the stiffness can be ignored; however, if the length of flanges and valves is a significant portion of the span length, then the OM-3 method may not be valid.

- If the pipe has two different diameters (i.e., has a reducer) in the span, the pipe no longer has a constant cross section and stiffness and, therefore, violates the assumptions used to derive β. There may be a way to use an effective diameter to calculate the allowable vibration velocity; however, this is not covered in OM-3. Also, if only a very small section of the pipe is a different diameter, it could be justified to use the OM-3 method.

The approach discussed here applies to bending-mode type (non-shell mode) which is often 1-20 Hz in mainline piping systems and 10-300 Hz for small-bore cantilevered piping subject to mainline motion. These methods do not apply to shell-mode vibration, i.e., ovalization of the pipe cross-section. The authors recommend that any application of these concepts be under the supervision of a qualified vibration technician or engineer capable of identifying these differences.

In-depth Review of the ASME OM-3 Velocity Limit: Endurance Limit and Stress Indices

The endurance limit of a material represents the maximum cyclic stress that can occur without cracking or failure. Taking the ASME smooth bar endurance limit (S_{el}) and dividing by the secondary and local stress indices (C_2 and K_2) results in the endurance limit used for the OM-3 vibration limit (Eq. (16)).

The ASME smooth bar fatigue curves represent the fatigue of a smooth and unwelded bar, while the stress indices (C_2 and K_2) represent the ratio of the stress in a specific component (elbow, tee, etc.) to the stress in a straight unwelded pipe.

$$S_{e,OM3} = \frac{S_{el}}{C_2 K_2} \tag{16}$$

When evaluating ASME B31.3 [6] piping for vibration using the OM-3 methods, it can be convenient to modify the definition of the fatigue endurance limit and the stress indices. In ASME B31.3, it refers to ASME B31J [7], which defines SIFs for different piping components such as elbows, tees, olets, threads, etc. The SIF is related to the fatigue strength of a component relative to a straight butt-welded pipe; therefore, it can be convenient to define the endurance limit relative to a welded component (S_{weld}) instead of a smooth bar component. The endurance limit for a ASME 31.3 component can then be calculated by taking the welded component endurance limit (S_{weld}) and diving by the SIF; see Eq. (17) for details.

$$S_{e,B31J} = \frac{S_{weld}}{SIF} \tag{17}$$

A fatigue curve that is commonly used for fatigue of welded components is from BS-7608 [8], which is a British standard. For a butt-welded component (class D), the base endurance limit is defined as 3844 psi 0-pk. This value can be adjusted based off the specific material elastic modulus (which is a function of temperature), the ratio of the membrane and bending stress, and whether the loading is constant or variable.

Now that β and the endurance limit have been redefined, the expanded equation for calculating the allowable vibration limit is shown in Eq. (19). Also, the vibration limit equation from OM-3 (Eq. (18)) is shown for reference and for comparison. It should be noted that the alternative equation uses the same underlying method as OM-3; the only difference is that some of the variables are redefined to be more accurate and be more relevant to B31.3 piping.

$$V_{allow.OM3} = \frac{C_1 C_4}{C_3 C_5} \frac{\beta(S_{el})}{C_2 K_2} \quad (18)$$

$$V_{allow.OM3.expanded} = \frac{C_1 C_4}{C_3 C_5} \cdot \frac{0.397}{\sqrt{\rho E}} \cdot \sqrt{1 + \left(\frac{D - 2t}{D}\right)^2} \cdot \frac{S_{weld}}{SIF} \quad (19)$$

CASE STUDY #1:

Relief valve piping near a hydrogen reciprocating compressor had a history of high vibrations. The piping relieves a knock-out drum on the suction side of the compressor and forms a dead-leg with the closed valve. Reciprocating compressor relief valve piping is a common area of high piping vibration due to ¼-wave acoustic resonance conditions that can form.

Single-point vibration measurements with a handheld data collector were taken at different locations along the piping and compared to the screening vibration limits per PVP2018-84847 [9] (see Figure 1 for details). The maximum measured vibration was 2.66 ips RMS, which was larger than the allowable screening vibration limit of 1.0 ips RMS.

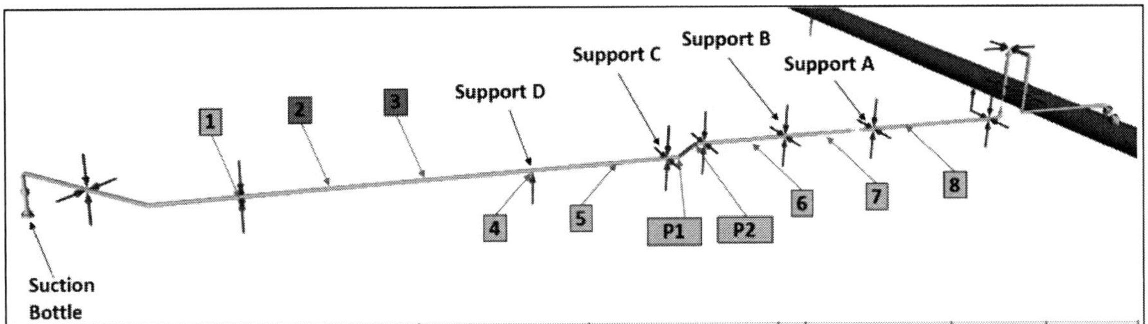

Location	measured vibration velocity: horizontal (ips RMS)	measured vibration velocity: vertical (ips RMS)	measured vibration velocity: axial(1) (ips RMS)	measured vibration velocity: resultant(2) (ips RMS)	Allowable velocity(3) (ips RMS)	allowable ratio	PASS/FAIL
1	0.524	0.537	0.190	0.77	1.0	0.77	PASS
2	**1.260**	**2.336**	**0.190**	**2.66**	**1.0**	**2.66**	FAIL
3	**0.929**	**1.914**	**0.190**	**2.14**	**1.0**	**2.14**	FAIL
4	0.163	0.081	0.190	0.26	1.0	0.26	PASS
5	0.332	0.474	0.190	0.61	1.0	0.61	PASS
6	0.051	0.035	0.306	0.31	1.0	0.31	PASS
7	0.125	0.126	0.306	0.35	0.707	0.50	PASS
8	0.095	0.086	0.306	0.33	1.0	0.33	PASS
P1	0.051	0.064	0.289	0.30	1.0	0.30	PASS
P2	0.073	0.052	0.245	0.26	1.0	0.26	PASS

Notes:
(1) highlighted yellow: the axial vibration was not measured. Assumed to have the same axial vibration magnitude as the inline piping
(2) vector magnitude of the vibration velocity direction magnitudes
(3) Allowable vibration screening limits per PVP2018-84847. Class A, periodic vibration. #7 is the only location with unsupported valve

Figure 1: Case Study #1, Measured Vibration Data Compared to Vibration Screening Limits

Since the screening vibration limits were exceeded, vibration limits were calculated using the expanded OM-3 method (Eq. (19)). A summary of the required inputs and the resulting allowable vibration velocity for this case study are shown in Table 1.

For this scenario, the piping properties were NPS 2 SCH 160, material of A106-B, and a temperature of 175°F. These properties are used to determine the diameter, thickness, elastic modulus, density, and endurance limit of the piping. The limiting (or largest) SCF for the piping span that had the high vibrations was the 45° LR elbows, which had an SIF of 1.0, per B31J.

There were no concentrated weights on the pipe span of interest, and because of this, the concentrated weight factor (C_1) was 1.0.

There was no insulation, and the fluid specific gravity was assumed to be 0.01, which along with the pipe density was used to determine the distributed weight factor (C_3).

The piping layout factor (C_4) was assumed to be for a straight pipe with fixed ends.

The measured dominant frequency of the vibration was 10.0 Hz. Also, the measured crest factor (CF) of the vibration was 1.92. The CF is the ratio of the 0-pk to the RMS vibration. A perfect sine or cosine vibration has a CF of 1.414, while a Gaussian random vibration will have a CF of ~4.5. Therefore, the closer the CF is to 1.414, the more sinusoidal the vibration is. The first natural frequency of the piping span was estimated to be 6.0 Hz.

Using methods based off OM-3 vibration limits, an allowable vibration limit of 6.43 ips RMS was calculated (using the measured CF of 1.92). The maximum measured vibration of 2.66 ips RMS is well below the calculated vibration limit (41% of the calculated vibration limit, to be exact).

Listed below is a summary of the takeaways from this case study:

- For some situations, the vibration screening methods can be very conservative. The OM-3 method can be used to reduce conservatism in the allowable vibration velocity without the need for advanced vibration measurement techniques or advanced analysis.
- An important and one of the most difficult parts of the OM-3 method is determining the first natural frequency of the piping. Determining the natural frequency of piping if it is currently vibrating can be difficult, especially for cases where pressure pulsations are the underlying issue. If the piping can be impact tested at rest, the natural frequency can easily be determined; however, for many applications that is not possible to shut down the equipment. There are methods to impact test while the piping is still vibrating; however, this requires specialized software or post-processing of the vibration data, or the ability to impact the system to create a response higher than the underlying background response. Alternatively, the natural frequency can be estimated with hand calculations.
- Even though the measured vibration was only 41% of the allowable vibration limit, it was unknown if the worst-case vibration was captured. The reciprocating compressors can operate at many different load cases, and sometimes variable speeds, which can affect the vibration amplitude of the attached piping. For this assessment, vibration measurements were only taken during one load case. It is always recommended to gather vibration measurements during all credible operating load cases to ensure that the maximum piping response is assessed.
- When adding piping supports for piping near reciprocating compressors, vibration measurements should be taken before and after the supports are installed to ensure the

additional supports reduce the vibration. If the new piping supports move one of the natural frequencies closer to one of the integer multiples of running speeds, the vibration can actually worsen. Either field testing or calculations to ensure adequate detuning of mechanical natural frequencies are advised prior to adding supports to piping attached to positive displacement machinery. This includes all piping, upstream and downstream of the machinery to the next major vessel.

- For piping that experiences significant vibration, any location where the pressure boundary is contacting another surface should be inspected for fretting damage. This is especially true for piping near reciprocating compressors because of the large pulsation forces.

Table 1: OM-3 Calculations for Case Study #1

Description	Variable	Input/ Output	Notes
Pipe size		NPS 2 SCH 160	
Pipe diameter (in)	D	2.375	
Pipe thickness (in)	t	0.344	
Elastic modulus (psi)	E	28.9×10^6	Based off A106-B @ 175°F
Pipe density (lb/in³)	ρ	0.283	Based off A106-B @ 175°F
Endurance limit (psi 0-pk)	S_{weld}	3690	From BS-7608; accounts for material and elastic modulus
Maximum SIF	SIF	1.0	SIF of NPS 2 SCH 160 LR elbow
Span length (ft)		34	
Insulation thickness (in)		0	
Insulation density (lb/ft³)		n/a	
Fluid specific gravity		0.01	Specific gravity relative to water
Concentrated weight (lb)		0	
Vibration frequency (Hz)		10.0	
1st natural frequency (Hz)		6.0	
Crest factor	CF	1.92	Ratio of 0-pk to RMS vibration
Concentrated weight factor	C_1	1.0	Per Fig. 10 of ASME OM-3. Function of the ratio of the concentrated weight to the unconcentrated weight of the piping span.
Distributed weight factor	C_3	1.001	Per section 5.1.2.4 of ASME OM-3. Function of the ratio of the distributed weight from insulation and fluid to the weight of the pipe.
Piping layout factor	C_4	1.0	Per section 5.1.2.4 of ASME OM-3. Assumes a straight pipe with fixed ends.
Frequency factor	C_5	1.0	Per section 5.1.2.4 of ASME OM-3. Function of the ratio of the vibration frequency to the first natural frequency.
Allowable vibration limit (ips RMS)		6.43	Using the expanded method in Eq. 19

CASE STUDY #2:
A NPS 3 boiler feedwater (BFW) piping was experiencing large vibrations. The piping was downstream of control valves and operated at 940 psig at 287°F. Single-point vibration measurements were taken with a handheld data collector at different locations along the piping and compared to the screening vibration limits per PVP2018-84847. The maximum measured vibration was 4.44 ips RMS, which was significantly larger than the allowable vibration screening limit of 1.0 ips RMS (see Figure 2 for details).

Figure 3 shows a spectrum and waveform of the vibration data taken at location 3B in the east/west direction. This location is where the maximum vibration occurs and is near the midspan of the vertical section of the piping. The dominate frequency of the vibration is 14.6 Hz, but there are other frequency components in the signal. The measured CF of this vibration was 2.47.

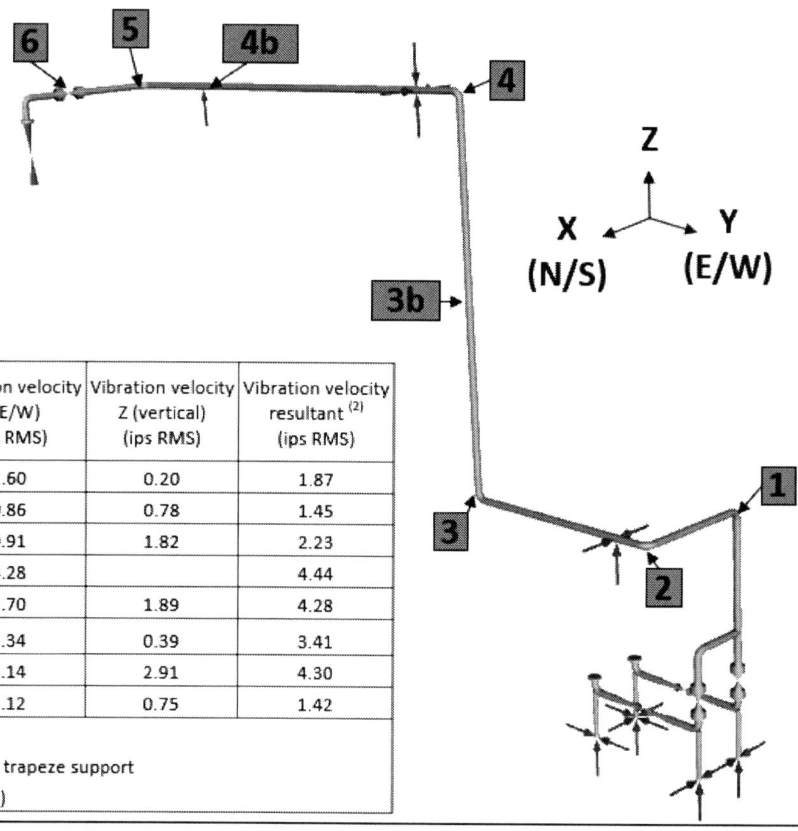

location	Vibration velocity X (N/S) (ips RMS)	Vibration velocity Y (E/W) (ips RMS)	Vibration velocity Z (vertical) (ips RMS)	Vibration velocity resultant (2) (ips RMS)
1	0.96	1.60	0.20	1.87
2	0.87	0.86	0.78	1.45
3	0.91	0.91	1.82	2.23
3b	1.16	4.28		4.44
4	1.02	3.70	1.89	4.28
4b (1)	0.56	3.34	0.39	3.41
5	0.45	3.14	2.91	4.30
6	0.45	1.12	0.75	1.42

Notes:
(1) location 4b was measured on the trapeze support
(2) resultant = sqrt(Vx^2+Vy^2+Vz^2)

Figure 2: Case Study #2, Summary of Measured Vibration Data

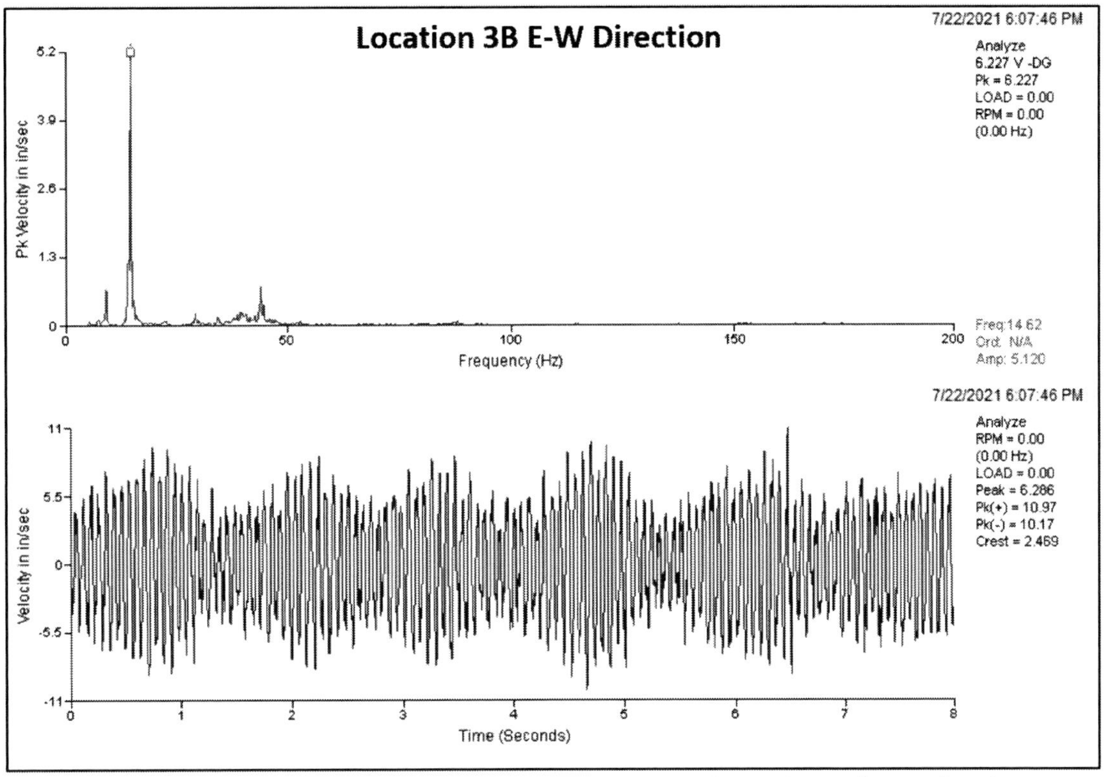

Figure 3: Case Study #2, Measured Vibration Data at Location 3B in East/West Direction

Since the screening vibration limits were exceeded, vibration limits were calculated using the expanded OM-3 method (Eq. (19)). A summary of the required input and the resulting allowable vibration velocity for this case study are shown in Table 2.

For this scenario, the piping properties were NPS 3 STD, material of A106-B, and a temperature of 287°F. These properties are used to determine the diameter, thickness, elastic modulus, density, and endurance limit of the piping. The limiting (or largest) SCF for the piping span that had the high vibrations was the 90° LR elbows, which had an SIF of 1.77, per B31J.

There were no concentrated weights on the pipe span of interest, and because of this, the concentrated weight factor (C_1) was 1.0.

There was no insulation, and the fluid specific gravity was assumed to be 1.0, which along with the pipe density was used to determine the distributed weight factor (C_3).

The piping layout factor (C_4) was assumed to be for a Z-bend.

The measured dominant frequency of the vibration was 14.6 Hz. Also, the measured CF of the vibration was 2.47. The CF is the ratio of the 0-pk to the RMS vibration. A perfect sine or cosine vibration has a CF of 1.414, while a Gaussian random vibration will have a CF of ~4.5. Therefore, the closer the CF is to 1.414, the more sinusoidal the vibration is. The first natural frequency of the piping span was estimated to be 12.0 Hz.

Table 2: OM-3 Calculations for Case Study #2

Description	Variable	Input/ Output	Notes
Pipe size		NPS 3 STD	
Pipe diameter (in)	D	3.5	
Pipe thickness (in)	t	0.216	
Elastic modulus (psi)	E	28.4x10^6	Based off A106-B @ 287°F
Pipe density (lb/in^3)	ρ	0.283	Based off A106-B @ 287°F
Endurance limit (psi 0-pk)	S_{weld}	3630	From BS-7608; accounts for material and elastic modulus
Maximum SIF	SIF	1.77	NPS 3 STD LR elbow
Span length (ft)		25	
Insulation thickness (in)		0	
Insulation density (lb/ft^3)		n/a	
Fluid specific gravity		1.0	Specific gravity relative to water
Concentrated weight (lb)		0	
Vibration frequency (Hz)		14.6	
1st natural frequency (Hz)		12.0	
Crest factor	CF	2.47	Ratio of 0-pk to RMS vibration
Concentrated weight factor	C_1	1.0	Per Fig. 10 of ASME OM-3. Function of the ratio of the concentrated weight to the unconcentrated weight of the piping span.
Distributed weight factor	C_3	1.193	Per section 5.1.2.4 of ASME OM-3. Function of the ratio of the distributed weight from insulation and fluid to the weight of the pipe.
Piping layout factor	C_4	0.74	Per section 5.1.2.4 of ASME OM-3. Assumes a Z-bend configuration.
Frequency factor	C_5	1.0	Per section 5.1.2.4 of ASME OM-3. Function of the ratio of the vibration frequency to the first natural frequency.
Allowable vibration limit (ips RMS)		1.88	Using the expanded method in Eq. 19

Using methods based off OM-3 vibration limits, an allowable vibration limit of 1.88 ips RMS was calculated (assuming a CF of 2.47). The maximum measured vibration of 4.44 ips RMS is significantly larger than the vibration limit; therefore, more action was required.

Rather than pursuing dynamic FEA to determine less conservative vibration limits, it was decided to design additional piping supports to reduce the vibration of the piping to acceptable levels. Figure 4 shows a summary of the details of the suggested piping supports. First, it was recommended to replace the upper and lower pipe guides with hold-down clamps to stiffen the piping. Secondly, it was recommended to add struts at two locations on the vertical piping span. There was existing structural steel at these heights which would make installation of the struts easier. Upon completing the thermal stress analysis, it was found that adding rigid sway struts at these locations over-restrains the piping and causes high stresses from thermal expansion. For this reason, the upper struts were recommended to be sway braces. A sway brace is similar to a rigid sway strut but consists of a spring compressed between two plates. Until the preload of the spring is reached, the sway brace acts as a rigid strut. When the force on the sway brace exceeds the preload, the spring starts to further compress and allows for movement of the piping. This is beneficial from a vibration point of view because as long as the vibration forces are less than the preload, the sway brace is rigid and should adequately restrain the vibration. This is beneficial from a thermal expansion stress point of view because once the preload of the spring is reached, the spring will compress and allow the pipe the move, which will relieve some of the thermal stresses in the piping. With these piping supports, the modeled stresses from thermal expansion were within code limits.

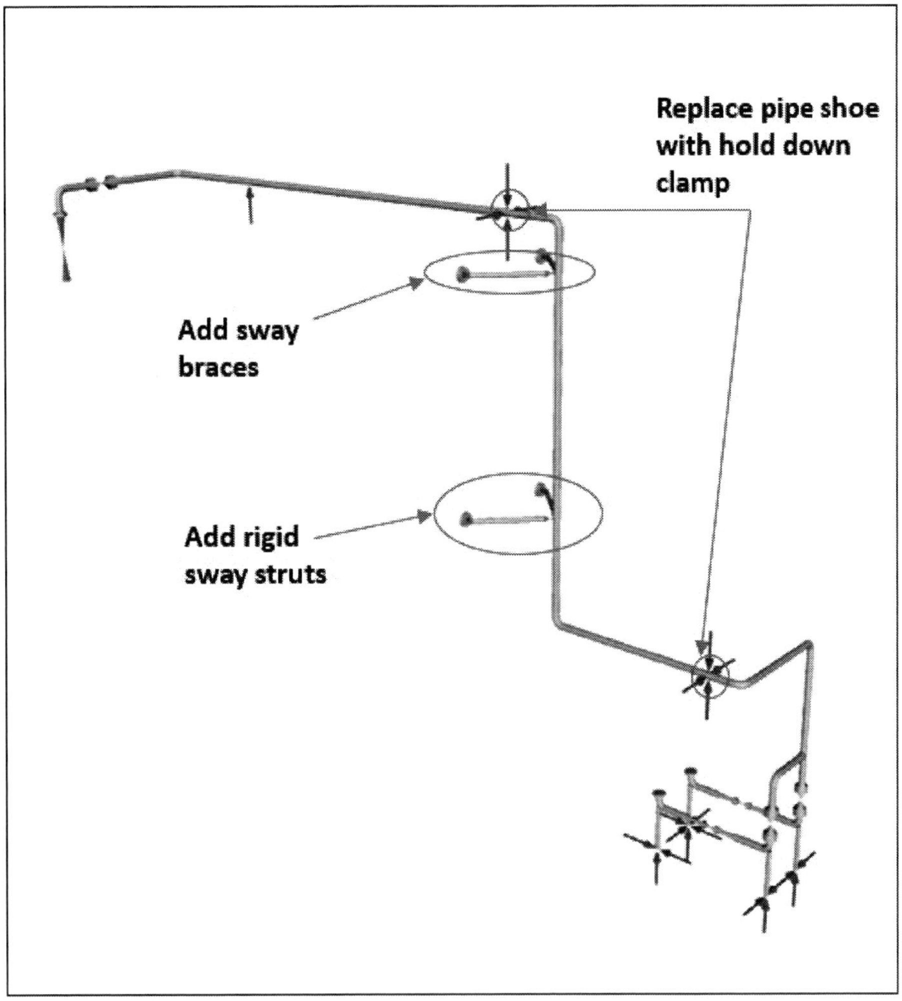

Figure 4: Case Study #2, Recommended Supports

Listed below is a summary of the takeaways from this case study:

- The layout factor is one of the more challenging values to determine for real piping systems. For this specific example, using the least conservative layout factor available in OM-3 of 1.33 for cantilever and simply supported pipe spans, versus the most conservative layout factor of 0.74 for Z-bends does not change the outcome of the assessment, indicating the need for mitigation and/or further advanced assessment (FEA).
- Typical pipe guides have gaps and, therefore, do not effectively restrain piping vibration unless the guide is in contact with the pipe shoe and the vibration force cannot overcome the friction. For cases where the guides do not adequately restrain the vibration, the guides can be replaced by hold-down clamps. It should be noted that, ideally, the hold-down clamps should have a liner to avoid fretting damage from metal-to-metal contact on the pressure boundary.
- For piping with a high design temperature, it may be difficult to add piping supports without causing high stresses by restraining thermal expansion. For cases like this, sway braces can be used.
- An alternative option to sway braces are hydraulic snubbers. For many piping vibration applications, sway braces are preferred to hydraulic snubbers. At higher frequencies, the hydraulic snubbers are not effective at restraining vibration. Hydraulic snubbers are best

used for applications with large transient vibrations such as water hammer events. Also, hydraulic snubbers require regular maintenance to be effective.

- For some piping vibration applications, it may be easier and more cost effective to add additional piping supports to reduce the vibration rather than performing a dynamic FEA. For this application, there was existing structure near the piping that made it relatively easy to attach new sway struts and sway braces. If there was not existing structure there, it may have been more cost effective to perform dynamic FEA rather than build an entire new structure to attach new piping supports.

REFERENCES:

[1] Energy Institute, Guidelines for the avoidance of vibration induced fatigue failure in process pipework, 2nd edition, 2008.

[2] J. C. Wachel, S. J. Morton and K. E. Atkins, "Piping Vibration Analysis," in *19th Turbomachinery Symposium*, College Station, TX, 1990.

[3] ASME, "Operation and Maintenance of Nuclear Power Plants (ASME OM)," ASME, 2022, pp. 119-121.

[4] ASME Boiler and Pressure Vessel Code, Section III: Rules for Construction of Nuclear Facility Components | Appendices (ASME BPVC.III.A), ASME, 2019.

[5] ASME Boiler and Pressure Vessel Code, Section III: Rules for Construction of Nuclear Facility Components | Division 1 - Subsection NB: Class 1 Components (ASME BPVC.III.1.NB), ASME, 2019.

[6] ASME, ASME B31.3: Process Piping, ASME, 2022.

[7] ASME, ASME B31J: Stress Intensification Factors (i-Factors), Flexibility Factors (k-factors), and Their Determination for Metallic Piping Components, ASME, 2017.

[8] The British Standards Institution (BSI), BS 7608: Guide to fatigue design and assessment of steel products, BSI Standards Limited, 2014+A1:2015.

[9] M. F. Bifano, L. Breaux, A. J. Feller and R. Brodzinski, "New Fatigue Screening Criteria for the Fitness-For-Service Assessment of In-Service Process Piping Vibrations (PVP2018-84847)," in *Pressure Vessels and Piping Conference (PVP)*, Prague, Czech Republic, 2018.

VIATC TRAINING CONFERENCE & EXPO
NIAGARA FALLS, NEW YORK JULY 26 -28, 2023

Fundamentals of Digital Signal Processing (Filtering and FFT)

John Hiatt
DEWESoft LLC.
John.hiatt@dewesoft.com

John Hiatt is a Business Development Engineer for DEWESoft LLC. He has 40 years of work experience. A B.S. in ME from the Military Academy and M.S. in ME From University of Texas at Austin. Work experience includes 10 years in the US Army serving in the first Gulf War in Armor Branch, 13 years at Ford Motor Company doing Powertrain NVH, 17 years as an application engineer in the areas of noise, vibration, durability, and environmental testing with LMS, Siemens and DEWESoft LLC.

ABSTRACT:
This paper will cover some common digital signal processing techniques, transforming time data to the frequency domain with Fourier Transform and filtering techniques with the infinite and finite Impulse response filter types. The main purpose of this paper is to inform the reader about how these mathematical processes impact the data. Also, understanding why the same process is done with two different engineering tools can yield different answers.

KEY WORDS: Fast Fourier Transform (FFT), Filter, Infinite Impulse Response, Finite Impulse Response, Convolution, Windows, Scaling, Frequency, Spectrum,

Transforming Data from Time to Frequency Domain:
Transforming data from the time to frequency domain is done with a mathematical process called the Fourier transform. This operation is reversible meaning going from time to frequency back to time yields the original time data. The discrete FFT (used on digitally sampled data) for both transformations are below:

Time to Frequency $$X_k = \sum_{n=0}^{n-1} x_n e^{-i2\pi\frac{k}{N}n}$$ Frequency to Time $$x_n = \frac{1}{N}\sum_{k=0}^{N-1} X_k e^{i2\pi\frac{k}{n}N}$$

Where:
X_k: Time Series
x_n: Frequency Spectrum
N: Number of Samples/Lines
n: Index Number Sample in time domain
k: Index Number Sample in Frequency domain

The Fourier Transform decomposes a time signal into sine waves with an amplitude and phase that allows the analyst to quickly detect the frequency of vibration and link the vibration to a source. We will not go into details of the math the following link provides more information on the transform with animations: https://training.dewesoft.com/online/course/fft-spectral-analysis. What we want to discuss is the assumptions of the transform, how these assumptions impact your data and what are the questions a typical analyzer will ask for the user to process. These questions will impact the results and if not done the same on two different systems will create different results.

The Fourier transform assumptions: 1. There is a constant time between samples, 2. The signal is infinite in duration, 3. The signal is periodic over the samples the transform is calculated on. A periodic signal can be described as a signal that starts and ends at 0 or if the signal is appended has no discontinuities.

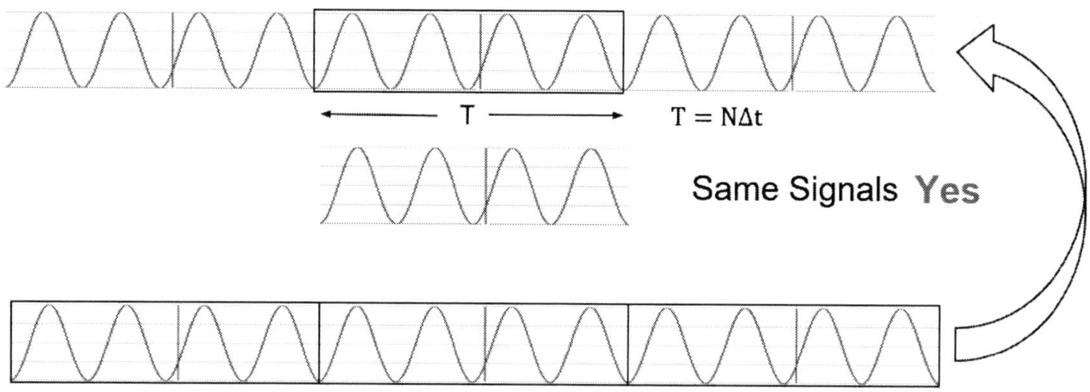

Figure 1: Periodic Signal

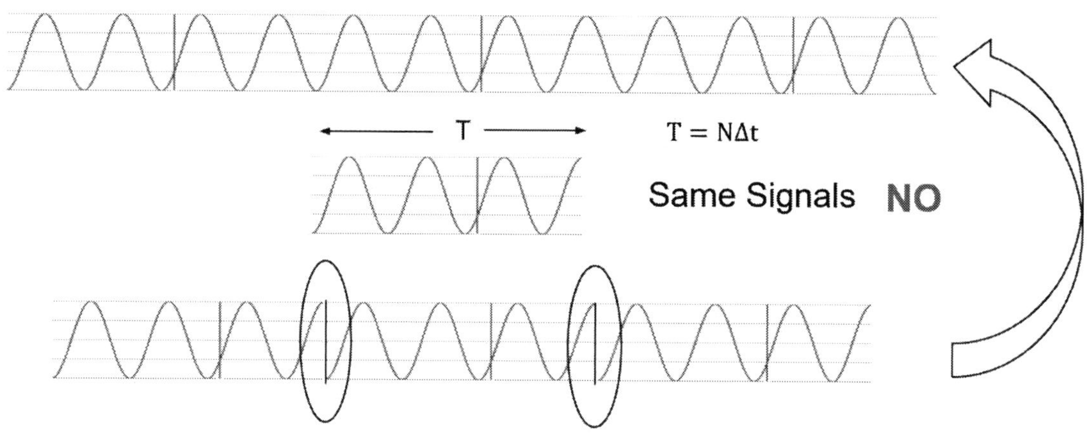

Figure 2: Non-Periodic Signal

When a Fourier transform is calculated on a non-periodic signal leakage will occur. The transformation of this discontinuity is a broadband spectrum in the frequency domain. This creates smearing of energy across the spectrum as illustrated in Figure 3 where the transformation of a spectrum with and without leakage is illustrated. Leakage commonly occurs when there is frequency content in a signal that does not fall on a spectral line, for example transforming a 100.5 Hz sine wave with a 1 Hz resolution. Although leakage can significantly affect the amplitude of a signal the overall energy (RMS Value) remains the same as shown in figure 3.

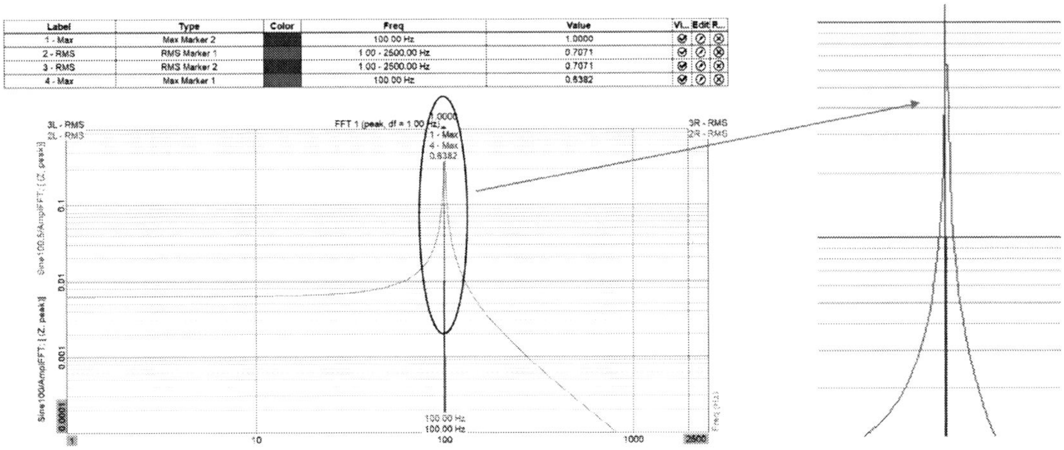

Figure 3: Example of leakage

The Leakage effect can be mitigated using a proper window. Windows simply performs a time weighting function that starts and ends at 0 (typically bell-shaped) that forces periodicity in the data processed. There are many types of windows used in FFT processing. A few of the more common window functions used in frequency analysis are Hanning, Hamming, Blackman, Blackman-Harris, Flat-top, Rectangular/Uniform/No Window. Figure 4 below shows the window function in multiplied by the original time series to create the windowed time domain output that is transformed to the frequency domain.

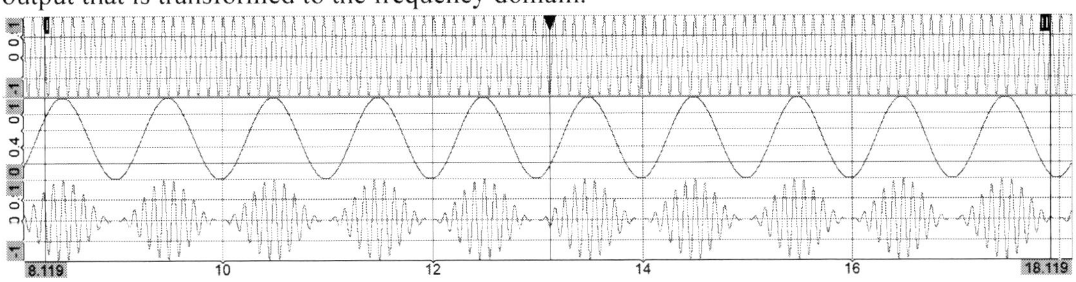

Figure 4: Hanning window applied to sine wave.

If no window is applied the maximum amplitude error possible would be 36%. In figure 5 a FFT is calculated with 1 Hz resolution on a 1 volt, 100 Hz (blue) and 100.5 Hz (red) sine wave with and without a Hanning Window. For periodic signal (blue), both windowed and unwindowed signals yield the correct amplitude. While non-periodic signal (red) the windowed signal as significantly less error due to the mitigation of leakage.

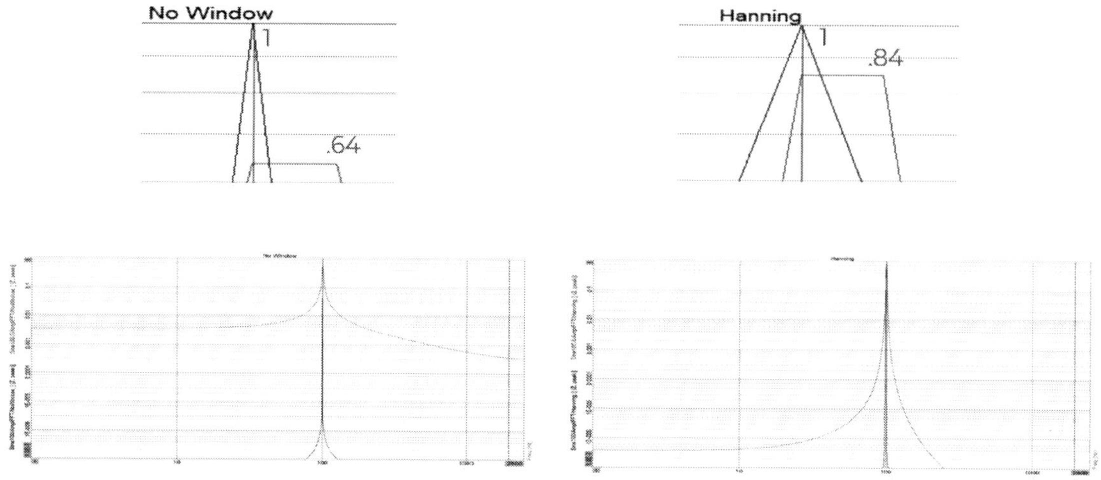

Figure 5: FFT of Hanning windowed an unwindowed sine wave.

Although windows mitigate leakage, they have two negative effects they smear the frequency resolution (noise bandwidth), and they alter the signal shape. The signal alteration can be seen in the time domain in Figure 4. Frequency smearing is visible in the frequency domain when observing the main lobe width of the window functions FFT.

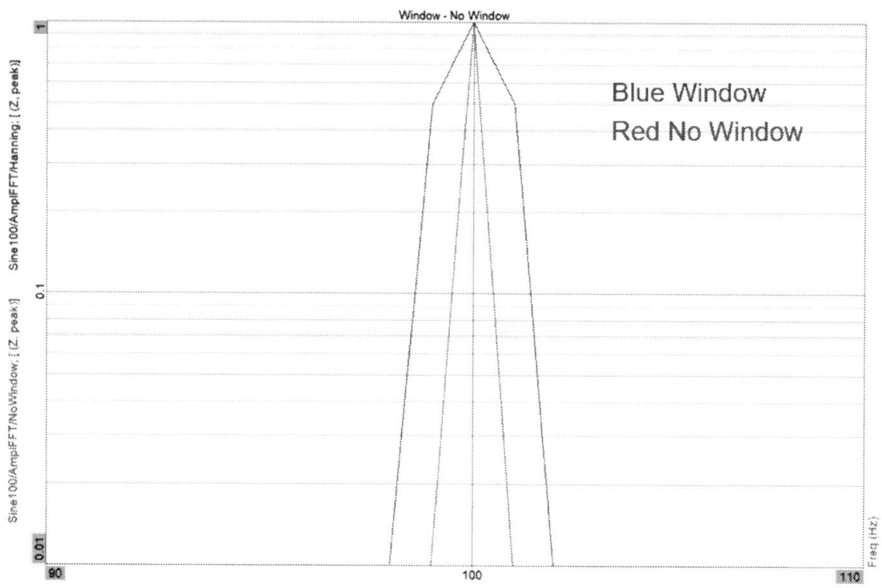

Figure 6: Frequency Smearing

Most analysis software automatically applies window correction factors. When looking at the spectrum overall level an energy correction is used. When creating a 2D graph, the amplitude correction is used. Keep in mind even with the amplitude correction there can be a 16% error observed, see Figure 5 but better than a 36% error with no window. The frequency smearing cannot be corrected. Table 1 gives a comparison of window effects note the Flattop window has a large noise bandwidth but a very minimal amplitude attenuation this window is ideal for calibration.

Window Type	Noise bandwidth/ Frequency Smearing	Max Amplitude Error (dB)	Amplitude Correction	Energy Correction
Rectangular	1	3.9	1	1
Hanning	1.5	1.4	2	1.63
Hamming	1.36	1.8	1.85	1.59
Blackman	2	1.1	2.8	1.97
Flattop	3.43	<0.01	4.18	2.26

Table 1: Window Comparisons for Frequency Smearing and Max Amplitude Error

Analysts need to understand if window correction factors are applied and what window was used in the analysis before comparing results from different analysis packages. Other factors that affect the output of the FFT and will be inputs to analyzers are spectrum scaling, spectrum type, frequency resolution (depends on data analyzed and spectrum type), averaging type, overlap, and double-sided spectrum.

FFT scaling should be labeled as either Peak, RMS, or Peak to Peak. Typically scaling creates differences of a factor of 2 or SQRT(2) (3dB or 1.414 or 30%).

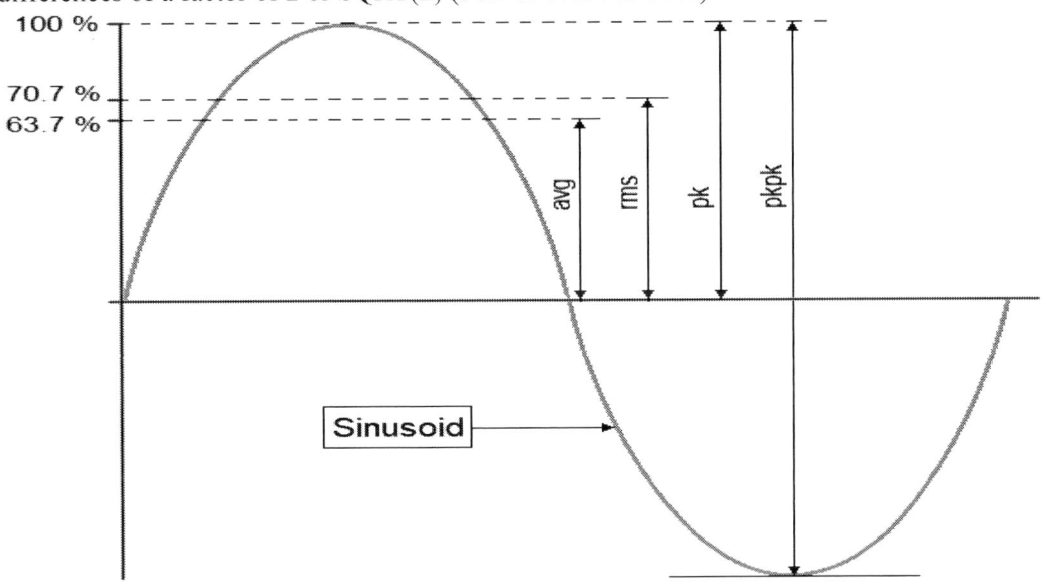

Figure 7: Spectrum Scaling Relationship to a Sine Wave

Spectrum types/outputs of the transform include spectrum, auto-power-spectrum, power spectral density (PSD), auto spectral density (ASD), and energy spectral density (ESD). We will not cover the applications of each of these functions but want to remind the reader that we must compare the same function types. The easiest way to recognize these functions are by the units see Table 2 below. If comparing spectrums, make surer the Sspectrum types/units are the same.

Spectrum Type	Output	Output Type sample unit g
Spectrum	Complex	g
AutoPower Power	Real	g^2
AutoPower Linear	Real	g
Power Spectral Density	Real	g^2/Hz
Auto Spectral Density	Real	g/sqrt(Hz)
Energy Spectral Density	Real	g^2-sec/Hz

Table 2: Spectrum Types calculated from the Fourier Transform

Frequency resolution can make a difference in the spectrum amplitude but depends on both the spectrum type and input signal type analyzed. We generally have three types of signals random (cannot be defined by a math function), deterministic (can be defined by a math function), and a combination of both. Two of the more common signal types, the PSD and Auto Power Linear have units of g^2/Hz and g respectively. The PSD is normalized by the frequency resolution whereas the Auto Power linear (simply the sqrt of the AutoPower Power) is not. We will take a 100Hz sine wave and a 2500Hz bandwidth white noise signal and process the functions at three different frequency resolutions 1, 2, and 4 Hz.

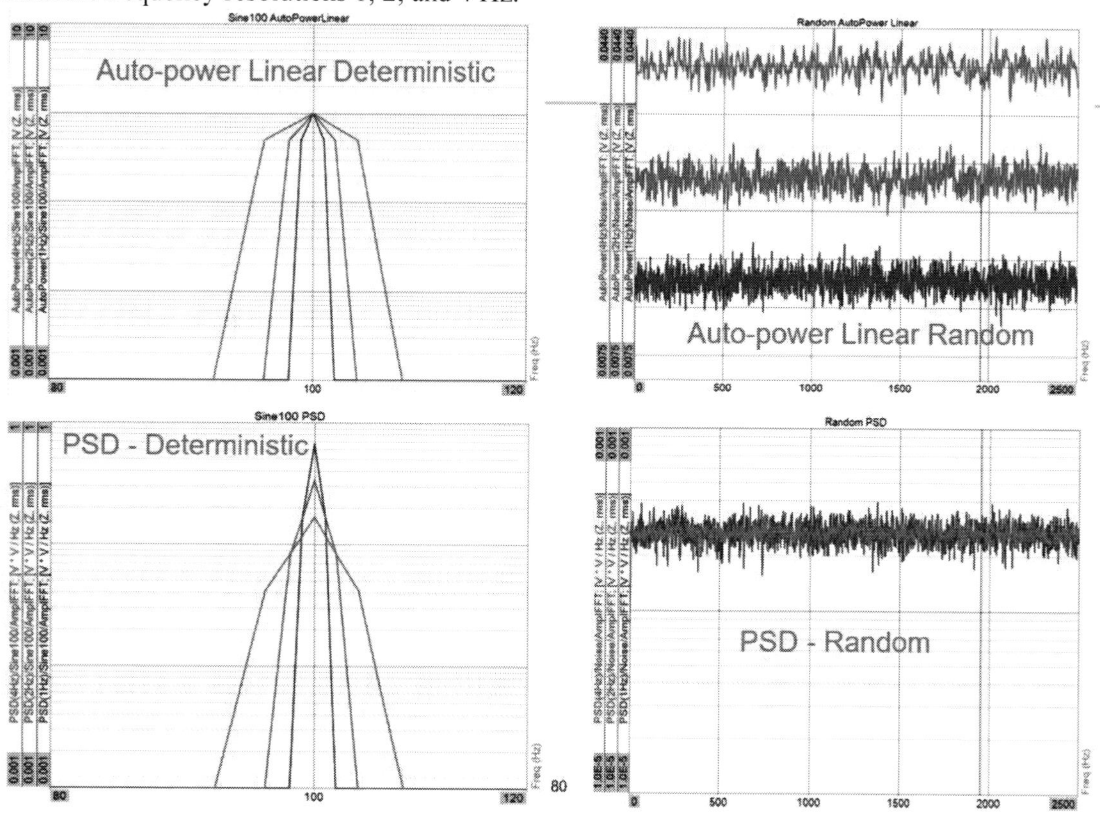

Figure 8: Processing effects of spectrum type on various inputs.

	Periodic (Sinusoidal)	Random
Auto-power Linear g	No Amplitude Change	Amplitude Change
PSD g^2/Hz	Amplitude Change	No Amplitude Change

Table 3: Comparison Spectrum type processed on various input types.

The main takeaway is frequency resolution should always be documented. PSD should typically be used on random data. If deterministic (lots of rotating components) use Auto-power and always document frequency resolution.

Averaging typically on stationary data the FFT is averaged. The most common average types are linear, Power (averaging squared units), exponential, and sometimes peak hold is considered an average. Make sure the same average type is used and if exponential averaging is selected there will be an additional time constant factor to set. This is all straightforward until we average the spectrum (complex output) here we have both the magnitude and the phase. If the phase is not referenced to any other signal, the phase of each spectral line is derived from the start of the time block the FFT is calculated on. This results in a different phase for a given spectral line from one FFT block to the next causing the amplitude to average out.

This can be mitigated in some systems by calculating a phase reference spectrum which simply subtracts the phase of the reference channel from the phase analyzed channel. Other common references are one pulse per rev signals on rotating shafts. Averaging reference channels is a clever way to remove components in the signal that do not have a constant phase relationship with the reference.

Overlap simply sets the overlap of the FFT block typical values 30-70%. This can mitigate some of the information lost when applying FFT windows and should be documented.

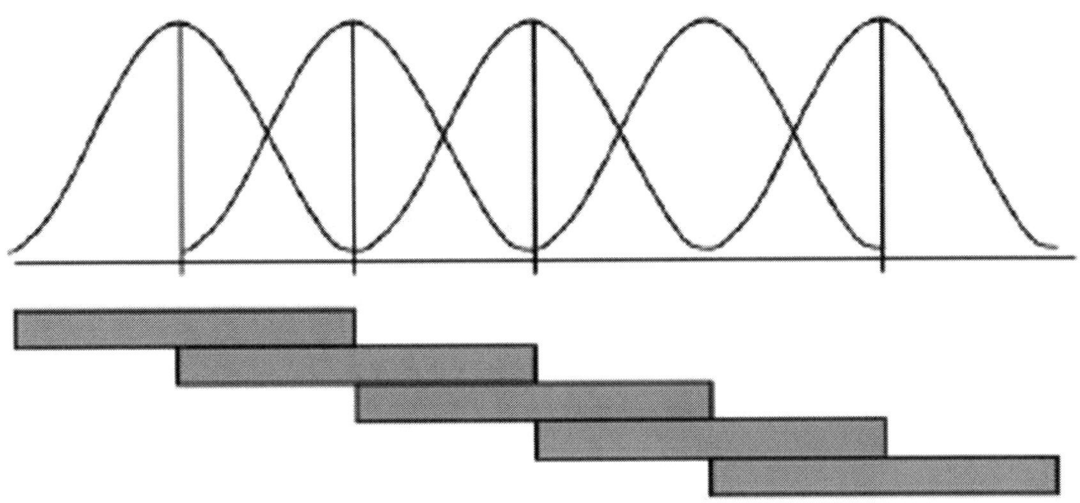

Figure 9 FFT Block Overlap

A Double-Sided Spectrum is the output of an FFT not to be confused with the full spectrum. Unlike the Full spectrum where analysts use the negative frequencies to make engineering decisions the negative frequencies on the FFT spectrum output are typically folded over the 0 Hz line and multiplied by 2. All analyzers I have seen do this correctly but if you are writing your own code keep this in mind.

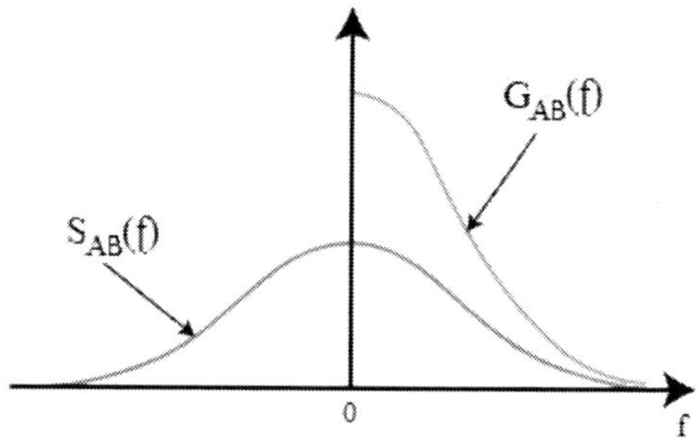

$$G_{AB}(f) \begin{cases} 0 & \text{for } f < 0 \\ S_{AB}(f) & \text{for } f = 0 \\ 2 \cdot S_{AB}(f) & \text{for } f > 0 \end{cases}$$

Figure 10: Double Sided Spectrum

Finite and Infinite Impulse Response Filter Affects (FIR and IIR respectively):

There are many filters in the data acquisition process. Filters in the process analog anti-aliasing, digital anti-aliasing (resampling), AC coupling (removes DC), User selected filters to remove unwanted frequencies, and weighting filters. The first three filters are typically done on the hardware. Some of which the user cannot set or manipulate. These filters are both analog and digital, IIR and FIR filters. When performing the filtering process on a signal there are typically trade-offs (anomalies). The pros are usually obvious, it's the cons (unwanted effects) that we will examine in this section.

Common filter types are Low Pass, High Pass, Band Pass, and Band Stop. FIR filters parameters require setting the number of taps (proportional to roll-off) and window type where IIR filters have prototypes (Butterworth, Bessel, Chebyshev) and a filter order (proportional to roll-off).

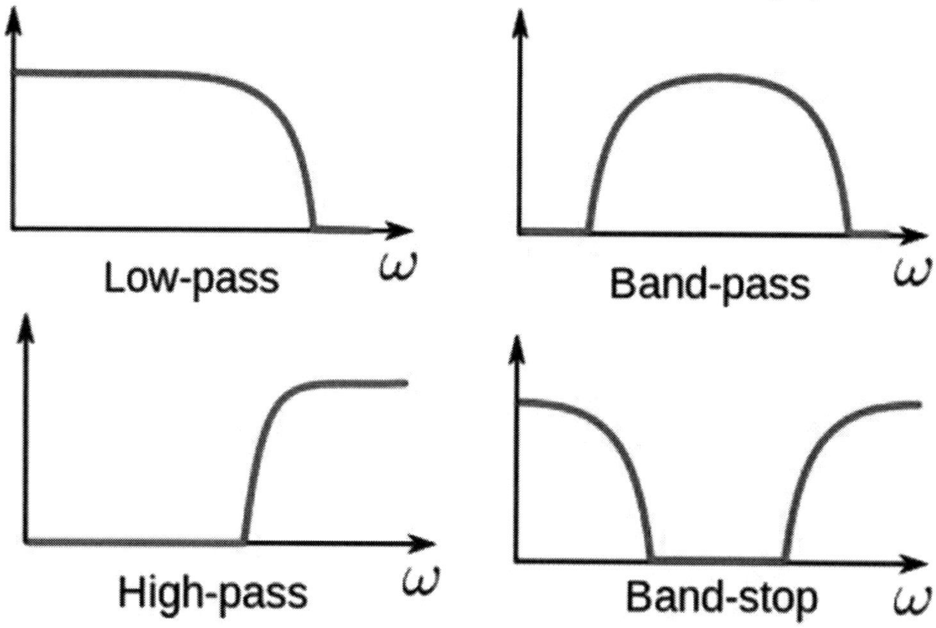

Figure 11: Filter types

Generalized filter process, we define the filter in the frequency domain perform inverse FFT obtain the IRF (Filter Impulse Response Function) and convolve the IRF with the time domain signal to get filtered results. The IIR has a similar flow but uses filter prototypes to calculate coefficients and recursive convolution which yields similar results with fewer coefficients with some trade-offs which we will discuss.

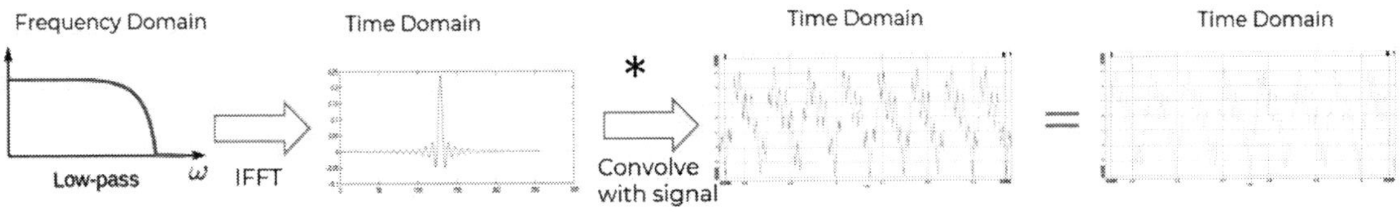

Figure 12: Filter Process

As mentioned, the mathematical filter process creates anomalies in the data which the analyst must be aware of. These anomalies include roll-off in the passband, distortion in the passband, group delay, phase delay, and distortion at beginning of the signal. Before we cover the anomalies, let's compare the FIR and IIR filters. This will give us some insight into the filter effects.

IIR	FIR
can be unstable	always stable
nonlinear phase response	linear phase response
faster	slower

Table 4: IIR and FIR Comparison

Roll-off in the passband is created because it is mathematically impossible to design a brick wall filter (no transition band). The specified frequency of the filter is typical -3dB (30%) at Nyquist frequency (0.5* Sample Frequency). In practice, this means there is signal attenuation in the top 20% of the bandwidth (BW) so be aware when comparing frequencies on a spectrum in the top 20% of the BW or if time domain analysis with significant frequency content in the top 20% of BW. For frequency domain analysis always sample at least 2.5*max frequency. Time domain rule thumb s 10x max frequency. Note some systems have more aggressive resampling filters. Simply use a function generator and these filter characteristics can be easily determined by inputting a sine wave at frequencies near the Nyquist frequency and recording the amplitude.

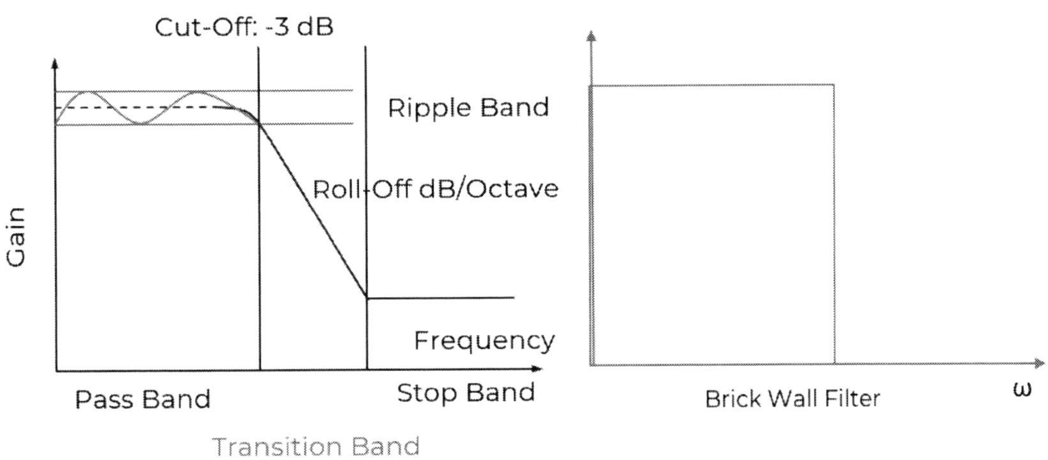

Figure 12: Filter Characteristics and Brick Wall Filter Diagram

Distortion in passband typically occurs in filters with very sharp roll-off (closer approximation of brick wall filter). The IIR Chebyshev filter is used because of the sharp roll-off (short transition band) the trade-off being ripple (see Figure 12 for ripple illustration) in the pass band. If using the filter most implementations will require you to input acceptable ripple in the pass band (distortion) more ripple more roll-off. The Bessel filter is the other extreme with lower roll-off rate for same order but very flat in the passband band while the Butterworth is a good compromise between the 2.

Group Delay and Phase Delay These filter properties are easily confused. Group delay is created because filters consist of coefficients that are convolved with a signal at the start of the signal as the filter passes through the signal there are no data points that align with the filter point causing the signal to be shifted in time. The greater the filter coefficients the longer the group delay. Since FIR filters require many more coefficients than IIR for the same roll-off, the group delay is much larger (this is why we say IIR filters are faster more efficient mathematical operations). Phase delay deals with how the phase of the individual frequency line is affected. This will be shown on a 2D graph with phase delay as function of frequency. FIR filters have a linear delay which is a constant delay as a function of frequency. For FIR filters the group delay and phase delay are the same. IIR filters have a nonlinear delay which means the phase delay is a function of frequency. See green line Figure 13.

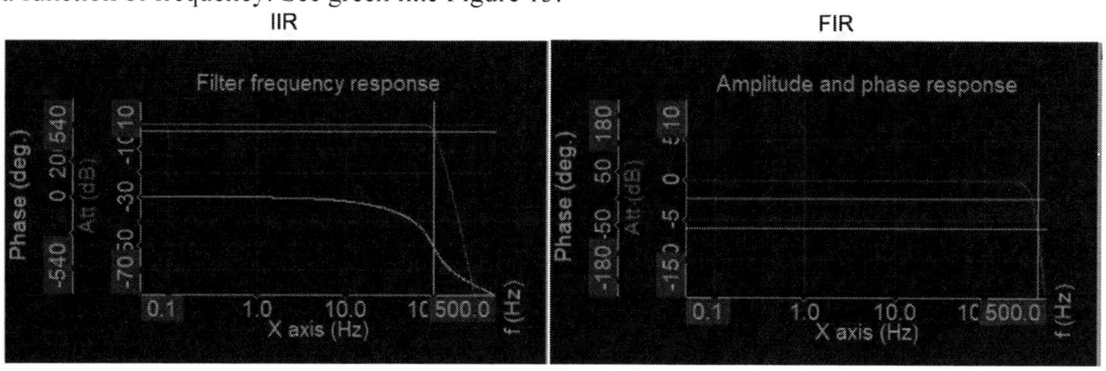

Figure 13: IIR and FIR Filter Frequency Response

The FIR filter group/phase delay can be corrected online and offline because it is a known constant quantity. The IIR delays can only be corrected offline typically specified with a 0 phase or multi-pass option. The system passes the IIR through the filter front to back and back to front. Check your analysis package for these options they can be critical when comparing the phase between channels in the frequency domain or the alignment of peaks in the time domain. Be aware that when channels have different filtering schemes it can affect your analysis. Example 1 pulse per rev tachometer/speed signal DC coupled and IEPE accelerometer AC coupled users will get wrong phase relationship between tachometer pulse and accelerometer order content this could create greater uncertainty when balancing.

Distortion at the beginning of the signal occurs because the filter has not had time to settle. Online this is typically not an issue since many systems start in a scope mode allowing filters to settle. Be aware of very low-frequency filters (like human body vibration or 0.05Hz ac coupling filters) these can take on the order of a minute to settle. Offline processing is where we run into more issues, especially on shock events where we sample very high (long duration large files) and are interested in a very short duration event (usually trigger acquisition). Often, we measure acceleration but are interested in velocity and displacement as well. When integrating we apply a high pass filter to remove the DC (otherwise integration will not yield a meaningful result) please note the distortion at beginning of the signal. The best practice is always acquire a little more data than you need.

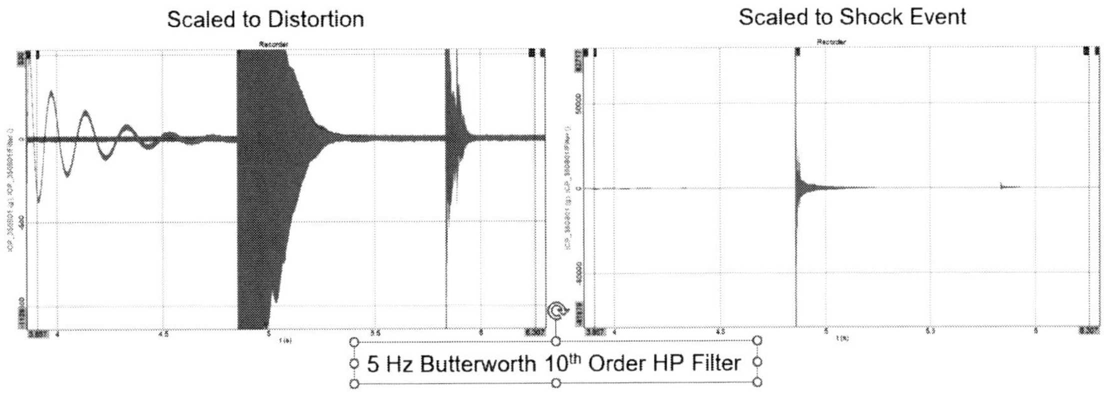

Figure 14: Filter Distortion at beginning of signal

Closing the loop Table 4 filter stability. It is possible for an IIR filter to be unstable, typically commercial code will not allow you to create an unstable IIR filter. However, if the software allows the user to generate their own IIR filter coefficients the user is responsible to check the stability of the filters the tools to do this are beyond the scope of this paper.

Summary:

The objective of this paper is to inform the reader on considerations of the Fourier transform both possible errors based on assumptions of the transform and how the multiple settings (inputs in the software) affect the output and the ability to compare results from different software packages. Hopefully, the reader also understands what parameters need to be specified when calculating the transform. Parameters include frequency resolution, amplitude scaling, amplitude function, window type, averaging type and number of averages, and overlap.

In addition, I wanted to make the reader aware of how filters affect data. The primary purpose of a filter is obvious, the issues created by the mathematical implementation are not so much. These artifacts/anomalies include roll-off in the passband, distortion in the passband, group, and phase delay and distortion at the beginning of the signal.

REFERENCES:

DEWESoft Training Material "Signal filtering, Signal suppression, Signal processing", https://training.dewesoft.com/online/course/filters

DEWESoft Training Material "FFT Spectrum Analysis (Fast Fourier Transform)", https://training.dewesoft.com/online/course/fft-spectral-analysis

Søren Linnet Gjelstrup "FFT Analysis: The Ultimate Guide", https://dewesoft.com/daq/guide-to-fft-analysis

VIATC TRAINING CONFERENCE & EXPO

NIAGARA FALLS, NEW YORK JULY 26 -28, 2023

BALANCING TECHNIQUES FORGOTTEN (OR LESSONS WE HAD TO RE-LEARN THE HARD WAY)

Thomas B. Lantrip, PE
Consulting Engineer – Southern Company Services
tblantri@southernco.com

Tom Lantrip is a consulting engineer with Southern Company Generation and works in the Performance and Reliability Department - Balancing and Vibration Analysis group. Tom has worked at Southern Company for 41 years, primarily in the field of balancing and vibration analysis, with an emphasis on turbines, fans, and pump components. He has a background in structural dynamics and in structural and seismic design, vibration analysis, and balancing of rotating machinery. Tom graduated from the University of Alabama with a B.S. in Civil Engineering, attended the University of California at Berkeley, and has an M.S. in Civil Engineering from the University of Alabama at Birmingham. Tom is a member of ASCE and is a professional engineer in the state of Alabama and holds a Category IV certification from the institute.

ABSTRACT:
Practice that becomes routine or schedule driven influences our ability to use proper procedures. The fundamental rules and prerequisites of machine balancing sometimes get lost in our many years of repetition and practice. Over time, the divergence from the fundamentals and basics of our balance training creates opportunities for errors and carelessness to affect or alter our performance and success in balancing of machines.

Instituting short-cuts and assumptions sometimes gives rise to allowing less than ideal outcomes for our work. This paper reminds us of those pitfalls and reconfirms the correct balance process to prevent errors in the future and reinforce the basics.

KEY WORDS: Balance, Phase Reference, Repeatability, Runout Compensation, Steady-State

FUNDAMENTALS OF BALANCING:

Phase Reference for Balancing

Besides the vibration analysis and preparation for machine balancing, the most valuable tool and installation is the phase reference sensor and signal acquisition. The phase reference (sometimes known as Tachometer, Speed Reference, Keyphasor ©, Trigger, etc.) supplies the once-per-rev signal of the rotating shaft that provides the timing relationship to the vibration response at once-per-rev frequency.

The phase reference can be acquired using many sensor types, which may include an optical-laser type sensor, magnetic sensor, proximity (non-contact) sensor or strobe-light sensor.

In permanent installations, the magnetic or non-contact proximity probe is typically used where in for temporary or short-term use an optical-laser type sensor is used with strobe-light as in-frequent technology of the modern practice.

Several scenarios have occurred in the personal experience of the author that have challenged the success of collecting accurate and valuable phase reference data.

The fundamental waveform plot of the vibration signal synchronized with the phase reference signal as shown in Figure 1. That same data similarly presented in the polar plot format that is shown in Figure 2 is key to the entire balance process and quality of the work. Calculation of the phase lag angle is by definition "the time interval between the phase reference pulse and the first positive peak of the vibration signal "for the respective measurement location divided by the period of the synchronous speed or time between two phase pulses (period).

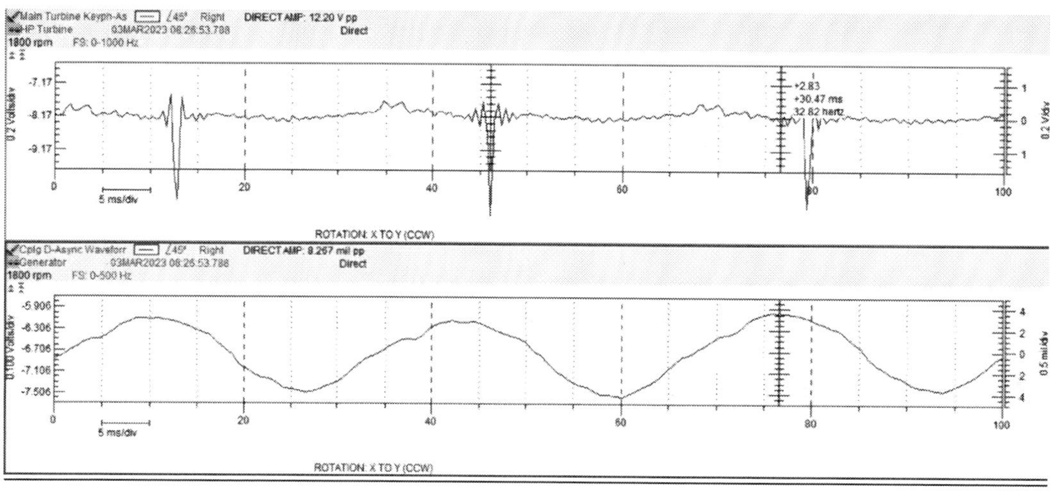

Figure 1 : Phase Reference Signal and Time Waveform of the Vibration Signal

In the overall time waveform plot above the "time to the first positive peak ~ 30.47 msec for a period of 33.3 msec (1/1800 rpm (30 Hz). This will yield a phase angle of ~329 degrees lag when manually calculated for the time waveform shown. Note, the 1X filtered waveform will provide a slightly different phase as shown in the polar plot in Figure 2 displaying 323 degrees lag.

It is imperative that the phase reference signal and the responding vibration data is collected in a consistent and repeatable manner for success to be achieved.

Using the vector data, the balance procedure of making one balance mass change, measuring the corresponding effect, and adjusting to close on the balance solution is both art and science.

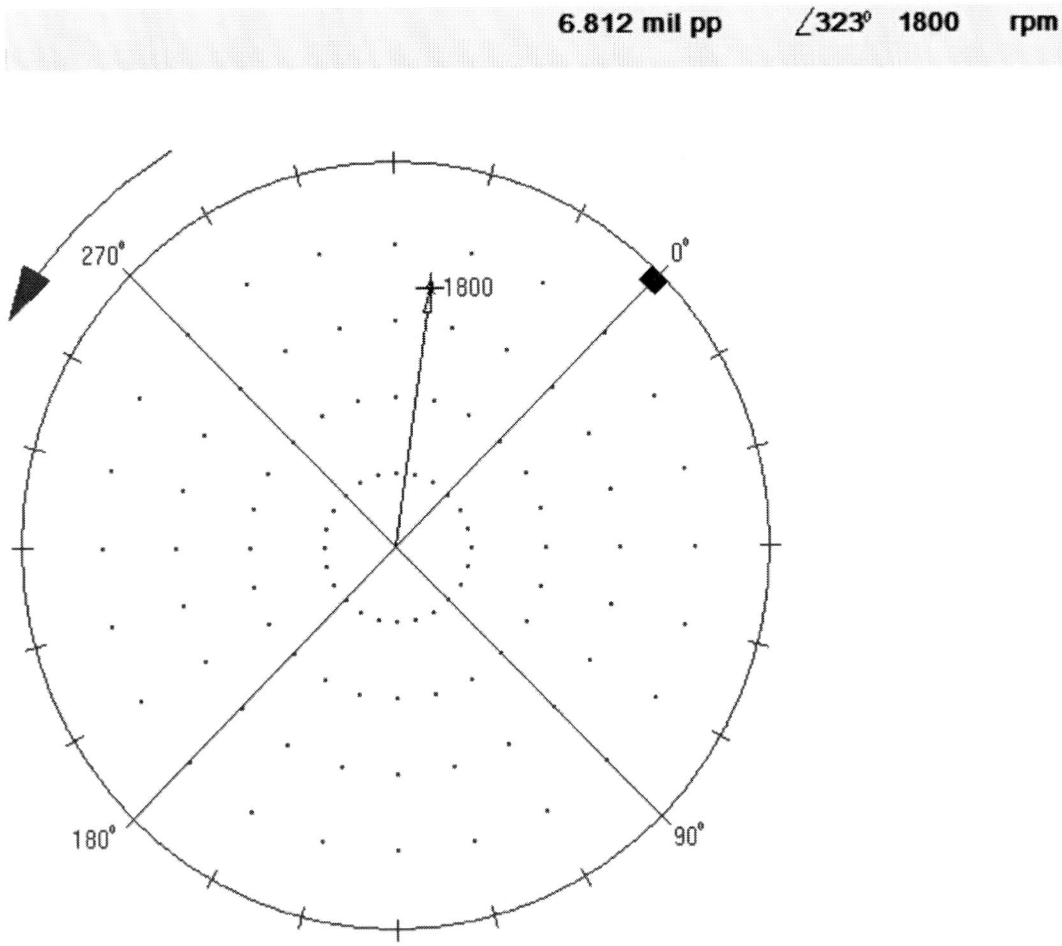

Figure 2 : Polar Plot of Typical Vector Amplitude and Phase from Time Waveform

The vector data is dependent on the accuracy and precision by which the data shown in the two above figures.

The total balance procedure which not only includes the vector interpretation of the data acquisition but also the physical location of the balance weight as it relates to the polar plot figure or balance map, the amount of the balance weight or weights, their mounting technique, including weldment weights, removal of fouling debris or dust, and management of trial weights, permanent weights, and weight consolidation all make up a very complex set of accurate steps to achieve a successful balance campaign.

Included are stumbles and reminders that plague our opportunity to achieve our target balance improvement. Mistakes or misinterpreted conditions, lack of knowledge of the sensor, hastily obtained or incomplete data, assumptions about the condition of the machine to be balanced all contribute to needed rework or minimal improvement from the balance performance.

STEAM TURBINE BALANCE ISSUE:

Steam Turbine Balance Issue

Turbine Phase Reference Startup Changes

During a balance iteration on a coal-fired power steam turbine, vibration data was collected during the operational cycle. This vibration data was recorded and analyzed using a remote monitoring and diagnostic on-line system, which has become more common with advances in network and internet applications. As part of the normal process for balancing of a steam turbine, recording and capture of the vibration data for each bearing of the turbine/generator and later analysis with the expectation that the fundamental frequency of vibration would be the operating frequency (1X).

The captured data showed unacceptable levels for continuous long-term operation. The fundamental fault (1X operating speed frequency) indicated in the review of the data suggested an unbalance was occurring. A balance solution was prepared after analysis of the detailed data suggested a balance solution was possible to reduce the vibration levels.

The turbine was a "sister unit" of another common machine where balance solutions were very predictable and past success had been achievable. As is common in balancing practice, knowledge of a machine's sensitivity to unbalance or balance vector response is a vital and important asset to the balance practitioner.

At the next outage opportunity, the balance solution was transmitted to the plant staff and installed according to the known phase reference location (notch) and proper rotor-phase angles. The determination procedure of the phase reference location which includes the use of the voltmeter or physical identification verification of the phase reference notch and alignment with the phase reference probe.

As is common on turbine-generator managed components, the operator was expecting a singular "solution" typically not a "trial run analysis ". This solution is commonly denoted as a "one-shot" approach, where the weight location and amount should be successfully located correctly without the need for rework or second trial or correction balance weight.

In this case, the balance weights were installed during the outage, and upon return to service the vibration levels were higher than previous vibration levels for the similar operating conditions. In addition, the vector analysis revealed that the vector was higher than before the balance with similar phase angles **(this typically is the case when the weight is added to the heavy spot, not good!)**

Post-Mortem of Failed Balance Success

As this situation was a remote analysis and balance solution, the information therefore for the balance weight installation relied upon the plant staff to properly locate the phase reference and install the balance weights with written instructions.

As is common, the blame would be that someone else made a mistake, not me! Subsequently, the instructions provided to the plant staff were revisited and appeared to be technically correct. In addition, the plant staff was interviewed prescriptively, and the results suggested the balance solution was correct, the plant staff actions were confirmed step-by-step as proper and correct. **What had gone wrong?**

During the next outage (several months later), a plant visit was in order. This visit was enabled to remove the balance weights that had been installed apparently in the "absolutely wrong" location on top of the heavy spot. The weight locations were verified, and the plant staff had followed the advised locations for the balance weights according to my instructions. **There was no blame or error on the part of the plant personnel.**

The unit was restarted up after the balance weights were removed with this time, a couple local analyzers were "piggy-back" to the remote system buffered outputs to acquire incredibly detailed information for the pre-balance condition.

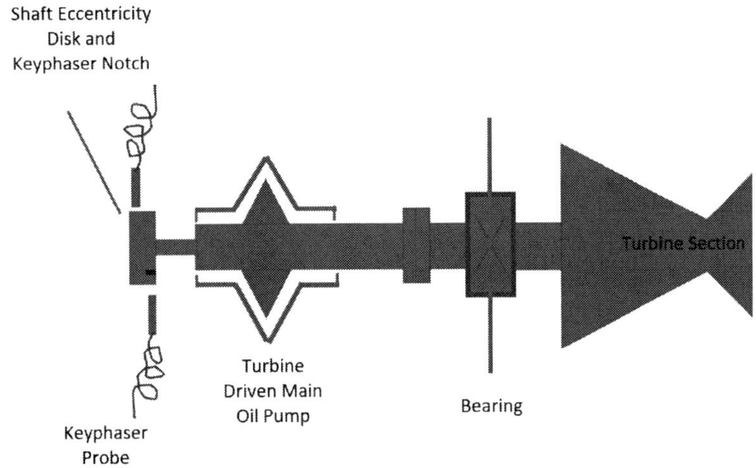

Figure 3 : Plan View Schematic of Steam Turbine Phase Reference Location

Figure 3 provides the layout of the unit with the phase reference (keyphaser) located on the eccentricity wheel on the end of the turbine main oil pump shaft. The eccentricity disk is a feature on steam turbines that helps the operator to recognize thermal bows in turbine rotors during startup conditions. Figures 4 provides the cross-sectional details of the phase reference (keyphaser probe and eccentricity probe).

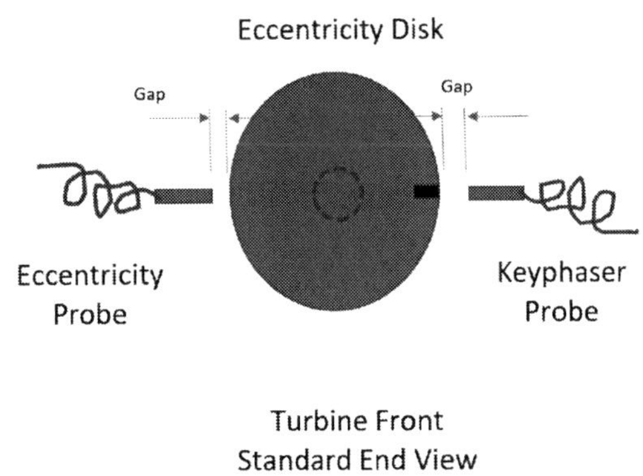

Figure 4 : Elevation View of Steam Turbine Phase Reference System and Location

Figures 5 provides the time waveform trace of the phase reference trace respectively during slow roll turning gear condition and startup.

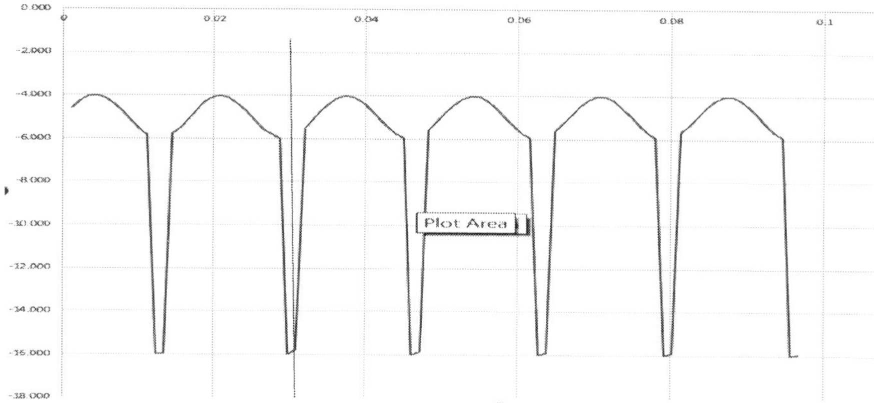

Figure 5 : Time Waveform Trace of Steam Turbine Phase Reference Signal including Pulse and Vibration on Eccentricity Disk

During the speed ascension process, the vibration data captured from the "piggy-back" vibration analyzers revealed that the gap voltage (DC) component of a non-contact Keyphaser © signal was shifting as a function of speed, with the apparent conclusion that the front standard of the turbine was moving (migrating) away from the eccentricity disk as viewed from the Keyphaser probe. As seen in Figure 4 the DC gap migrated to such a high-level that the "notch" pulse was apparently truncated to a flat line with the resulting positive portion of the vibration sine wave creating a "new pulse". The phase reference instrumentation transitioned from the reference phase "pulse" shown in Figure 5 to the high-speed reference phase "pulse" shown in Figure 6.

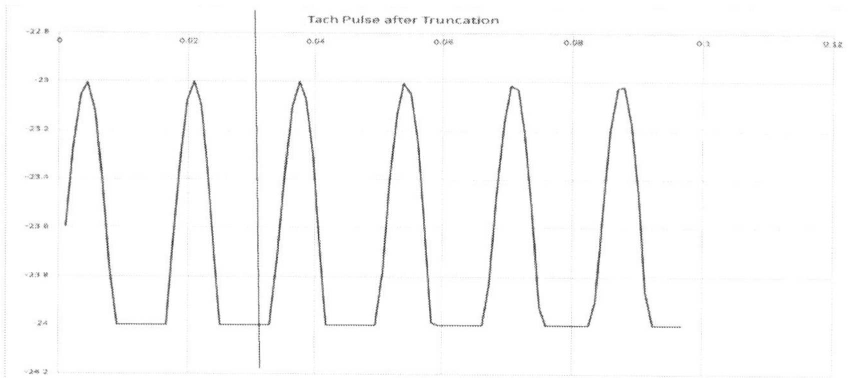

Figure 6 : Phase Reference Pulse after Signal Truncation

(Phase Reference Pulse is approximately ~180 from pulse in Figure 3)

Problem: Keyphaser position (out-of-range) during transient conditions (heat-up evolution)

Lesson Learned: Always review the phase reference signal and validate the phase calculation manually to determine that the phase angle used for balancing makes sense.

KNOW YOUR SENSOR!

Velocity Phase Rolling Due to Wiring

When collecting vibration signals using velocity sensors, it is important to understand the type of sensor being used. The coil-style velometers will flip phase due to wiring, check the polarity or check with alternative sensor. This is not a fatal flaw when using the velocity sensor for trending amplitude or some types of analysis as the amplitudes will be the same with either wiring.

When using the coil-type velocity sensor for single balancing, it is clear to see that the positive peak if that is desired could be flipped, thus changing the phase angle by 180 degrees. If using that data for first trial or balance correction, placing the weight on the heavy spot is not the most desirable outcome.

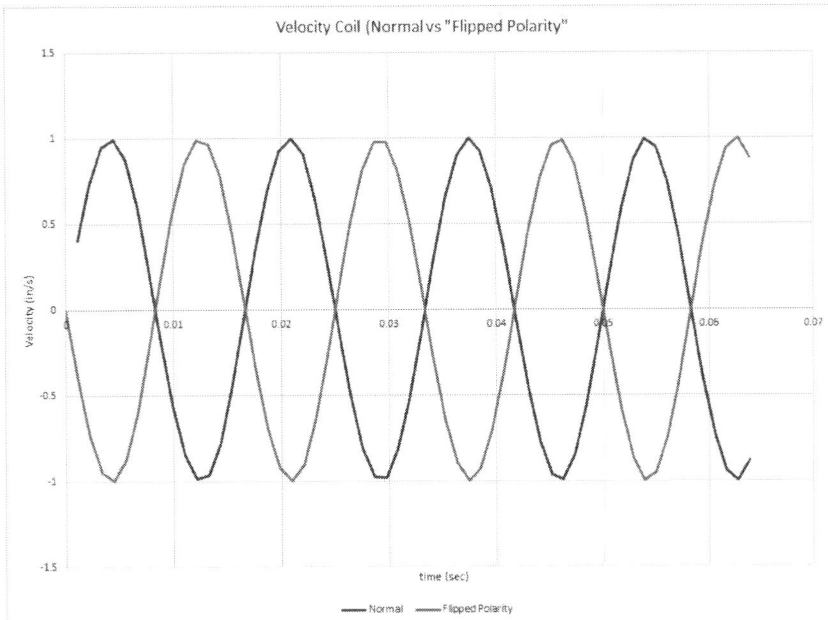

Figure 7 : Velocity Signal Polarity "Flipped" Depending on the Wiring

<u>Accelerometer vs Velocity Transducer</u>

When collecting vibration signals using accelerometer or piezo-velocity sensors, it is important to understand the type of sensor being used. If the sensor is an accelerometer instead of a velometer the phase angle will be different by 90 degrees.

Using previous balance influence data or trying to determine the 1st trial weight amount and location could be obviously compromised by mistaking one sensor for another as the solid-state sensors look similar as shown in Figure 8 below.

If a single shot balance solution is the desired choice (in my opinion this is always the inferred approach) which utilizes the initial collection of the vibration data at the balance speed (once-per-revolution) frequency and a trial weight amount and rotor angle is calculated and placed accordingly. After which, the machine is returned to service with no additional balance weight adjustment needed and a significant reduction in the vibration levels (reduction by a factor of 3-10 times reduced) is achieved. This approach would require absolute knowledge of the sensor polarity and/or sensor type.

In another case, such as the most general case, the iterative balance solution that results in multiple runs using the vector response and correction related to the original condition, with trial weights added and removed until the reduction of the vibration to the prescribed level. The knowledge of the sensor polarity/type would not be immediately necessary to achieve success as the solution calculation is a relative movement and proportioning of the balance weights.

For the added benefit of collecting and cataloging balance influence information, it is vital that the polarity/type of the sensor, proper units and calibration of the sensor, phase reference relationship, exact amount of balance weight, including radius and exact angular attachment locations.

Figure 8 : Photo of Accelerometer and Velocity Sensor Side-by-Side

(Which one is the accelerometer?)

Problem: Use sensors and configurations that you can validate yourself.

Lesson Learned: Know your sensors and what they tell you.

RUNOUT (SLOW ROLL) COMPENSATION CONCERNS

When using shaft relative probes, always perform slow roll compensation, particularly if the runout levels are significant. Shaft relative probe vibration data is unfortunately biased by the inherent shaft surface condition of the probe projection view while looking at a finite band of the rotating shaft. Slow roll compensation involves the subtraction or compensation of the slow speed filtered amplitude and phase vector from the high-speed service condition of vibration in a polar or complex form.

Table 1 : Table of Balance Results for Uncompensated Vibration, Bearing Housing and Final Compensated Shaft Vibration

	Location	Position	Original 11/25/2009 1X (in/s pk)	Phase (deg lag)	Post Trial 1 11/25/2009 1X (in/s pk)	Phase (deg lag)	Post Trial 2 11/25/2009 1X (in/s pk)	Phase (deg lag)		
7B	Fan Inbd Vert	0	0.007	0	0	0	0	0		
Fan	Fan Inbd Horiz	90 R	0.1	72	0.094	13	0.019	297		
900 rpm	Fan Outbd Vert	0	0.002	0	0	0	0	0		
	Fan Outbd Horiz	90 L	0.126	235	0.106	162	0.036	66		
									Final Compensation	
			(mils p-p)*	(deg lag)	(mils p-p)*	(deg lag)	(mils p-p)*	(deg lag)	(mils p-p)**	(deg lag)
	Fan Inbd Y Shaft	45 L	2.85	58	1.26	48	1.78	85	0.17	193
	Fan Inbd X Shaft	45 R	2.96	168	2.33	150	2.01	163	0.15	29
	Fan Outbd Y Shaft	45 L	3.64	324	2.68	318	1.31	338	0.77	180
	Fan Outbd X Shaft	45 R	2.3	98	2.37	78	1.66	73	0.15	1
					Added 11 ounces at 210 deg lag		Removed Trial 1 Added 8.5 ounces at 255 deg lag			

* Shaft Displacement Uncompensated

** Shaft Displacement Compensated

Table 2 : Table of Balance Results for Compensated Shaft Vibration

	Location	Position	Original (PreBalance) 11/25/2009 1X (mils p-p)	Phase (deg lag)	Slow Roll 11/25/2009 1X mils p-p	Phase (deg lag)	Compensated Original 11/25/2009 1X (mils p-p)	Phase (deg lag)	Uncompensated Levels: Original (PreBalance) 11/25/2009 1X (mils p-p)	Phase (deg lag)
7B	Fan IB Y 45 Left	Vert	2.85	58	1.83	79.9	1.339	27	2.85	58
Fan	Fan IB X 45 Right	Horiz	2.96	168	2.17	165	0.801	176	2.96	168
900 rpm	Fan OB Y 45 Left	Vert	3.64	324	2.04	346	1.908	300	3.64	324
	Fan OB X 45 Right	Horiz	2.3	98	1.62	78	0.955	133	2.3	98
	Location	Position	Post Trial 1 11/25/2009 1X (mils p-p)	Phase (deg lag)	Slow Roll 11/25/2009 1X mils p-p	Phase (deg lag)	Compensated Post Trial 1 11/25/2009 1X (mils p-p)	Phase (deg lag)	Post Trial 1 11/25/2009 1X (mils p-p)	Phase (deg lag)
7B	Fan IB Y 45 Left	Vert	1.26	48	1.83	79.9	1.011	301	1.26	48
Fan	Fan IB X 45 Right	Horiz	2.33	150	2.17	165	0.608	83	2.33	150
900 rpm	Fan OB Y 45 Left	Vert	2.68	318	2.04	346	1.300	271	2.68	318
	Fan OB X 45 Right	Horiz	2.37	78	1.62	78	0.750	78	2.37	78
	Location	Position	Post Trial 2 11/25/2009 1X (mils p-p)	Phase (deg lag)	Slow Roll 11/25/2009 1X mils p-p	Phase (deg lag)	Compensated Post Trial 2 11/25/2009 1X (mils p-p)	Phase (deg lag)	Post Trial 2 11/25/2009 1X (mils p-p)	Phase (deg lag)
7B	Fan IB Y 45 Left	Vert	1.78	85	1.83	79.9	0.168	190	1.78	85
Fan	Fan IB X 45 Right	Horiz	2.01	163	2.17	165	0.176	9	2.01	163
900 rpm	Fan OB Y 45 Left	Vert	1.31	338	2.04	346	0.765	180	1.31	338
	Fan OB X 45 Right	Horiz	1.66	73	1.62	78	0.149	1	1.66	73

When acquiring shaft relative data for the balance calculation, always collect slow roll data (filtered vibration 1X amplitude and phase) to use as a compensation base line, otherwise errors can occur in the final solution. Table 1 provides data from a balance campaign for a large Induced Draft (ID) fan at an electric power generating plant. The data was collected on the bearing housing of the fan bearings consisting of seismic data and shaft relative vibration data.

The values tabulated show shaft relative levels at 3.64 mils p-p for the initial condition of vibration and 1.5-2.0 mils p-p for the final (Post Trial 2) condition.

Normally, an acceptable shaft reduction from 3+ mils p-p to 2.0 mils p-p is not a significant improvement, however if the compensated levels are reviewed, the results are presented for the final compensation shaft levels at or below 0.77 mils as compared to the original shaft level as shown in Table 2 and Figure 9 for the compensated shaft relative vibration.

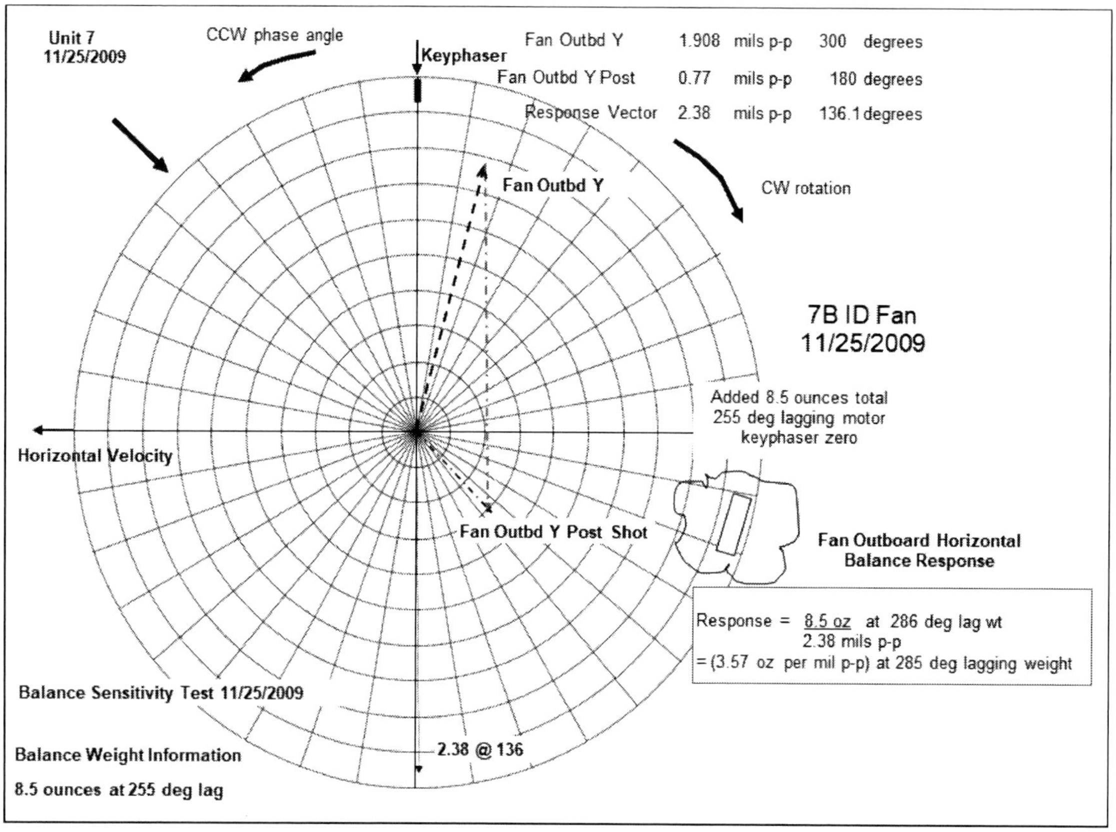

Figure 9 : Shaft Relative Compensated Vibration Response on Vector Plot (Fan Outboard Y Probe Pre and Post Balance)

Also, the seismic vibration levels do not require the same base line vector adjustment as their response is not related to the surface condition of the proximity probe target area.

Problem: If using shaft relative proximity probes, make sure to acquire slow roll runout data to validate the compensated shaft vibration levels.

Lessons Learned: Utilize shaft slow roll compensation methods for balance solutions, prepare machine shaft surfaces during maintenance periods and new machine procurement, backup shaft data with seismic sensors if applicable.

PHASE REFERENCE MISTAKEN IDENTITY

Location of Phase Reference Unknown on Machine

Redundant phase reference can change the results significantly if it is not known which location is being used. In some cases, for permanently installed monitoring systems, a permanent phase reference is preferred, however if redundant probes are sometimes utilized as shown in Figure 10 (above and below each other), the user must be careful to verify which phase reference probe is being used when balancing a machine. A visual check and/or phase reference verification is recommended. When using an optical phase reference this is typically more controlled, but errors can still exist if multiple optical reference marks are used.

Figure 10 : Redundant Phase Reference Probes (Looking at Notch)

Redundant Reflective Tape (Two Pieces of Reflective Tape /Abandoned Reflective Tape)

Inherently, the condition of a machine to be balanced does not have a good permanent phase reference system. In this case, a temporary method is needed by adding a reflective tape mark to the surface of the rotating element. With this, care should be taken to make sure the machine is safe to access and place a small part of reflective tape on the shaft and secure it properly.

It may be possible that someone has balanced the same machine before and installed a piece of reflective tape in a similar location band on the shaft. If two pieces of reflective tape are encountered on a rotating shaft with the phase reference optical sensor, two discrete events will occur on the vibration acquired signals, thus creating the following likely scenarios:

1. The speed indication will be twice as high as expected.
2. The phase angle could be very erratic and inconsistent.
3. The phase angle will reflect the twice frequency phase angle (Figure 11)
4. The phase angle will be consistent and stable and at the proper speed (Figure 12), which reference tape is the one related to the vibration amplitudes and phase? **When you stop the machine which tape is the one to use?**

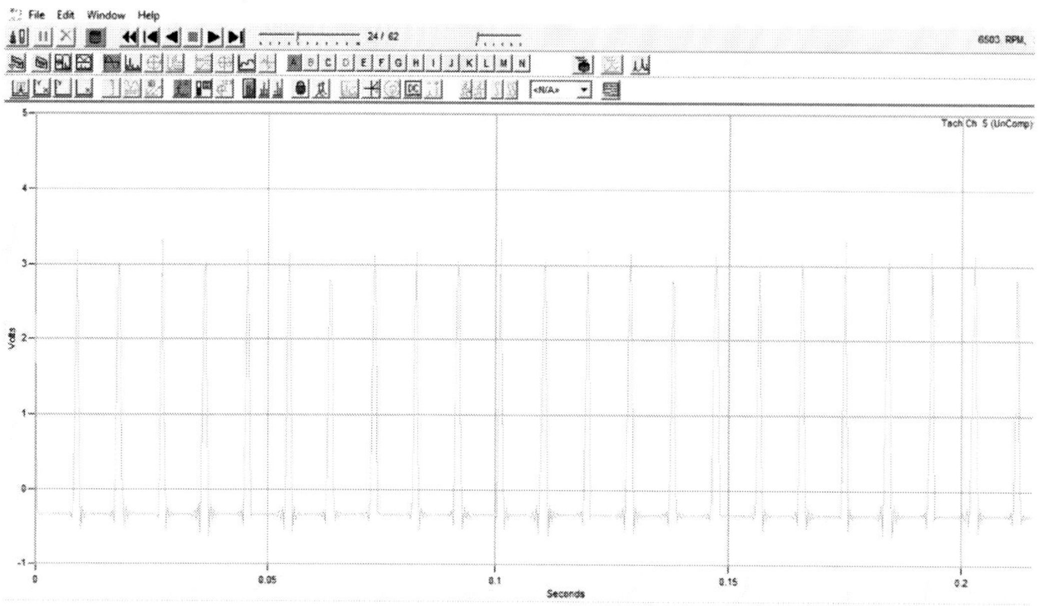

Figure 11 : Phase Reference Pulse twice the Speed Rate of the Machine

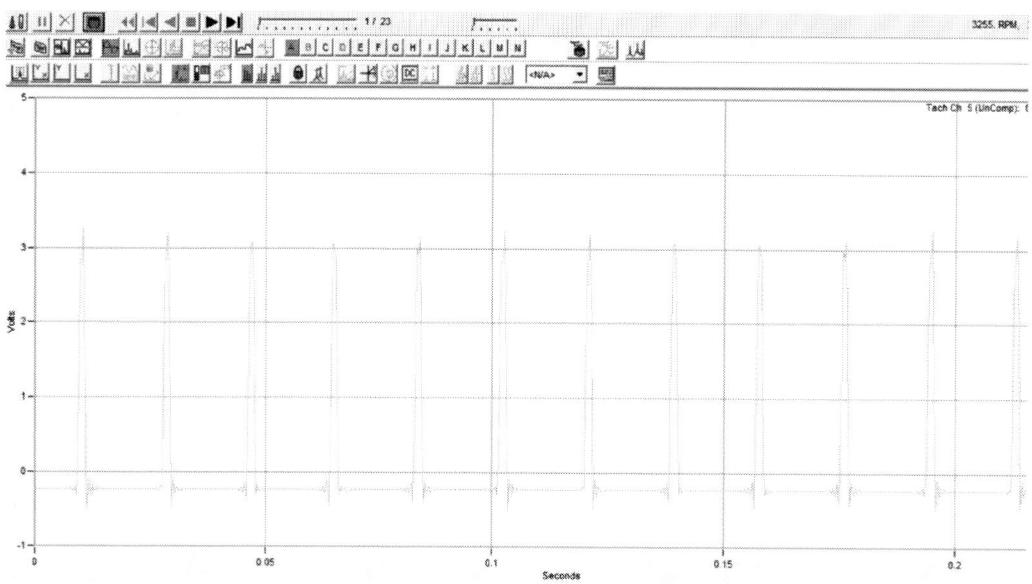

Figure 12 : Phase Reference Pulse at the Correct Speed Rate of the Machine

Problem: Inconsistent phase data, and/or waisted starts, stops, delays, incorrect phase locations. Which phase reference tape or mark is the one used by the vibration analyzer.

Lessons Learned: Examine the shaft with it stopped or using a strobe light. Two pieces of reflective tape (accidentally or not visualizing) and cleaning the shaft surface.

REPEATABILITY OF BALANCE CONDITION

Gas turbines by their nature would not be repeatable during the initial 30-60 minutes of operation. Gas turbines experience significant temperature changes during their operational cycle particularly as the unit load is advanced quickly.

The temperatures create thermal bows that alter the unbalance or mass center of the rotor during the startup sequence.

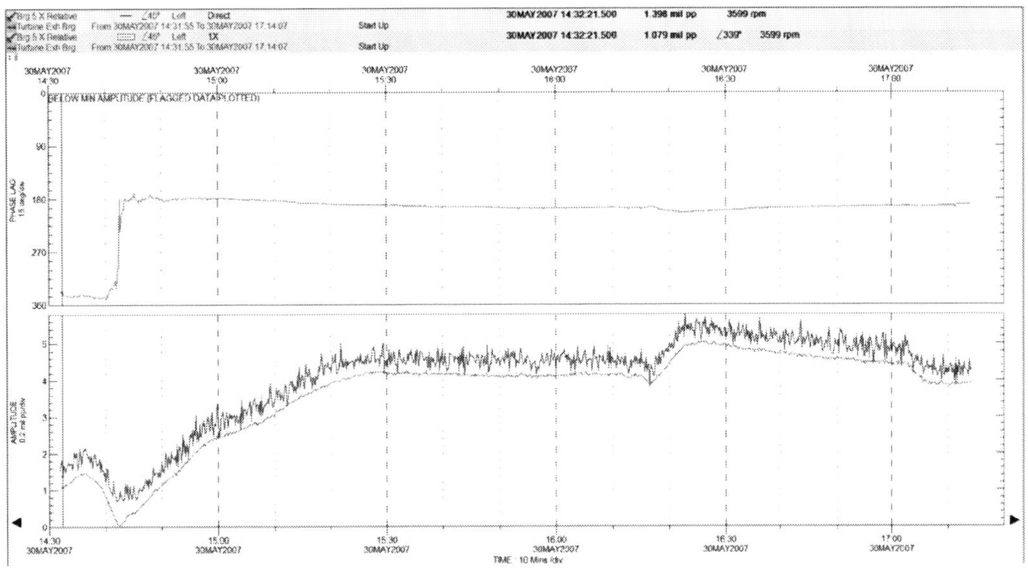

Figure 13 : Gas Turbine Vibration Trends Over a Typical Warm Startup

Figure 13 represents a typical vibration vs time trend for a gas turbine operational run over a multiple hour duration. The typical gas turbine balance approach uses the fundamental requirement to collect vibration data for the unique operating conditions at one or more of the following conditions or combinations of the following conditions including Full Speed No Load condition (FSNL), Minimum Load, Maximum Load, Cold Rotor, Hot Rotor, Warm Rotor.

A fundamental tenant of machine balancing assumes that the only change between the unbalanced (original state of unbalance) and the post-balanced condition (or after adding a balance weight) is that the treatment to the rotor from the addition of the balance weight?

With that understanding, a gas turbine was initially evaluated and balanced using that approach as summarized in the polar plot below. Vibration data was collected initially for the original state and plotted on the polar plot (Figure 14) as $4A_1$ as 6.6 mils p-p at 230 degrees on the plot. A balance weight was then added (Shot #1) which was calculated and determined utilizing previous balance sensitivity data on a similar or unique gas turbine.

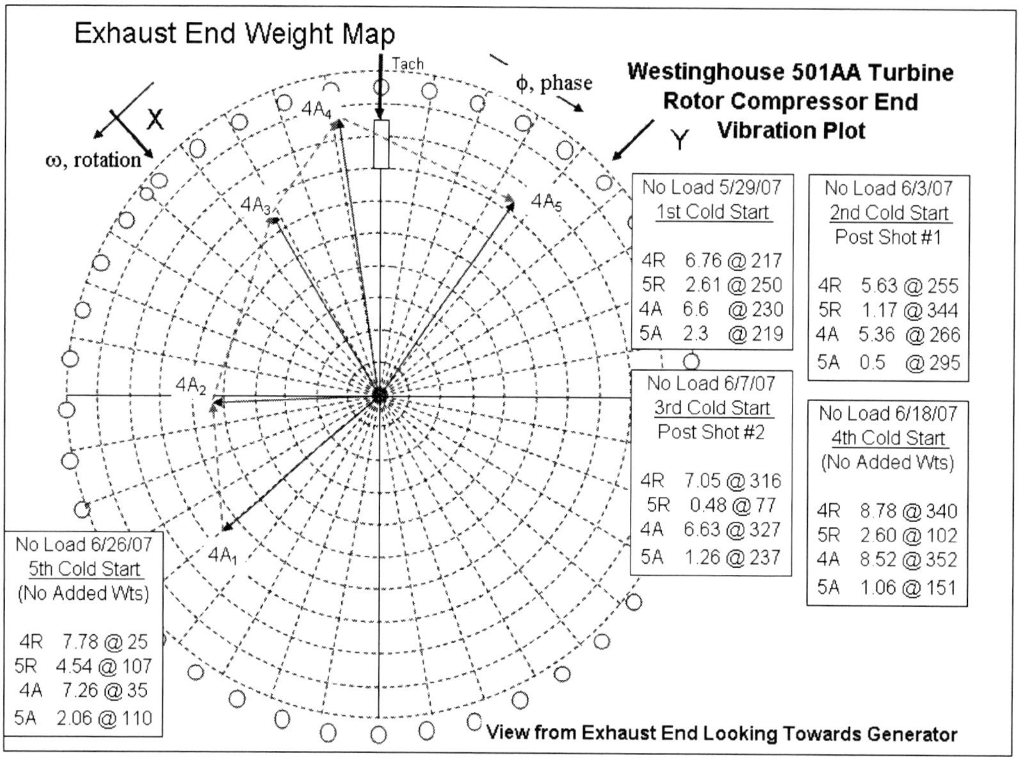

Figure 14 : Polar Plot Showing Various Conditions During Balance Iteration

The unit was restarted for the same condition as the original condition data (Cold) which required waiting for 4-5 days for the unit to cool for weight installation and to recover the same operating conditions as had been collected previously. The resulting data as shown as $4A_2$ at 5.36 mils p-p at 266 degrees which resulted in an improvement as compared to the original but not totally acceptable as a final correction. The (Shot #1) weight was removed and recalculated for another location as would normally be conducted according to balance solution regimen and placed as (Shot #2). With this placement (Shot #2) and rerun 4 days later, $4A_3$ at 6.63 mils p-p at 327 degrees the amplitude was similar to the levels collected originally, even though a proper balance procedure was followed. This now appeared curious as to the unanticipated results and experience with this style of gas turbine unit.

To reconcile this question, an additional startup run was performed on the unit with all balance weights removed including (Shot #1) and (Shot #2) to repeat the original unbalanced condition as denoted by the $4A_1$ condition. The results were the $4A_4$ vector location which is no where close to the $4A_1$ condition. Again, another run was made $4A_5$ original condition with no repeat, as a result the difference between the original vector $4A_1$ condition and $4A_5$ condition indicates approximately 160-170 degree from the original vector?

The OEM (Original Equipment Manufacturer) was promptly contacted for insight and comment, including other industry contacts as to the behavior. The OEM supplied additional maintenance knowledge that some components after extensive run time and startup cycles would experience "slippage" of the rotor disk sections.

The unit was placed into an outage, to inspect and remedy the condition if possible.

Figures 15 provides a photograph of one inspection of an adjacent disk stage to another disk with a match mark (drilled hole shared between separate stages) used as an index mark.

Figure 16 provides another adjacent disk index hole which has moved relative to its adjacent index hole. The relocated adjacent disk was finally found to be "spun" or reindexed approximately 160-170 degrees away from the mating adjacent disk.

In summary, the compressor disk had "spun" on the shaft element which created a moving unbalance condition. The original assumption when the balance project had begun was that our balance condition was repeatable for the same conditions. We had not checked or verified our assumption, and it led us down a tortuous path.

Balance Conditions are Repeatable and Steady

1. Check for repeatability on gas turbines if possible.
2. Check gas turbines during exact sequences of time and load.
3. Allow for sufficient cool down, turning gear time, etc. and compare cold, hot and warm startup conditions.

Problem: When data doesn't add up?

Lesson Learned: Don't run before you walk! Make sure you have good repeatable data otherwise you introduce too many variables.

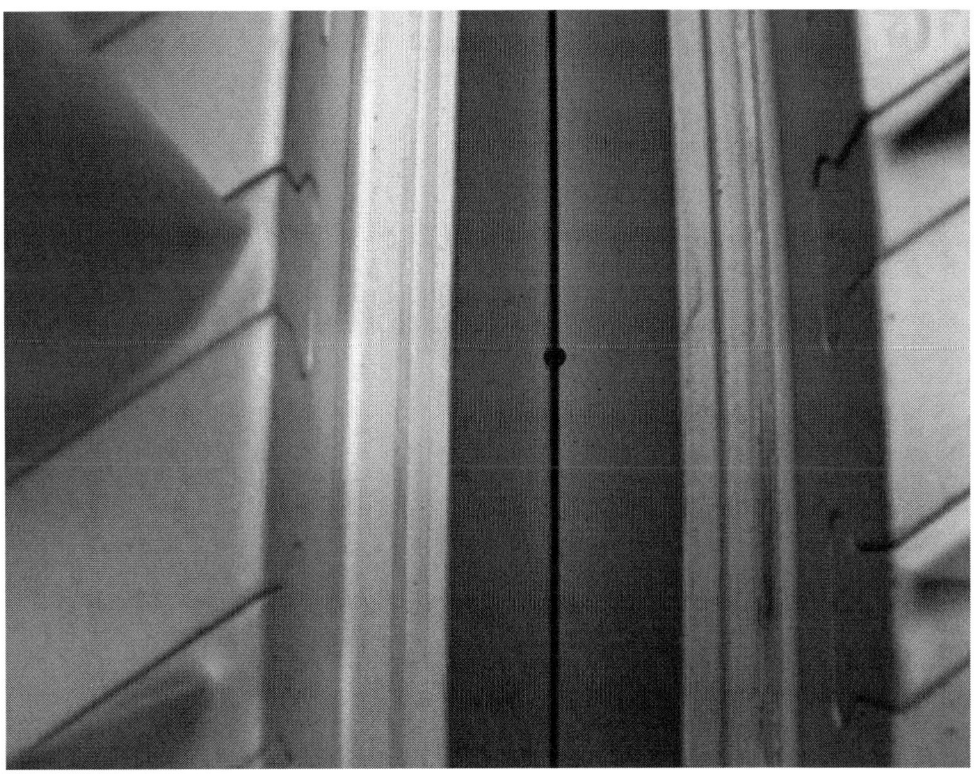

Figure 15 : Marking on Adjacent Disks of Gas Turbine Rotor

Figure 16 : Marking on Adjacent Disks of Gas Turbine Rotor after Slipping between Adjacent Disks

It is important to have repeatable vibration amplitude and phase at the operating speed frequency of the machine. Some machines including fans, motors, turbines, and generators have the potential to change with respect to the operating conditions of the machine and the degraded condition of the machine. Figure 17 provides a polar plot presentation history of this type condition.

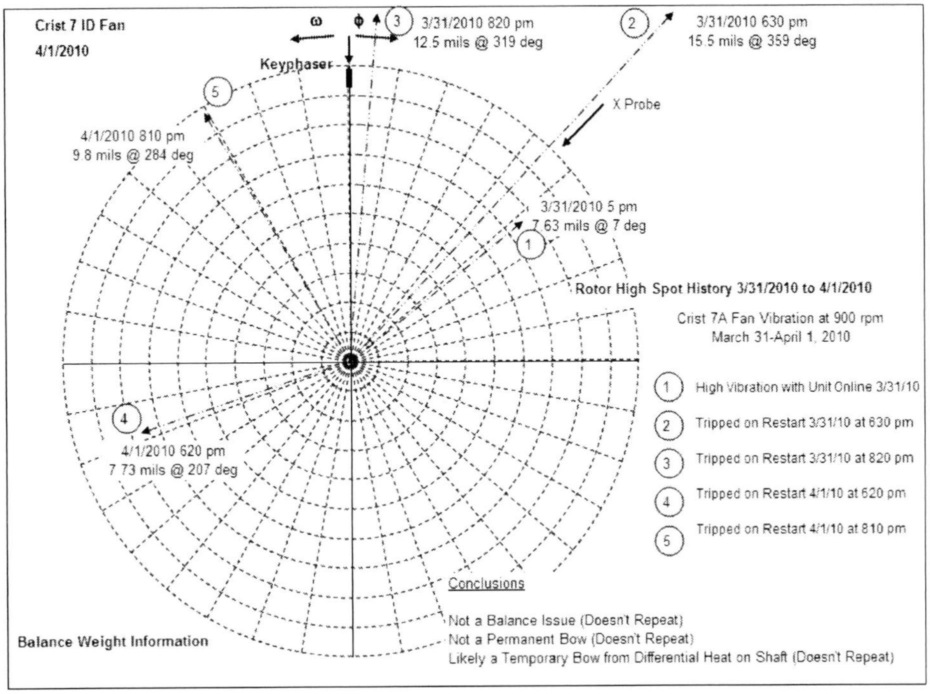

Figure 17 : Vector Location for Multiple Starts during Various Conditions

In this case, the fan was shut down and restarted after the fan was intentionally stopped and parked after the unit came out of service. A restart was attempted with vibration levels excessive at 15.5 mils p-p as shown as (Location 2). The fan was taken out of service again and restarted a day later with levels at 7.73 mils p-p as shown with (Location 4). The fan was restarted again 2 hours later with 9.8 mils (Location 5).

Upon examination of the polar plot figure, the vector location is in drastically different quadrants every start and restart.

A balance request was obtained after the 1st restart (Location 2) and additional data was collected because of the unusual levels and vector position.

This condition was considered to be due to a thermally sensitive fan rotor, which was exposed to a heat thermal gradient in the surrounding duct work that would create a condition that required the fan to be rolled out like a turning gear prior to restarting. Another choice was to shut the boiler off and allow the fan to cool completely prior to starting.

Problem: When data does not repeat within a reasonable quadrant of the polar plot or if you have suspicions?

Lesson Learned: Repeat the run on the machine, you cannot balance a moving target!

VALIDATE DATA FROM REMOTE VIBRATION SOURCES

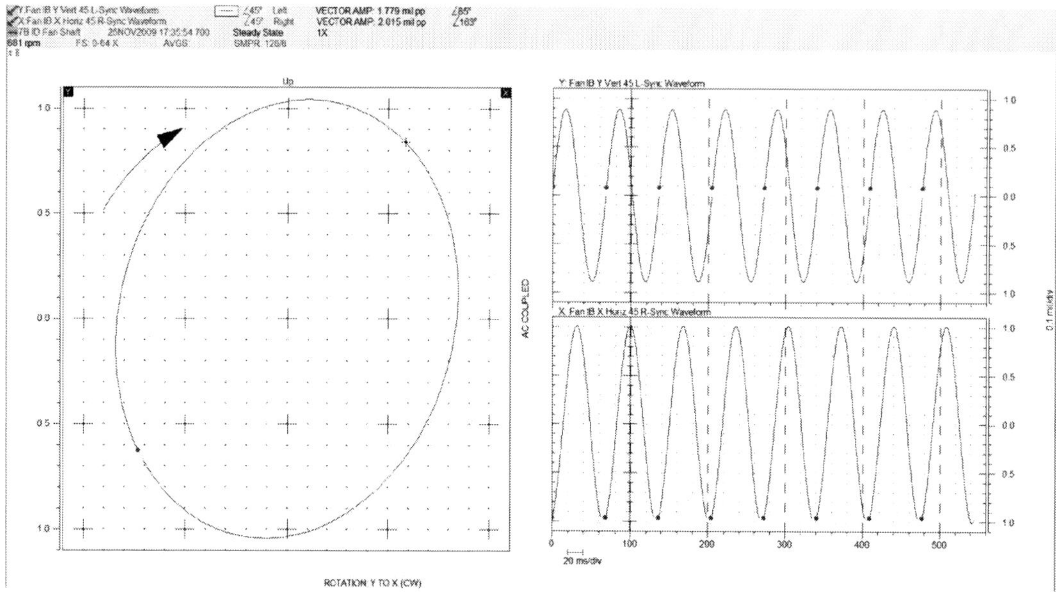

Figure 18 : Once-per-Rev (1X) Vibration of two Orthogonal Probes on Fan Bearing

(Orbit Presentation and Time Waveform)

When reviewing data, always check the precession of the probe waveform to make sure the location of each probe (orthogonal probes as shown in Figure 18) is proper. If the probes are swapped or the location is misinterpreted, major problems can occur with the balance solution, balance calculation, and final results.

CONCLUSIONS AND LESSONS LEARNED

Balance success is dependent on many factors.

The factors include the proper data acquisition, the proper balance procedure and analysis of the conditions required to perform the balance campaign.

Consideration with respect to the phase reference accuracy, the vibration measurement and proper sensor recognition all contribute to the acceptable balance product.

Acknowledgement of the proper repeatable condition and analysis are required for machines that exhibit varying vibration with respect to time and condition.

Knowledge and experience that supplements the appropriate assumptions that can prevent mistakes and eliminate errors and missteps.

Always review the phase reference signal and validate the phase calculation manually to determine that the phase angle used for balancing makes sense.

Know your sensors and what they tell you.

Utilize shaft slow roll compensation methods for balance solutions, prepare machine shaft surfaces during maintenance periods and new machine procurement, backup shaft data with seismic sensors if applicable.

Examine the shaft under the phase reference with it stopped or using a strobe light, redundant reflective tape or redundant phase reference probe can alter the accuracy of the phase measurement or weight placement process.

Make sure you have good repeatable data otherwise you introduce too many variables when balancing.

If the data isn't repeatable for the same conditions, a balance process is not possible.

When performing reviews and using remote vibration data sets, validate your probe locations using the review of the orthogonal probes, otherwise the vector location would be in error.

BIBLIOGRAPHY:

Jackson, Charlie "Single Plane Balancing" – Presented at the 8th International Pump Users Symposium- February 1991 Published in the Proceedings of the Texas A&M Turbomachinery Lab, Department of Mechanical Engineering - Texas A&M University, College Station, TX

Balancing of Rigid and Flexible Rotors (Chapter 3) Dr. Edgar J. Gunter and Charles Jackson. (Notes from VI Class – Rotordynamics and Balancing – Graves Mountain Syria, VA)

Rotor Dynamics and Balancing Course Notes – Syria, Virginia 1992 (Jackson, Gunter, Eshleman)

Vibration and Balance Problems in Fossil Plants: Industry Case Histories – Electric Power Research Institute EPRI Report CS-2726 November 1982 – Prepared by Public Service Company of Indiana (N.L Baxter)

VIATC TRAINING CONFERENCE & EXPO
NIAGARA FALLS, NEW YORK JULY 26 – 28, 2023

MULTIPLANE BALANCING USING COMMONLY AVAILABLE MODERN TOOLS

Michael J. Titone, PE
Machinery Diagnostic Services Technical Leader, Bently Nevada
Michael.Titone@bakerhughes.com

Michael J. Titone, P.E. is presently a Machinery Diagnostic Services Technical Leader for Bently Nevada. He joined the company in 2011. In prior job assignments, he worked for a power generation company and a vibration consulting firm in various roles. Since 1989, Michael has been a practicing engineer in the areas of vibration analysis, balancing, predictive maintenance/monitoring, mechanical system design/analysis, and I&C system design.

Michael earned a BS in mechanical engineering from Louisiana Tech University and an MBA from Mississippi College. He also earned 18 hours of graduate credit in mechanical engineering classes. He is a licensed professional engineer in the state of Mississippi and certified as a Category III Vibration Analyst through the Vibration Institute.

Steven P. Ciesla, PE
Machinery Diagnostic Services Technical Leader, Bently Nevada
Steven.Ciesla@bakerhughes.com

Steven P. Ciesla, P.E. is presently a Machinery Diagnostic Services Technical Leader for Bently Nevada. He re-joined the company in 2013. Prior to that he worked as a predictive maintenance engineer at a nuclear power point where he was responsible for the vibration and lubrication programs. His work history includes being as a field engineer for both Bently Nevada and Westinghouse Process Control. He has over 20 years of hands-on, vibration analysis experience in the power generation industry.

Steve earned a BS in mechanical engineering from California State Polytechnic University, Pomona and an MSE in mechanical engineering from the University of Nevada, Las Vegas. He is a licensed professional engineer in the states of Nevada and California and certified as a Category IV Vibration Analyst through Mobius Institute.

ABSTRACT:

Balancing complex machine trains is often performed using expensive commercial software or proprietary software. In both cases, the programs may not be readily available for some users. Regardless of whether the user has access to software, an understanding of the calculation methodology is required to use the computer codes properly. The current paper reviews the calculation techniques and algorithms used in multiplane balancing and how they can be implemented in common programming languages, such as VBA and/or Python. The authors discuss the development of example spreadsheets and discuss the techniques used to develop the tools (using VBA). The results of the spreadsheet calculations and Python code are compared to the results of a commercial multiplane balance code.

KEY WORDS: Balancing, Multiplane, Python, Excel

INTRODUCTION:

Field balancing of complex machine trains may be accomplished using a variety of techniques including single plane, static-couple, or multiplane. Single plane and static-couple techniques provide simple methodologies for performing hand calculations and may be sufficient to achieve a satisfactory overall balance condition for a complex machine train. However, based on the authors' experiences, full consideration of the machine train is a better approach in many cases.

For the purposes of this paper, a typical complex machine train is a steam or gas turbine generator with rigid couplings, three to twelve bearings, multiple critical speeds, and multiple operating conditions. The techniques discussed, however, are not exclusive for complex machines. The techniques are equally usable on simpler machines with flexible coupling such as fans. Also, for the paper, a well-balanced machine is the goal for a balancing program with well-balanced meaning all bearings below about 3.0 mils pk-pk of 1X vibration at normal operating conditions. The vibration through critical speeds and off-normal operating conditions should be below alarm levels. The measured vibration could be shaft vibration relative to the bearings or absolute shaft vibration.

The most effective approach to achieving the goals stated in the previous paragraph is often to use a variety of balancing approaches with a multiplane balancing program playing a critical role in the analysis. Multiplane balancing programs are available from some commercial sources. However, the programs do not appear to be common. A quick internet search resulted in only two or three products that appear to be currently available and to contain the needed features. Additionally, users of a commercial program should understand most of what the program is doing under the hood in order get the best results.

This paper reviews the calculation techniques and algorithms used in multiplane balancing and how they can be implemented in common programming languages and tools, such as Excel, Excel with VBA, and Python. The authors discuss the development two example spreadsheets and the development of a simple Python program. The results of the developed spreadsheet and Python calculations are compared to the results of a commercial multiplane balance code.

The authors are not computer programming experts. If the reader is looking for elegant, professional coding, a search elsewhere is probably in order. The goal of the paper is to show that two seasoned engineers can quickly duplicate some of the key features of a commercial multiplane balancing tool without an exceptional amount of programming aptitude or experience. Both authors did graduate from engineering school when computer programs were used and written by students, and both have used spreadsheet products as personal computers have evolved from DOS based machines to more user-friendly GUI based machines.

MULTIPLANE BALANCING:

Rotor balancing has been the subject of a great deal of research. Foiles et al. summarized some of this work in their paper from 1998, which listed one hundred and sixty-six references. [1] Goodman's paper [2] from 1964 is considered to be the introduction of the least squares balancing procedure. Darlow cataloged a generous amount of useful information on balancing in his 1989 text. [3] The reader should consult these sources and the plethora of sources available online as a starting point for additional information on the details of rotor balancing.

Multiplane (MP) balancing is a broad term. It is used to describe many techniques that are used to balance a complex machine train in multiple planes. In the context of this paper, MP balancing is any balance campaign where influence coefficients are calculated based on calibration runs involving a weight change in a single balance plane. Static-couple (SC) balancing is conveniently integrated with MP balancing by using a virtual single plane for static pairs and a virtual single plane for couple pairs. The MP approach employed in the field also involves knowledge of the machine modes and incorporates aspects of modal balancing.

Multiplane balancing may be performed in shop and field environments. In a shop, the typical balancing process is limited to one rotor with two bearings in a balance machine. However, a multiplane, modal approach may be used for at-speed balancing in the shop. In the field, the machine is coupled, and access to balance planes is limited to accessible planes. In the shop, the goal is to balance to a standard that is likely well-defined. Multiple runs in a shop environment are expected and are not as costly as multiple runs in the field. In the field, the goal is to balance so that the vibration is acceptable, but the limits are not as clearly defined, and "good enough" is often defined by the machine owner.

The techniques described herein are based on practical field balancing. The following points should be considered:

- One goal is to balance the machine so that all probes are reading below the alarm setpoints.
- Another goal is to balance the machine so that the machine is considered well-balanced using OEM criteria. This level of balance could be significantly better than simply balancing to levels below alarms.
- The balancing effort should take as few runs as possible.
- Only field accessible balance planes are available.
- Coupling vibrations and balance planes should be considered, depending on the machine design and operating characteristics.
- Multiple speeds and operating conditions must be considered.
- Machine train designs vary greatly, and the design plays a major role in determining the approach to field balancing.

The overall goal of the multiplane field balancing process is to achieve an acceptable balance condition in the shortest possible time. The balance campaign owner (balance engineer) uses influence coefficients, machine design knowledge, and other factors in pursuit of this goal. The goals of the process do not include determining the influence coefficient for all balances planes or achieving a perfect balance.

MACHINE TRAIN DESIGN CONSIDERATIONS:

The machine train design for a complex machine such as a large steam turbine generator must be considered when designing a tool for multiplane balancing. The tool should allow for a variety of machine train configurations and a large number of measurements. Each "measurement" for the purposes of a generic tool is simply a point to minimize. The measurements could be multiple measurements from the same point taken at different load or speed points. If the tool is adequately designed for complex machines, handling of simpler machines should not be an issue.

There is no standard turbine generator machine train design. The authors have balanced a variety of machines from smaller "three-bearing" machines typically found in paper mills to twelve-bearing designs with six rotors and two bearings per rotor. Some machines have "shared" bearings, where a bearing is used to support two adjacent rotors.

The machine train design also allows for operation at speeds that involve up to three critical speeds. Smaller, stiffer rotors tend to operate above their first critical speed and below their second critical speed. Some low pressure (LP) rotors operate near their second critical speed. Some large, flexible generator rotors operate near their third critical speed.

In addition to the turbine and generator rotors, the couplings and jack shafts can play a significant role in the machine response. In a rigidly coupled machine, these parts act as small rotors.

The locations and numbers of balance planes are also significant. Some rotors have field balance planes at both ends only, some have them only at one end, and some have field accessible midspan balance planes. Also, some rigid couplings include a field balancing plane.

MULTIPLANE FIELD BALANCING SCENARIO:

A multiplane balance tool (MPBT) can be used throughout the field balancing process for the evaluation of potential and actual balance shots. The MPBT helps the balance engineer to understand the effect of balance shots on the full machine train by automating the calculations. The MPBT does not fully automate the process and does not replace critical thinking and engineering judgement.

The first step in field balancing is to determine when to start. Under ideal circumstances, the balance campaign should not be started until the machine has reached stable conditions at all applicable loads. Circumstances are often less than ideal and field balancing may be required to enable the machine to reach full speed conditions. Regardless, the decision makers should consider relevant factors and come to an agreement that the time is right to begin balancing.

The repeatability of the system should be considered throughout the balancing process. Before starting, if there are concerns about the repeatability, then additional testing and data collection may be required. For example, a thermally sensitive generator should be tested to ensure repeatability prior to balancing.

Linearity of the system is another condition that should be met for balancing. If the data shows symptoms of rubbing or looseness, these malfunctions introduce non-linearities that may prevent successful balancing.

Next, a decision is required as to the type of balance shot and the location of the shot in the machine train. A complex machine train can easily offer an overwhelming number of options. Consider a high pressure (HP) turbine in a twelve bearing machine train. The options for addressing a balance problem in the HP turbine may include (1) a single plane shot in the bearing one plane, (2) a single plane shot in the midspan, (3) a single plane shot in the bearing two plane, (4) a static pair shot in the end planes, (5) a couple pair shot in the end planes, and (6) a shot in the coupling between the HP and the intermediate pressure (IP) turbine.

To decide the initial shot in that HP, the balance engineer should consider a variety of factors. If the engineer has access to balancing data on the machine in question or on a similar machine, the MPBT can become a valuable tool at this point in the process. Consider the six options identified in the previous paragraph. All of the options could have significant effects on the IP rotor, and possible minor effects on other rotors in the train. Assuming that the vibration is monitored with X-Y shaft relative and seismic probes at each bearing and each rotor is supported by two bearings, the affected vibration readings could include relative, seismic, and shaft absolute data at two locations per bearing for a total of twenty-four measurements. If all six options for a balance shot were to be included, there would be a potential of one hundred and twenty-four evaluations. Obviously, the problem could be simplified. The relative and seismic readings could be ignored since the shaft absolute contains contributions from both measurements, and one probe per bearing could be considered. This would decrease the complexity to six balance options with four measurements, but the manual calculations required would remain unwieldly.

Even if the path forward seems obvious to the balance engineer for the HP rotor, the actual machine condition may involve additional complications. For example, another rotor, assume the A low pressure (LPA) turbine is operating with vibrations similar to the HP rotor. Assuming prior influence coefficient data is available, the MPBT can help the balance engineer to model what will happen if LPA is balanced before, after, or at the same time as the HP rotor. The wisdom of balancing two rotors in the same run may be questionable, but the MPBT allows predicting what should happen, and these predictions can help to guide the decision on the initial balance shot. For example, if the initial shot is made on the HP-IP coupling, the predicted effects on LPA might be enough to eliminate the need for further balancing of LPA.

Eventually, the decision on the first balance shot will be made, and the balance campaign begun. After installation of the first balance shot, the machine should be started and operated until it reaches operating conditions that the balance engineer judges to be sufficient for evaluating the effect of the shot. At this point, the balance engineer must evaluate the effect of the first shot and determine the path forward. If the shot was somewhat successful, optimizing the shot might be advisable. If the shot was good enough, then moving to another rotor might be the best decision. Often, the shot is good enough, and further shots are not required, although additional balancing may be performed to fine tune the results in a future planned outage.

MULTIPLANE BALANCE TOOL FEATURES:

Based on the previous discussions, the following features could be defined as important in an MPBT:

- Allow for multiple balance planes
- Allow for multiple measurements
- Allow for multiple shots to be installed in multiple planes
- Allow the user to evaluate the effect of a specific shot

In the authors' experience, commercial MPBT programs typically provide the features listed above. However, the intent of this paper is to illustrate the implementation of these features using readily available tools. The authors considered alternatives including MATLAB and open source MATLAB clones such as Scilab and GNU Octave. The authors determined that Excel would be a good option since practically all computers run Excel. The authors also selected Python instead of MATLAB or a clone as an option since Python is a general purpose, open source programming language that is used extensively and since it includes technical libraries (also called modules) that make it suitable for scientific and engineering programs. Another reason the authors selected Python is that it can be compiled, and the authors were not aware of a simple method to compile programs written in the MATLAB clones. Finally, Python can be run on most current computer operating systems, including Windows, Mac OS X, Linux, etc. There is even a version of Python named Pythonista for Apple's iOS.

MULTIPLANE FIELD BALANCE PROCEDURE:

The balancing process is well-described in a variety of references, including the cited references for this paper. The field balancing procedure described above involves the following high level steps regardless of the number of measurements or balance planes:

- Run 0 – Obtain the reference or as-found data {O}
- Shot 1 – Install a calibration shot in a single balance plane. This "single" plane shot might actually be a static or a couple pair. {Wcal}
- Run 1 – Operate the machine at conditions similar to those when the reference data was collected. {O+C}
- Evaluate the effect of the calibration shot {C} and calculate influence coefficients for all measurement locations and conditions. {H}
- Using Run 1 as a reference decide whether to optimize Shot 1, whether to install a second shot in a different location, or whether to conclude the balance campaign.

The balancing process described above is typically continued until the as-left vibration is acceptable. Figure 1 provides a flow chart illustrating a typical balancing campaign.

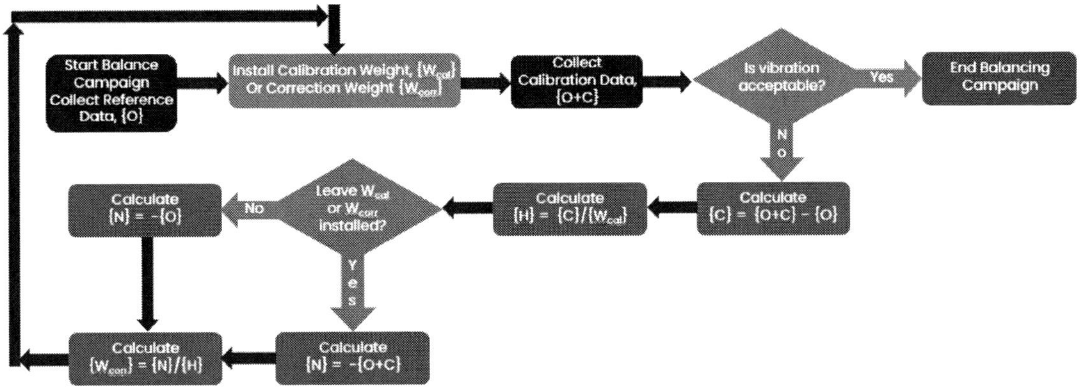

Figure 1, Balancing campaign flow chart

The balancing process requires conventions and notation. The authors use the conventions adopted by their employer, Bently Nevada. Goodman [2] used different conventions, and the current paper is based largely on implementing the process as described by Goodman. Table 1 provides a comparison of the nomenclature used by Bently Nevada and in the Goodman paper.

Table 1, Nomenclature comparison

Variable description	Goodman	Bently Nevada
Original vibration vector (Zero rotor data)	A	O
Vibration after adding calibration (trial) weight	B	O+C
Vibration due to calibration weight		C
Influence coefficients (response coefficients)	α	H
Number of measurements	M	M
Number of balance planes	N	N
Number of operating conditions	K	
Probe (pickup) locations	L	
Residual vibration	ε	R

A convention for displaying influence coefficient angles is also required. The calculation process does not depend on the adoption of any specific convention for influence coefficient angles. The process does require that the influence coefficient convention is consistent when sharing influence coefficients from one tool to another. Similarly, comparing results between the tools used for the analysis requires an understanding of the conventions used by the tools.

The Bently Nevada convention is that all calculations are performed with phase angles based on the probes. The location of the Keyphasor is not required for these calculations. Zero is located on the rotor at the point in time when the Keyphasor fires for the probe of interest. The zero position for X and Y probes are therefore inherently different physical locations on the rotor.

Another convention could be to convert all phase angles such that zero degrees would be the Keyphasor notch. This convention requires that the Keyphasor probe location is known and input accurately.

One commercial MBPT gives the user an option of how to display the influence coefficients (IC's). One option is that the IC's are displayed relative to a reference transducer for a given balance plane, and the user can select the reference transducer. Another option is that the IC's are displayed relative to the location of the calibration weight. This results in the IC's reflecting the force-response lag angle.

The authors' advice on this subject is to always try to obtain the data required to calculate influence coefficients. Then the influence coefficients can be calculated in the user's standard convention.

IMPLEMENTING THE PROCEDURE – REQUIREMENTS FOR EXCEL OR PYTHON:

Unbalance in a rotor manifests as a vibration occurring at the running speed frequency, commonly referred to as 1X. The 1X vibration is a vector quantity, composed of both an amplitude and direction. The amplitude is measured in applicable engineering units (EU) such as mil pp, um pp, in/s pk, etc. The direction measurement is called phase lag and is measured in angular degrees, typically 0-360, or possibly -180 to +180. The phase lag measurement is a timing measurement between two events. For the critical turbomachinery described in this paper, it is normal to have permanently installed vibration instrumentation including a Keyphasor probe that provides a once-per-turn reference for the rotor speed measurement and for the phase measurement. The Keyphasor observes a once-per-turn target on the shaft, such as a notch, hole, or key. The elapsed time from when the measurement system recognizes the Keyphasor pulse (when the Keyphasor fires) until the 1X filtered vibration waveform reaches its first positive peak is divided by the time for one revolution of the shaft (also known as the period). This time ratio is multiplied by 360 degrees to simplify the analysis. Using degrees allows analysts to deal with phase angles rather than a decimal representing the percentage of a rotation, i.e., 231 degrees is equivalent to 64.2% (0.642) of a rotation of the shaft. The use of angles also makes the process of locating a physical spot on the rotor more intuitive for analysis and for locating weights in the field.

Often less critical machines such as fans do not have permanent instrumentation. Phase measurements can be made on these machines. The process is similar, but often an optical or laser sensor is used to generate the once per revolution reference signal. Historically strobe lights were used, but due to the advancement of temporary optical and laser sensors, strobe lights are no longer commonly used for phase measurements.

It is natural to describe 1X vibration in polar coordinates (r, θ) as it naturally matches the amplitude and directional nature of vectors. However, in order to perform the balancing calculations in Excel it is necessary to convert the vectors from polar coordinates to cartesian coordinates (x, y). The (x, y) components could also be represented in the Real and Imaginary planes as a complex number $(x + yi)$ shown in Figure 2.

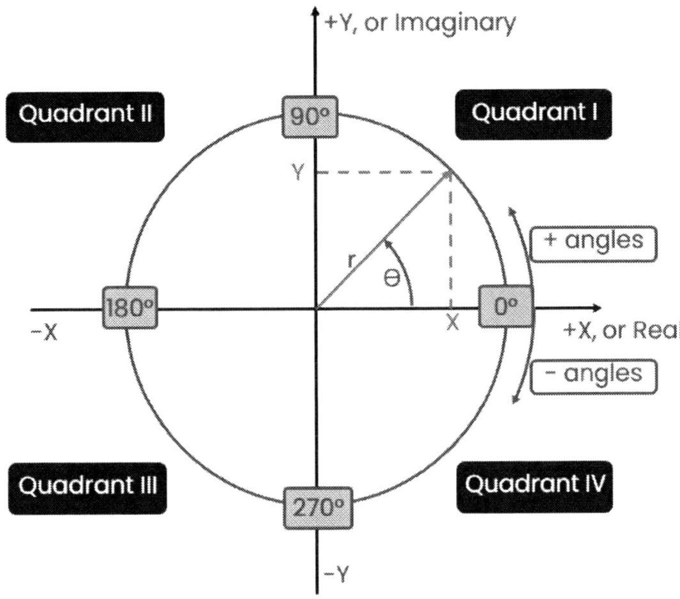

Figure 2, Polar, x-y, and real-imaginary coordinates

The following equations convert the vector from polar coordinates to rectangular components.

$$X = r\cos(\theta) \quad (1)$$

$$Y = r\sin(\theta \quad (2)$$

Python has the capability of performing calculations directly using complex numbers ($z = a + bj$) such that conversion to the rectangular components is not necessary. Note that complex numbers are typically written as "$z = a + bi$" in mathematical context, where $i^2 = -1$. Python uses a "bj" instead of "bi." The use of "j" is consistent with notation in the electrical engineering field where "i" is used for current. The advantage of using complex numbers in Python is that a single complex number is treated as one value or one quantity where other notations for vectors such as (r, θ) or (x, y) are treated as two quantities, resulting in additional complexity in handling arrays. A full discussion of complex numbers is beyond the scope of this paper, but many on-line resources are available.

Goodman describes the balancing problem as, "Mathematically the problem of balancing rotating machinery may be stated as follows: Given N balance planes (locations for correction masses) at M = K x L vibration readings, taken at L different pickup locations at each of K different operating conditions of speed and load, find the optimum magnitudes and angular locations of correction masses in the N balance planes to minimize the amplitudes of the M vibration readings." [2]

The mathematics do not distinguish between the vibration probes (pickup locations L) or operating conditions of load and speed (K and L). This allows the balancing engineer to consider all of the vibration probes on a rotor (not just the Y probes or the X probes), all of the vibration probes on the machine train (all bearings) and multiple speed and load conditions (near resonance, full speed no load, and different load points) dependent on how the end-user operates their machine.

The balance planes used for solutions on typical large turbine generators include the following:

- End planes – designed balance planes at the ends of the rotors near the bearings
- Midspan planes – balance planes near the center of rotor span, available on some rotors
- Coupling balance planes – balance planes located on the rims of couplings
- Static or couple balance planes – planes used in calculations to account for installation of static or couple pairs

The balancing campaign may involve one or more balance planes. Any balance plane or any combination of balance planes can be used in a solution if the applicable influence coefficients are available. As with the measurements, the mathematics do not involve the physical locations or characteristics of the balance planes.

Influence coefficients can only be calculated based on a change in one balance plane with a reference run and a calibration run. Therefore, if a machine train contains twelve bearings with balance planes at the end plane of each rotor and on each coupling rim, the machine would have seventeen balance planes. Obtaining sufficient data to have reliable influence coefficients for all seventeen balance planes is only practically possible over a long period of time with multiple balance campaigns on a family of machines.

An exact solution to this problem can only be calculated if the number of balance planes N is equal to the number of vibration readings M. Mathematically, this system is described as "exactly-determined". As previously discussed, the balance engineer is typically collecting more measurements (vibration probes and operating conditions) than the balance planes that are used in the solution which results in an "over-determined" system. Additionally, the practical process of field balancing requires that the minimum number of balance planes be used to achieve an acceptable balance state. The goal is not to perfectly balance the machine. The goal is to balance the machine to a level that meets the customer requirements with the minimum number of runs. Typically, the balance engineer works with one balance plane per run. This "single" balance plane may actually involve two physical planes for static and couple pairs, but mathematically, the pairs are treated as a single balance plane. As such, this becomes an optimization problem. An "under-determined" system would result from having more balance planes than measurements, which in the authors' experience does not occur when field balancing the typical machines described in this paper.

IMPLEMENTING THE PROCEDURE:

The multi-plane balancing procedure is summarized in the flow chart shown in Figure 3. For most cases, the influence vector matrix is not square because the number of vibration measurements, M, is larger than the number of balance planes, N, used in the solution. In matrix notation, the Residual vibration vectors {R} are equal to the matrix product of the Influence Vector matrix [H] and the Correction Weight vector {Wcor} added with the Original vibration vectors {O}. In order to solve for the correction weights, the matrix pre-multiplying {Wcor} must be inverted, however that process is only valid for square matrices. Therefore, both sides of the equation are multiplied by the transpose of the influence vector matrix, $[H]^T$. The mathematics associated with minimizing the residuals results in the formation of the "normal" equations that are fundamental to least squares regression, hence the term Least Squares Balancing. [4]

$$\{R\} = [H]\{Wcor\} + \{O\} \tag{3}$$

Assume $\{R\} = 0$, substitute the Null vector $\{N\}$ for $\{-O\}$, and rearranging:

$$\{N\} = [H]\{Wcor\} \tag{4}$$

Transpose [H], which means switching the rows and columns and pre-multiply both sides of the equation.

$$[H]^T\{N\} = [H]^T[H]\{Wcor\} \tag{5}$$

Let $[Z] = [H]^T[H]$ which is now a square matrix and pre-multiply both sides by $[Z]^{-1}$.

$$[Z]^{-1}[H]^T\{N\} = [Z]^{-1}[Z]\{Wcor\} \tag{6}$$

The matrix product of the inverse of a matrix and the matrix itself, results in the identity matrix [I], which has values of 1 on the diagonal and 0 everywhere else.

$$[Z]^{-1}[H]^T\{N\} = [I]\{Wcor\} \tag{7}$$

$$\{Wcor\} = [Z]^{-1}[H]^T\{N\} \tag{8}$$

These matrix operations are used to solve for the correction weights that minimize the residuals in a least squares form. Although the resulting vibration amplitudes from one iteration will result in the minimum residual, the vibration amplitudes may not be evenly distributed. Goodman refers to the first iteration as Plain Least Squares (PLS). Therefore, an additional process, Weighted Least Squares (WLS), is required to minimize the maximum residual amplitude. In the weighted least squares procedure, each of the simultaneous equations from the previous iteration is weighted by the amplitude of the residual vibration. This process has been found to converge in three iterations in most cases, according to Goodman. Details are provided in the appendix to Goodman's landmark paper. [2]

The authors coded the example from Goodman using Python. This code and the resulting output are provided in Appendix A.

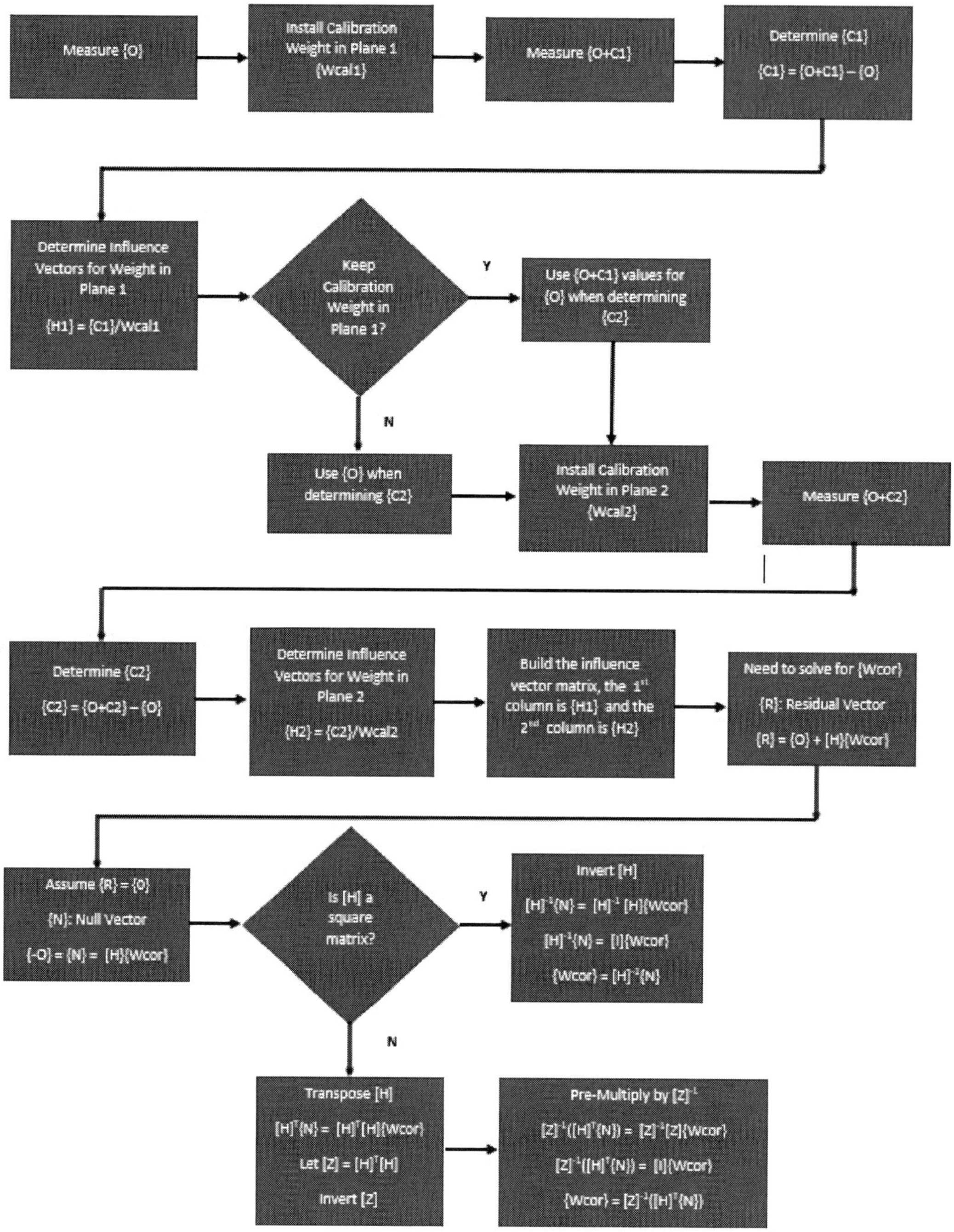

Figure 3, Balancing calculation flow chart

IMPLEMENTING THE PROCEDURE IN EXCEL USING VBA:

Excel natively possesses some complex number capabilities but not enough to easily perform the matrix calculations in a straight-forward manner in a spreadsheet. As such, the vectors are broken into cartesian (x, y) components and manipulated accordingly. The matrix calculations presented require multiple calculations between entries based upon row and column location. To accomplish this, Excel's Visual Basic for Applications (VBA) programming language was utilized. An Excel worksheet is utilized for data entry, calculations, and results presentation.

This worksheet provides for the input/output (I/O) to the VBA code. In VBA, the vibration vectors in the [H] matrix and {N} vector are populated as complex numbers from the cartesian components calculated in the worksheet.

The sequence of matrix calculations is summarized below:

1. Calculate the transpose of [H], denoted as [H_T]
2. Calculate [Z] = [H_T] [H]
3. Invert [Z], denoted as [Z_I]
4. Calculate {B} = [H_T] {N}
5. Calculate the correction weights, {Wcor} = [Z_I]{B}

Figure 4 provides an example VBA code for performing the necessary matrix operations described above to calculate the solution of a single plane balance problem with six measurements (M = 6).

```
'
'Calculate the transpose [H_T] = [H]^T
'
For K = 1 To 6
    H_T(1, K) = H(K, 1)
Next
'
'Calculate the product [Z] = [H_T] x [H]
'
Yankee = 0#
For K = 1 To 6
  Yankee = Application.WorksheetFunction.ImSum(Yankee, Application.WorksheetFunction.ImProduct(H_T(1, K), H(K, 1)))
Next
Z = Yankee
'
'Calculate the inverse [Z_I] = [Z]^-1
'Z_I = 1 / Z
Z_I = Application.WorksheetFunction.ImDiv(1#, Z)
'
'Calculate the product [B] = [H_T] x [N]
'
Yankee = 0#
For K = 1 To 6
  Yankee = Application.WorksheetFunction.ImSum(Yankee, Application.WorksheetFunction.ImProduct(H_T(1, K), N(K, 1)))
Next
B = Yankee
'
'Calculate the product [W] = [Z_I] x [B]
'
Yankee = 0#
  Yankee = Application.WorksheetFunction.ImProduct(Z_I, B)
W = Yankee
'
```

Figure 4: VBA code for performing matrix operations for a single plane solution with six measurements

The most difficult part of the VBA implementation is performing the matrix inversion calculation. For a single plane problem, the [Z] matrix is a single complex number, wherein the matrix inversion is performed via division with a complex number. However, when the number of balance planes increases, the size of the [Z] matrix increases. For two balance planes, the [Z] matrix is 2x2 and for three balance planes, the [Z] matrix is 3x3. Inversion of a 2x2 or 3x3 matrix in VBA requires the calculation and use of determinants and/or Matrix of Minors, Matrix of Cofactors and the Adjoint Matrix. The complexity of the matrix inversion of a 2x2 matrix is shown in Figure 5.

```
'
'Calculate the inverse [Z_I] = [Z]^-1
'
Yankee = 0#
Yankee = Application.WorksheetFunction.ImDiv(1, Application.WorksheetFunction.ImSub(Application.WorksheetFunction.ImProduct(Z(1, 1), Z(2, 2)), Application.WorksheetFunction.ImProduct(Z(1, 2), Z(2, 1))))
Z_I(1, 1) = Application.WorksheetFunction.ImProduct(Yankee, Z(2, 2))
Z_I(1, 2) = Application.WorksheetFunction.ImProduct(Yankee, -1, Z(1, 2))
Z_I(2, 1) = Application.WorksheetFunction.ImProduct(Yankee, -1, Z(2, 1))
Z_I(2, 2) = Application.WorksheetFunction.ImProduct(Yankee, Z(1, 1))
'
```

Figure 5: VBA code for performing 2X2 matrix inversion

The intent of balancing calculations is for the vibration response from the balance weights to cancel out the original vibration. However, an exact solution is only possible when the number of measurements (M) is equal to the number of balance planes (N). It is common to have more measurements than balance planes, which results in residual vibration. In matrix form, the residual vector is given by:

$$\{R\} = [H]\{Wcor\} + \{O\} \tag{9}$$

Goodman presents the following two terms:

1. S "sum of squared residuals" = $\Sigma(R_m)^2$
2. R "rms residual" = sqrt(S/M) where M is the number of measurements.

Both S and R are minimized using the least-squares methodology described by Goodman [2]. Other techniques for minimizing S and R could also be used, such as using the Excel Solver tool as described in the current paper.

IMPLEMENTING THE PROCEDURE IN EXCEL WITHOUT VBA:

As an alternative to programming in VBA (or even Python) one can take advantage of Excel's Solver tool to iteratively determine appropriate balance weights using influence vectors obtained from the installation of trial weights in the applicable balance planes. If the reader is unfamiliar with the Excel Solver tool, it is described my Microsoft as follows:

"Solver is a Microsoft Excel add-in program you can use for what-if analysis. Use Solver to find an optimal (maximum or minimum) value for a formula in one cell — called the objective cell — subject to constraints, or limits, on the values of other formula cells on a worksheet. Solver works with a group of cells, called decision variables or simply variable cells that are used in computing the formulas in the objective and constraint cells. Solver adjusts the values in the decision variable cells to satisfy the limits on constraint cells and produce the result you want for the objective cell." [5]

Link to Solver description: https://support.microsoft.com/en-us/office/define-and-solve-a-problem-by-using-solver-5d1a388f-079d-43ac-a7eb-f63e45925040#:~:text=Solver%20is%20a%20Microsoft%20Excel,formula%20cells%20on%20a%20worksheet

The Solver can be installed in Excel by following the steps listed below: [5]

1. In Excel 2010 and later, go to File > Options
2. Click Add-Ins, and then in the Manage box, select Excel Add-ins.
3. Click Go.
4. In the Add-Ins available box, select the Solver Add-in check box, and then click OK. Notes:
    - If the Solver Add-in is not listed in the Add-Ins available box, click Browse to locate the add-in.
    - If you get prompted that the Solver Add-in is not currently installed on your computer, click Yes to install it.
5. After you load the Solver Add-in, the Solver command is available in the Analysis group on the Data tab.

Link to Solver installation steps: https://support.microsoft.com/en-us/office/load-the-solver-add-in-in-excel-612926fc-d53b-46b4-872c-e24772f078ca

Native Excel functions can be used to calculate the balancing variables (C, H, N, Wcor) as shown in Figure 6. Note that cells with a blue background indicate user input for the Slow Roll Vector (if applicable), Original Response Vector, Calibration Weight, and Calibration Response Vector.

Measurement #	1	1X Amplitude (mil pp)	1X Phase (degrees)	x	y
Slow Roll Vector	SR1	0.00	313	0.000	0.000
Original Response Vector	O1	10.10	222	-7.506	-6.758
Compensated Original Response Vector	O1'	10.10	222	-7.506	-6.758
Calibration Weight	Wcal	36.50	295	15.426	-33.080
Calibration Response Vector	O1+C1	2.10	205	-1.903	-0.887
Compensated Calibration Response Vector	O1'+C1	2.10	205	-1.903	-0.887
Calibration Vector	C1	8.12	46	5.603	5.871
Influence Vector	H1	0.22	111	-0.081	0.207
Opposite Compensated Original Vector	N1'	10.10	42	7.506	6.758
Correction Weight	Wcor	45.43	291	16.029	-42.506

Figure 6: Single plane balance in Excel

Synchronous (1X) vibration vectors are typically described as an amplitude (in engineering units, such as mil pp, in/s pk, g's pk) and a phase lag (degrees) measurement, e.g., 5.0 mils pp @ 72°. The amplitude will be denoted as A and the phase as F. Excel works with the mathematical or trigonometric conventions of angles in four quadrants and positive and negative components described above. Since the phase angles used in vibration are between 0 and 360°, these measurements must be converted to a mathematical convention to use the trigonometric functions available in Excel. Therefore, in Excel, the vibration measurements will be transformed from polar coordinates between 0 and 360° to rectangular (X, Y) coordinates. The rectangular coordinates are denoted as X and Y.
Note for trigonometric functions in Excel, which are used for the coordinate transformation, angles are measured in radians instead of degrees.

The x- and y- components of a vector A @ F are given by:

$$X = A*COS(RADIANS(F)) \quad (10)$$

$$Y = A*SIN(RADIANS(F)) \quad (11)$$

Conversely, given the x- and y- components of a vector, one can calculate the amplitude (A) and phase (F) as follows:

$$A = SQRT(X^\wedge 2 + Y^\wedge 2) \quad (12)$$

$$F = MOD(DEGREES(ATAN2(X,Y)),360) \quad (13)$$

The phase calculation requires additional explanation. The ATAN2 function within Excel returns the angle in radians between -π and +π for the specified x- and y- coordinates. Note that -π radians is equivalent to -180 degrees and +π radians is equivalent to +180 degrees. Additional background information about the ATAN2 function and the use of radians as an angle unit is readily available from online sources.

The DEGREES function converts the angle measurement from radians to degrees.
Since the ATAN2 function will produce angles between -180 and +180, the MOD() function within Excel is used to convert the angle from 0 to 360. The MOD() function returns the remainder after a number is divided by a divisor (360 in this application).

To subtract vectors (which is used for slow roll compensation and determining the calibration (C) vector) the x and y components are subtracted independently. In this section, the apostrophe (') is used to denote quantities that have been slow roll compensated. For example, the Compensated Original vector components (O') are calculated as follows:

$$O'x = Ox - SRx \tag{14}$$

$$O'y = Oy - SRy \tag{15}$$

Note that the Calibration vector components are identical whether using compensated or uncompensated data, but compensated data will be used for completeness.
Example, the Calibration vector components (C) are calculated as follows:

$$C'x = Cx = (O+C)'x - O'x \tag{16}$$

$$C'y = Cy = (O+C)'y - O'y \tag{17}$$

The influence vector is calculated by dividing the Calibration vector (C) by the calibration weight (Wcal).

$$H = C / Wcal \tag{18}$$

When dividing two complex numbers (C and Wcal in this case), the X (Real) and Y (Imaginary) components are calculated by multiplying both the numerator and denominator by the complex conjugate of the denominator. For the complex number, a + bi, the complex conjugate is a – bi. In the denominator, the product of (a + bi)(a – bi) becomes $a^2 + b^2$. In the numerator, the distributive property is used to multiply the two complex numbers and the X (Real) and Y (Imaginary) components can be determined. Extending this principle to vector space results in Equations 19 and 20.

$$Hx = (Cx * Wcalx + Cy * Wcaly) / (Wcalx^2 + Wcaly^2) \tag{19}$$

$$Hy = (Cy * Wcalx - Cx * Wcaly) / (Wcalx^2 + Wcaly^2) \tag{20}$$

In order to calculate the correction weight, Wcor, the Null vector (N) is utilized. The Null vector (N) is a vector the same length as the Original (O) vector but in the opposite direction (phase angle different by 180 degrees), assuming the calibration weight has been removed. This is accomplished by multiplying the x and y coordinates by -1 as shown in Equations 21 and 22.

Note that the Null vector could also be defined as a vector that is created to cancel out any vibration present in the machine. For example, if the calibration weight remains in the machine, the null vector is equal and opposite to the O+C vector.

$$N'x = -1 * O'x \tag{21}$$

$$N'y = -1 * O'y \tag{22}$$

Ideally the vibration response created by the installation of the balance weight(s) will cancel out the original vibration vector.

$$\{O'\} + [H]\{Wcor\} = 0 \tag{23}$$

$$[H]\{Wcor\} = \{-O'\} \tag{24}$$

$$[H]\{Wcor\} = \{N'\} \tag{25}$$

$$\{Wcor\} = [H]^{-1}\{N'\} \tag{26}$$

In vector component space, the correction weight is calculated using Equations 27 and 28.

$$Wcorx = (N'x * Hx + N'y * Hy) / (Hx\text{^}2 + Hy\text{^}2) \tag{27}$$

$$Wcory = (N'y * Hx - N'x * Hy) / (Hx\text{^}2 + Hy\text{^}2) \tag{28}$$

These calculations can be easily copied in Excel to account for multiple measurement points (probes, speed, and/or load conditions) as well as expanded to use multiple balance planes. An example header of a spreadsheet with two balance planes and twelve measurement points is shown in Figure 7.

Minimize RMS Residual	0.978	19.95	144		
Minimize Maximum (Mini-Max) Amplitude	1.54	19.05	143		
Correction Weight	Wcor1	20.24	145	-16.659	11.493
Compensated Predicted Result Measurement #1	PR1'	0.78	178	-0.784	0.030
Compensated Predicted Result Measurement #2	PR2'	0.33	64	0.146	0.300
Compensated Predicted Result Measurement #3	PR3'	1.02	283	0.230	-0.991
Compensated Predicted Result Measurement #4	PR4'	1.54	174	-1.527	0.171
Compensated Predicted Result Measurement #5	PR5'	1.20	45	0.849	0.843
Compensated Predicted Result Measurement #6	PR6'	1.54	272	0.064	-1.535
Compensated Predicted Result Measurement #7	PR7'	0.57	134	-0.399	0.408
Compensated Predicted Result Measurement #8	PR8'	0.69	201	-0.649	-0.244
Compensated Predicted Result Measurement #9	PR9'	0.85	182	-0.847	-0.023
Compensated Predicted Result Measurement #10	PR10'	0.36	77	0.081	0.354
Compensated Predicted Result Measurement #11	PR11'	0.36	77	0.081	0.354
Compensated Predicted Result Measurement #12	PR12'	0.36	77	0.081	0.354

Minimize RMS Residual	0.978	6.81	117		
Minimize Maximum (Mini-Max) Amplitude	1.54	6.12	117		
Correction Weight	Wcor2	6.85	116	-3.031	6.141

Weight Plane #1

Measurement #	1	1X Amplitude (mil pp)	1X Phase (degrees)	x	y
Slow Roll Vector	SR1	0.00	[illegible]	0.000	0.000
Original Response Vector	O1	0.86	239	-0.440	-0.733
Compensated Original Response Vector	O1'	0.86	239	-0.440	-0.733
Calibration Weight	Wcal1	12.00	135	-8.485	8.485
Calibration Response Vector	O1+C11	0.45	186	-0.449	-0.047
Compensated Calibration Response Vector	O1'+C11	0.45	186	-0.449	-0.047
Calibration Vector	C11	0.69	91	-0.008	0.686
Influence Vector	H11	0.06	316	0.041	-0.040
Opposite Compensated Original Vector	N1'	0.86	59	0.440	0.733
Correction Weight	Wcor	14.96	103	-3.446	14.559

Weight Plane #2

Measurement #	1	1X Amplitude (mil pp)	1X Phase (degrees)	x	y
Slow Roll Vector	SR1	0.00	[illegible]	0.000	0.000
Original Response Vector	O1	0.86	239	-0.440	-0.733
Compensated Original Response Vector	O1'	0.86	239	-0.440	-0.733
Calibration Weight	Wcal	12.00	315	8.485	-8.485
Calibration Response Vector	O1+C12	0.45	186	-0.449	-0.047
Compensated Calibration Response Vector	O1'+C12	0.45	186	-0.449	-0.047
Calibration Vector	C12	0.69	91	-0.008	0.686
Influence Vector	H12	0.06	136	-0.041	0.040
Opposite Compensated Original Vector	N1'	0.86	59	0.440	0.733
Correction Weight	Wcor	14.96	285	3.446	-14.559

Figure 7: Two plane balance in Excel

Instead of calculating the correction weights directly, the predicted results (1X vibration) based upon the influence vectors and correction weight(s) are calculated. Then Excel's Solver function is used to minimize either the RMS Residual or the maximum amplitude (Mini-Max) of these results.

The predicted result from a single measurement (for a two plane solution) is calculated as follows:

$$PR1' = O1' + H11 * Wcor1 + H12 * Wcor2 \tag{29}$$

where:
PR1' = Compensated predicted vibration result for measurement 1
O1' = Compensated original vibration for measurement 1

H11 = Influence vector for measurement 1 corresponding to a weight placed in plane 1
H12 = Influence vector for measurement 1 corresponding to a weight placed in plane 2
Wcor1 = Correction Weight placed in balance plane #1
Wcor2 = Correction Weight placed in balance plane #2

These calculations would be made for each of the measurements and weight planes. This method is easily extendable to 3 or more planes if necessary.

The RMS Residual calculation is simply the square root of the sum of each of the measurement amplitudes squared divided by the number of measurements.

$$\text{RMS Residual} = \text{SQRT}(((PR1')^2 + (PR2')^2 + \ldots (PRN')^2)/M) \tag{30}$$

This solution method should result in equivalent results to the least squares method. An alternative solution is to minimize the maximum predicted amplitude of all of the measurements. This seems to be a more natural solution as machine owners and operators are typically only concerned with the maximum vibration amplitude observed.

$$\text{Mini-Max} = \text{MAX}(PR1'{:}PRN') \tag{31}$$

To use the Solver function in Excel, navigate to the Data Ribbon and select Solver from the Analyze tab as shown in Figure 8.

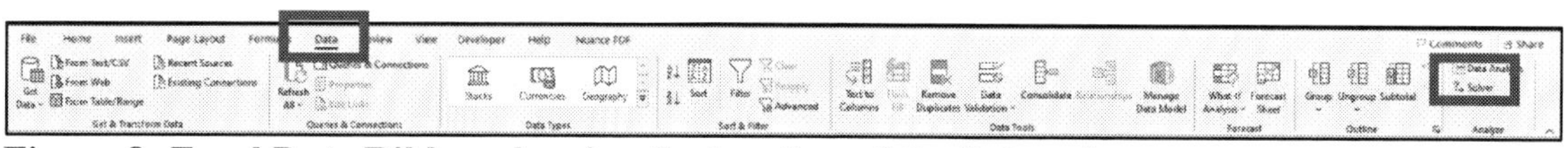

Figure 8: Excel Data Ribbon showing the location of the Solver feature

The Solver Parameters dialog box will allow the user to set up the problem to be solved.

Enter or click on the cell corresponding to either the RMS Residual or Mini-Max calculation for the "Set Objective" field (Figure 9).

Select the Min radio button to perform a minimization optimization (Figure 9).

The selection of the "By Changing Variable Cells" is dependent on how many balance planes that are configured in the spreadsheet. In this case, there are two planes, so the cells corresponding to the size and angle of Wcor1 and Wcor2 are selected (Figure 9).

The "Subject to the Constraints" entry is not required, but constraints can be added to ensure that the optimization generates positive values for the weights and phase angles below 360 degrees (Figure 9).

The Make Unconstrained Variables Non-Negative check box is selected as default as well as the GRG Nonlinear Solving Method.

In the authors' experience, it is possible to get unexpected or poor solutions from the Solver based upon the initial values input in the Wcor fields. It is recommended to use engineering judgement to select starting Wcor values based upon the results from the individual measurement values. Additionally, the Options button adjacent to the "Select a Solving Method" drop-down box can be configured to help Solver find a global minimum rather than a local minimum (Figure 9). Select the GRG Nonlinear tab (Figure 10).

The Converge field dictates the amount of relative change in the previous five iterations performed needed to stop the Solver. The Solver will require more time/iterations to converge as the Converge field value is decreased. No discernable differences were noticed when using Forward (default) or Central Derivatives, however theoretically central difference derivatives should offer a better (smaller) truncation error than a forward difference derivative but at the cost of additional calculation time.

Selecting the "Use Multistart" checkbox will run the GRG Nonlinear method repeatedly, beginning with different starting values for the Variable Cells (Decision Variables). This should help the Solver determine a better solution. The Population Size box dictates the number of different starting points. The minimum value is 10, even if left blank and the maximum value is 200. The larger the value results in more runs, thereby taking more time to calculate a solution. If the Random Seed box is populated, then the Solver will begin at the same starting point each time. A blank value will result in a random starting point.

Selecting Require Bounds on Variables will speed up the solution process. However, both lower and upper bounds are required for each Variable Cell (Decision Variable). Those bounds are entered in the Subject to Constraints field from the Solver Parameters dialog box. If the size of balance weights (i.e. plug weights or dovetail weights) are known, then the size of a single weight could be used for the lower bound.

The authors were able to repeatedly get good results using these values but encourage the user to make changes as necessary to achieve satisfactory results:

Convergence: 0.00001
Derivatives: Central
Multi-Start: Selected
Population Size: 20
Random Seed: Blank
Require Bounds: Not Selected

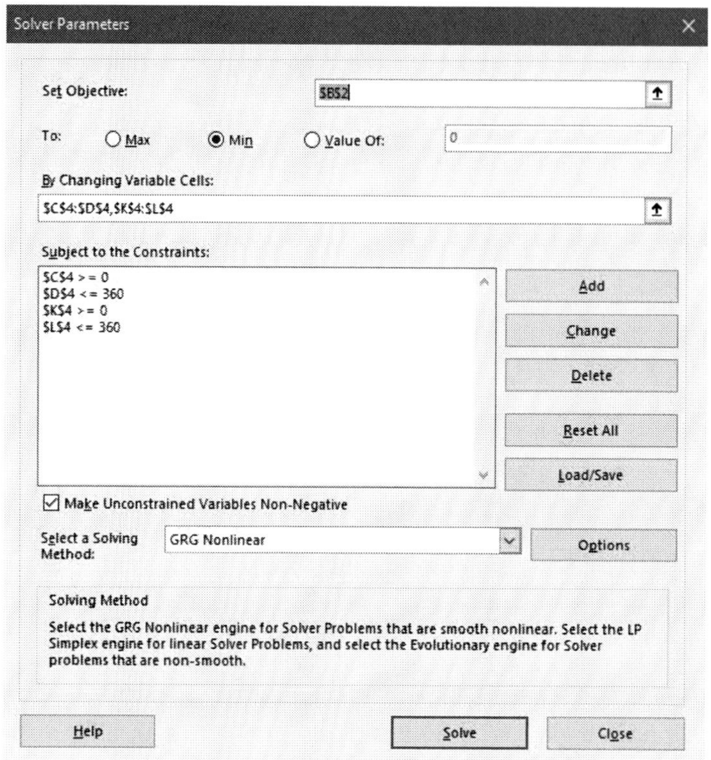

Figure 9: Solver parameters dialog box

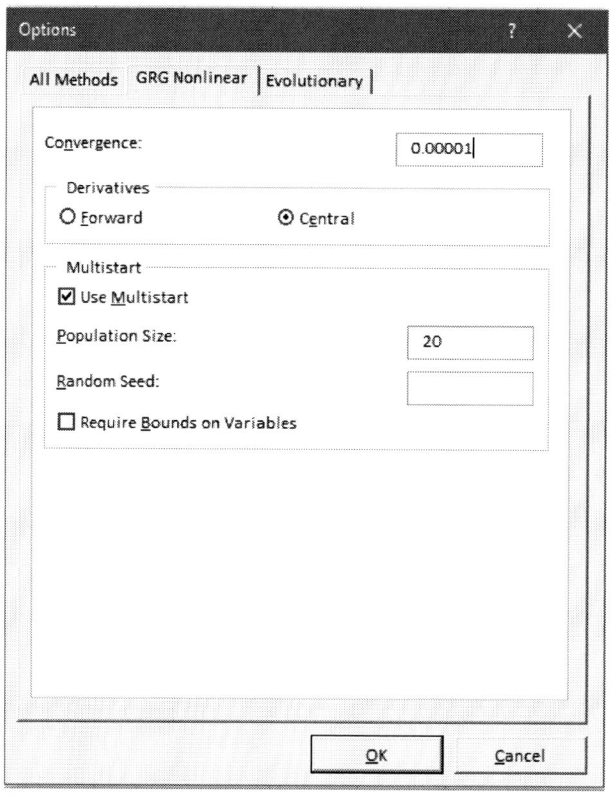

Figure 10: Select a Solving Method Options, GRG Nonlinear tab settings

Once all of the parameters are set, click on the Solve button of the dialog box and Excel will do the optimization and determine if a solution is found or not. When a solution is reached, Excel will display a message as shown in Figure 11. Select the OK button to accept the solution.

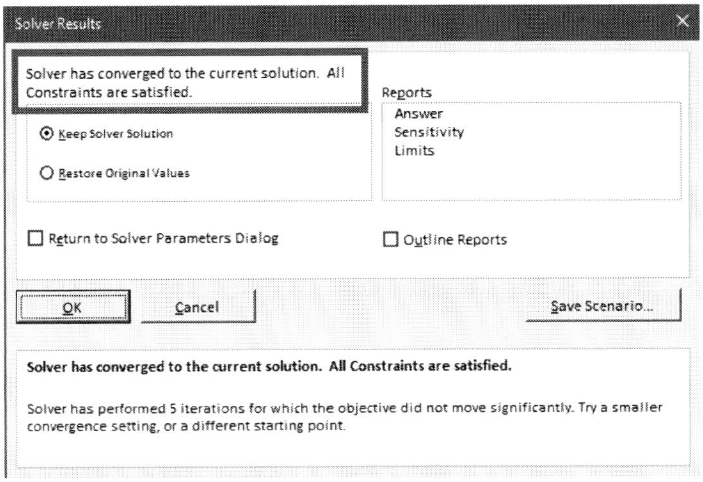

Figure 11: Solver results dialog box indicating that a solution has converged

PYTHON BASICS AND INSTALLATION:

The primary internet site that is the source for the Python language standards, license details, interpreter download, etc. is www.python.org. The reader that is not experienced with Python and interested in learning the language should start with a visit to this site.

Since Python is an interpreted language, compiling the code is not required for running programs written in Python. Compiling is possible but not required. Generally, a new user will start by installing the interpreter and interacting with it through a console that includes an editor. The user can enter commands directly in the Python Command Line for simple tasks and to become familiar with the language. Most practical uses of Python require writing programs that are executed by the interpreter.

The authors selected the Spyder environment for use in developing the codes written for this paper. The Spyder application provides a convenient graphical interface that allows the user to write programs and run the programs within the Spyder environment. There are many similar environments available, and the authors make no claims that Spyder is the best available.

Information about Spyder, including the program download is available from the following website:

https://www.spyder-ide.org/

While not used for the current paper, another Python development tool should be mentioned, the Python notebook. One popular implementation of a Python notebook is Jupyter. The notebook is a collection of short code snippets that can be used to perform parts of a larger task. For example, the user could create a notebook of balancing tools.

Many of the powerful capabilities of Python are accessible through libraries. The libraries can be available as free open source libraries, but some libraries are commercially licensed for a fee. An example of a free library is Tkinter, which supports the development of Graphical User

Interfaces. The Spyder environment includes a large number of libraries useful for technical programming, and this is another reason the authors chose to use Spyder. Two of the primary libraries used for technical programming are NumPy and Matplotlib, and both are installed with Spyder. To view the modules (libraries) installed the user can enter *help ('modules')* in the console command line.

IMPLEMENTING THE PROCEDURE IN PYTHON:

Python is capable of directly dealing with complex numbers and performing matrix calculations in a direct format rather than implementing nested do loops. In order to do this, the Python library NumPy is used. NumPy (Numerical Python) is the fundamental package for scientific computing in Python and allows the using of matrices, linear algebra, and complex numbers. Depending on the installation of Python used, one may need to install NumPy by entering the following command in the Python console:

>>>pip install numpy

With NumPy installed, accessing NumPy and its functions requires it to be imported in the Python code by including the following line at the beginning of the code:

Import numpy as np #import the numpy library

In this example of Python coding, the authors used an Excel spreadsheet to import the applicable data (Slow Roll, O, O+C, N and Wcal measurements) rather than manually inputting them. As such, the Pandas library is installed and imported similarly to NumPy, by entering *pip install pandas* in the Python console. Pandas is an open source Python package built on top of NumPy which simplifies working with data (Figure 12).

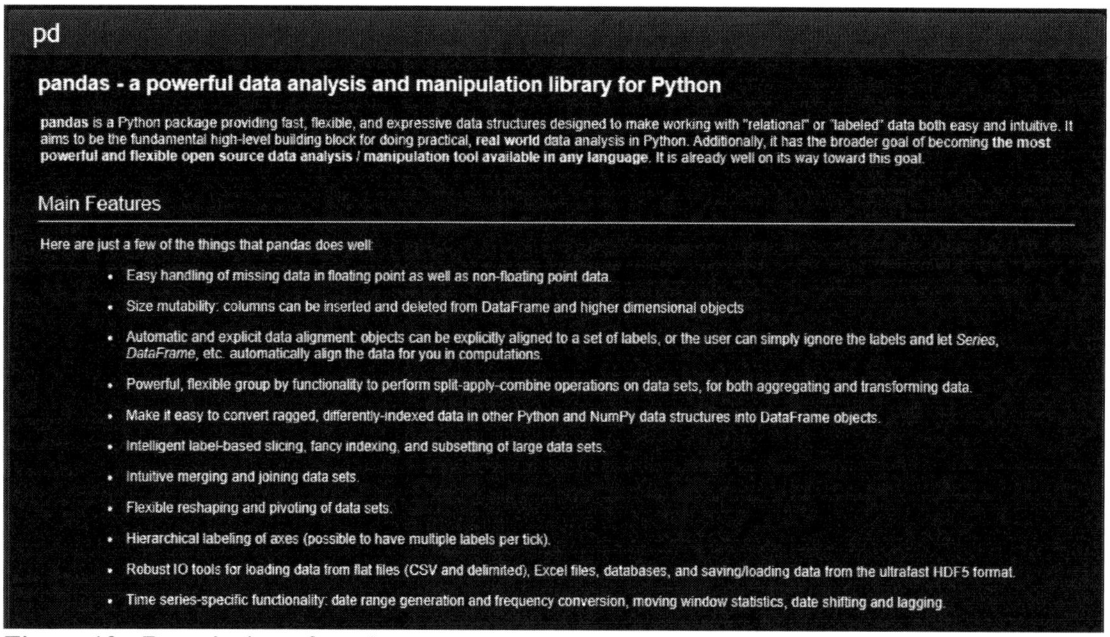

pd

pandas - a powerful data analysis and manipulation library for Python

pandas is a Python package providing fast, flexible, and expressive data structures designed to make working with "relational" or "labeled" data both easy and intuitive. It aims to be the fundamental high-level building block for doing practical, **real world** data analysis in Python. Additionally, it has the broader goal of becoming **the most powerful and flexible open source data analysis / manipulation tool available in any language**. It is already well on its way toward this goal.

Main Features

Here are just a few of the things that pandas does well:

- Easy handling of missing data in floating point as well as non-floating point data.
- Size mutability: columns can be inserted and deleted from DataFrame and higher dimensional objects
- Automatic and explicit data alignment: objects can be explicitly aligned to a set of labels, or the user can simply ignore the labels and let *Series*, *DataFrame*, etc. automatically align the data for you in computations.
- Powerful, flexible group by functionality to perform split-apply-combine operations on data sets, for both aggregating and transforming data.
- Make it easy to convert ragged, differently-indexed data in other Python and NumPy data structures into DataFrame objects.
- Intelligent label-based slicing, fancy indexing, and subsetting of large data sets.
- Intuitive merging and joining data sets.
- Flexible reshaping and pivoting of data sets.
- Hierarchical labeling of axes (possible to have multiple labels per tick).
- Robust IO tools for loading data from flat files (CSV and delimited), Excel files, databases, and saving/loading data from the ultrafast HDF5 format.
- Time series-specific functionality: date range generation and frequency conversion, moving window statistics, date shifting and lagging.

Figure 12: Description of pandas

Similar to NumPy, the code requires the *import pandas as pd* statement at the beginning of the program.

The O, O+C, and Wcal data (amplitude and phase angle) is simply entered in an Excel Worksheet in polar (r and ϕ) coordinates (Figure 13).

	A	B	C	D	E	F
1	O(r)	O(phi)	O+C(r)	O+C(phi)	Wcal(r)	Wcal(phi)
2	3.8	103	2.7	67	16.0	35
3	3.0	359	0.7	2		

Figure 13: Excel worksheet showing O, O+C, and Wcal data in polar coordinates

A second Excel Worksheet is used to convert from polar coordinates to cartesian coordinates. Additionally, the Null Vector values are calculated by multiplying the x and y values of O by -1 (Figure 14).

	A	B	C	D	E	F	G	H
1	O(x)	O(y)	O+C(x)	O+C(y)	Wcal(x)	Wcal(y)	N(x)	N(y)
2	-0.854814007	3.702606246	1.054974047	2.485363104	13.10643271	9.177222982	0.854814007	-3.702606246
3	2.999543085	-0.052357219	0.699573579	0.024429648			-2.999543085	0.052357219

Figure 14: Excel worksheet showing O, O+C, Wcal, and N data in rectangular coordinates

A third Excel Worksheet is used to convert the cartesian coordinates to complex numbers (a + bj) which is read by the Python program and used for the balancing calculations. This conversion utilizes Excel's COMPLEX function (Figure 15).

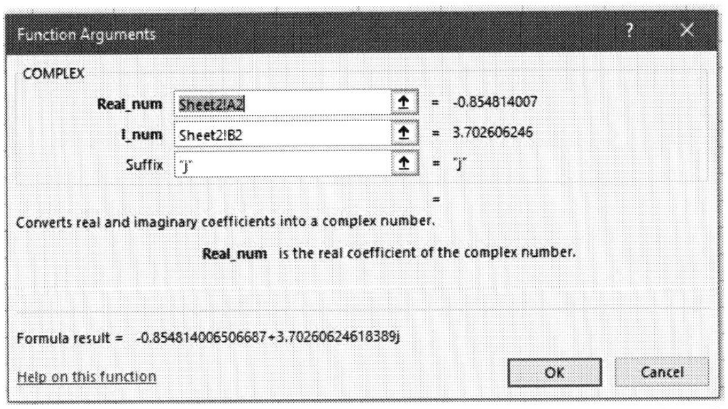

Figure 15: Excel function to create a complex number from rectangular coordinates

An Excel worksheet with the O, O+C, Wcal, and N data in complex number format is shown in Figure 16. Note that Slow Roll data was ignored for this exercise but could easily be implemented. The manipulation of the vector (amplitude and phase) measurements to x- and y-coordinates and then to complex number notation in Excel is not mandatory. Python has the capability of performing calculations and transformations that would mimic the work performed in Excel. However, since vibration data is commonly tabulated in Excel for presentation and reporting purposes, it felt natural to the authors to manipulate the data in Excel in complex number format and have Python read the data directly. Note that the Python implementation of the example problem from Goodman [2] provided in Appendix A demonstrates a technique to handle complex numbers and polar coordinates within Python code.

	A	B	C	D
1	O	O+C	Wcal	N
2	-0.854814006506687+3.70260624618389j	1.05497404692104+2.48536310432159j	13.1064327086239+9.17722298161674j	0.854814006506687-3.70260624618389j
3	2.99954308546917-0.0523572193118507j	0.699573578913367+0.0244296476917507j		-2.99954308546917+0.0523572193118507j

Figure 16: Excel worksheet showing O, O+C, Wcal, and N data in complex number format

In Python, the hashtag (#) is used for comments. The beginning of the following Python codes contains multiple comments about the libraries installed and descriptions of the variables used in the code (Figure 17). After importing NumPy and Pandas, the data is imported from Excel and the Plain Least Squares (PLS) is performed.

```
#Installed numpy, from command prompt, pip install numpy
#Installed pandas, from command prompt, pip install pandas
# A: Original Vibration vectors
# B: O+C Vibration vectors
# Wcal: Calibration (trial) weight
# Alpha: Influence coefficient vector
# Alpha_T: Transpose of Alpha
# W: Correction Weight, -{[Alpha]^T[Alpha]}^-1[Alpha]^T[A]}

import numpy as np
import pandas as pd

#file = "Turb_Pump_Data.xlsx"
#file = "ID_Fan_Data.xlsx"
file = "Metz.xlsx"
#file = "JPP_U4_Data.xlsx"
#file = "Metz_Xprobe.xlsx"
#file = "WC3_Generator_Data.xlsx"

xl = pd.ExcelFile(file)
df = xl.parse("Sheet3")
Temp = df.to_numpy()

A = Temp[:,0]
A = A.astype(complex)
A = A.reshape(-1,1)

B = Temp[:,1]
B = B.astype(complex)
B = B.reshape(-1,1)

Wcal = Temp[0,2]
Wcal = complex(Wcal)

Alpha = (B - A) / Wcal #Caciuates the influence vectors (Alpha)
Alpha_T = np.transpose(Alpha) #Perform the transpose on Alpha which is needed for the Least Squares methodology

W = -1. * (np.matmul(np.linalg.inv(np.matmul(Alpha_T,Alpha)),np.matmul(Alpha_T,A))) #Goodman equation 8
print("Plain Least Squares")
print("W", np.round(np.abs(W[0]),2), np.round(np.mod(np.angle(W[0], deg=True),360),0))
print(" ")
```

Figure 17: Python code implementation for Plain Least Squares

An implementation of the Weighted Least Squares (WLS) algorithm in Python is shown in Figure 18. The iterations are based on comparing the difference between the RMS residual from the current iteration versus the previous iteration and terminating if the difference gets below a specified tolerance or a specified maximum number of iterations is reached.

```
#Weighted Least Squares Iteration
Iter = 0
Iter_Max = 100 #Maximum number of iterations
diff = 1.0 #Initial variable for difference between iterations
tol = 0.0001 #Set tolerance for convergence
M = len(A)

Eps = np.matmul(Alpha,W) + A #Calculate the residual vectors, Epsilon, Goodman equation 9
Eps_Amp = np.abs(Eps) #Calculate the amplitude of the residual vectors
S = np.sum(np.square(Eps_Amp)) #Calculate S "Sum of Squared Residuals"
R = np.sqrt(S/M) #Calculate R "RMS Residual"
S_old = S
R_old = R

while diff > tol:
  Iter = Iter + 1
  if Iter > Iter_Max:
      break
  Eps = np.matmul(Alpha,W) + A #Calculate the residual vectors, Epsilon, Goodman equation 9
  Eps_Amp = np.abs(Eps) #Calculate the amplitude of the residual vectors
  S = np.sum(np.square(Eps_Amp)) #Calculate S "Sum of Squared Residuals"
  R = np.sqrt(S/M) #Calculate R "RMS Residual"
  print("Iteration", Iter)
  print("S", np.round(S,4))
  print("R", np.round(R,4))
  print(" ")
  Weight_Factor = np.abs(np.divide(Eps_Amp,R)) #Calculate the Weighting Factor to apply to Alpha_T
  Alphaprime_T = np.multiply(np.transpose(Weight_Factor),Alpha_T) #Calculate Alphaprime_T
  Wprime = -1. * (np.matmul(np.linalg.inv(np.matmul(Alphaprime_T,Alpha)),np.matmul(Alphaprime_T,A)))
  print("W", np.round(np.abs(Wprime[0]),2), np.round(np.mod(np.angle(Wprime[0], deg=True),360),0))
  print(" ")
  Alpha_T = Alphaprime_T
  W = Wprime
  Eps = np.matmul(Alpha,W) + A #Calculate the residual vectors, Epsilon, Goodman equation 9
  Eps_Amp = np.abs(Eps) #Calculate the amplitude of the residual vectors
  S = np.sum(np.square(Eps_Amp)) #Calculate S "Sum of Squared Residuals"
  R = np.sqrt(S/M) #Calculate R "RMS Residual"
  diff = np.abs(R - R_old)
  R_old = R

#print("W", np.round(np.abs(W[0]),2), np.round(np.mod(np.angle(W[0], deg=True),360),0))
#print(" ")
print("Eps_Amp")
print(np.round(Eps_Amp,2))
```

Figure 18: An implementation of Weighted Least Squares

TEST CASES:

Multiple test cases (using field balancing data from real machines) were solved utilizing the programs described in this paper. Three test cases are compared in Appendix B. These cases were solved using the Plain Least Squares (PLS) and Weighted Least Squares (WLS) methodologies implemented in Python, using the Excel Solver spreadsheet, minimizing on both the RMS Residual and Mini-Max Residual, and using a commercially available MPBT, Bently Balance. Sufficient data is provided in Appendix B for the reader to replicate the testing.

Test Case 1:8 vibration measurements and 1 balance plane on a 4 bearing machine train (Steam Turbine – Feedwater Pump)
Excellent agreement is observed between all 5 of the balancing results, with calculated correction weights between 2.81 – 3.14 oz and calculated correction weight angles between 344° – 359°.

Test Case 2:6 vibration measurements and 2 balance planes on a vane-axial fan (Overhung fan wheels mounted on each end of the motor shaft)
The results from the two Excel solutions and the commercial MPBT are very similar even though the Solver Mini-Max solution is trying to optimize a different parameter. The results

from the Python codes indicate fairly different values from the correction weight angles and weights, especially on the second balance plane.

Test Case 3:12 vibration measurements and 1 (static pair) balance plane on a 6 bearing machine train (HP/IP Steam Turbine – LP Steam Turbine – Generator)
The Excel Solver RMS solution is identical to the commercial MPBT solution, and the Excel Solver Mini-Max result appears reasonable, within 1 ounce and 3 degrees of the RMS solution. However, both least squares solutions implemented in Python show significant differences in the correction weight and angle. This suggests that the Python implementation may be suffering from an unknown deficiency at this time. As previously discussed, field balancing of turbomachinery requires the use of engineering judgement at all times and tools such as a MPBT program, whether self-made or commercially acquired, should not be utilized blindly.

CONCLUSIONS AND RECOMMENDATIONS FOR FUTURE WORK:

Based on the work performed for this paper, the authors concluded the following:

- Creating a multi-plane balancing tool (MPBT) program in a readily available spreadsheet package (Excel) or an open-source programming language (Python) is within the capability of everyday engineers and provides an alternative to the limited number of available commercial products.
- Using Excel with VBA was a valid method. However, this tool was the most difficult to implement, especially for multiple balance planes.
- Using Excel with the Solver was the easiest tool to implement and scale, and it gives the most consistent results.
- Using Python was convenient due to the use of complex numbers for the matrix math. However, the authors results were not as consistent as the results obtained with the Excel Solver.

The authors are planning to continue this work. The goal is to improve the implementation of the Least Squares Python code by solving the identified issue, adding a simple graphical user interface, and making the code more general. Additional ideas include adding some basic plots to the code.

REFERENCES:

[1] W. C. Foiles, P. E. Allaire and E. J. Gunter, "Review: Rotor balancing," *Shock and Vibration 5*, pp. 325-336, 1998.

[2] T. P. Goodman, "A Least-Squares Method for Computing Balance Corrections," *Journal of Engineering for Industry*, pp. 273-279, August 1964.

[3] M. S. Darlow, Balancing of High-Speed Machinery, New York: Springer-Verlag, 1989.

[4] S. C. Chapra and R. P. Canale, Numerical Methods for Engineers, Second Edition, McGraw-Hill, 1988.

[5] "Microsoft Support," [Online]. Available: support.microsoft.com.

APPENDIX A
Python Implementation of Goodman Example

```
S = np.sum(np.square(Epsilon)) #Calculate S "Sum of Squared Residuals"
R = np.sqrt(S/3.0) #Calculate R "RMS Residual"
print("\nS = ", round(np.real(S),3), " R = ", round(np.real(R),3), "\n")

# the following loop implements the Weighted Least Squares calculations from Goodman
# the procedure is essentially the same as the Plain Least Squares with the addition
# of the weighting factors
print("Begin Weighted Least Squares calculation\n")
diff = 1.0       #initial variable for difference between interations
toler = 0.001    #set tolerance for convergence if set to 0.1 covergence with two iterations
max_iter = 5     #set maximum number of iterations
iter = 0
Alpha_p = Alpha_complex
Epsilon_p = Epsilon
S_p = S
R_p = R
R_p_old = R
while diff>toler:
    iter = iter+1
    if iter>max_iter:
        break
    print("Begin iteration", iter+1, "\n")  # the first weighted iteration is actually the second iteration
    Alpha_p = Alpha_p*np.abs(Epsilon_p/R_p)
    Alpha_p_T = np.transpose(Alpha_p)
    W = np.matmul(np.linalg.inv(np.matmul(Alpha_p_T,Alpha_complex)), np.matmul(Alpha_p_T,N))
    Wmag = np.abs(W)                              # convert to polar amplitude
    Wphase = np.mod(np.angle(W, deg = True), 360)  # convert to degrees
    print("Correction Weights")
    for i in range(2):
        location = i+1
        print("Balance Plane ", location, np.round(Wmag[i],2), "at", np.round(Wphase[i], 1), " degrees")
    Epsilon_p = np.matmul(Alpha_complex, W)+Acomplex
    print("\nResidual (Epsilon) after iteration", iter+1)
    mag = np.abs(Epsilon_p)                               # convert to polar amplitude
    phase = np.mod(np.angle(Epsilon_p, deg = True), 360)  # convert to degrees
    for i in range(3):
        location = i+1
        print("Location ", location, np.round(mag[i], 4), "at", np.round(phase[i],1), " degrees")
    S_p = np.sum(np.square(Epsilon_p))
    R_p = np.sqrt(S_p/3)
    diff = np.abs(R_p_old-R_p)
    if diff < toler:
        print("\nConvergence tolerance achieved\n")
        print("S_p = ", round(np.real(S_p),3), " R_p = ", round(np.real(R_p),3), "\n")
        break
    else:
        R_p_old = R_p
    print("S_p = ", round(np.real(S_p),3), " R_p = ", round(np.real(R_p),3), "\n")

print("the program has ended")
```

Code Output in Python Console

```
Correction weight from first iteration (Plain Least Squares)

Balance Plane  1 [0.81] at [360.]  degrees
Balance Plane  2 [1.48] at [0.]  degrees

Residual (Epsilon) after first iteration

Location  1 [0.4762] at [0.]  degrees
Location  2 [0.0952] at [0.]  degrees
Location  3 [0.381] at [180.]  degrees

S =  0.381  R =  0.356

Begin Weighted Least Squares calculation

Begin iteration 2

Correction Weights
Balance Plane  1 [1.] at [360.]  degrees
Balance Plane  2 [1.8] at [0.]  degrees

Residual (Epsilon) after iteration 2
Location  1 [0.4] at [0.]  degrees
Location  2 [0.4] at [0.]  degrees
Location  3 [0.4] at [180.]  degrees
S_p =  0.48  R_p =  0.4

Begin iteration 3

Correction Weights
Balance Plane  1 [1.] at [360.]  degrees
Balance Plane  2 [1.8] at [0.]  degrees

Residual (Epsilon) after iteration 3
Location  1 [0.4] at [0.]  degrees
Location  2 [0.4] at [0.]  degrees
Location  3 [0.4] at [180.]  degrees

Convergence tolerance achieved

S_p =  0.48  R_p =  0.4

the program has ended
```

APPENDIX B
Comparison of Test Case Results

Test Case 1 – Turbine Driven Feedwater Pump
Number of measurements: 8
Number of balance planes: 1

O(r)	O(θ)	O+C(r)	O+C(θ)	Wcal(r)	Wcal(θ)
0.5	324	0.4	309	3.0	0
0.8	296	0.5	327		
0.2	246	0.3	246		
0.7	159	0.8	173		
0.7	43	0.5	65		
4.3	329	0.6	296		
0.6	127	0.6	74		
3.2	145	0.8	15		

Python (PLS):	3.00 oz @ 352^{0}
Python (WLS):	3.14 oz @ 344^{0} (21 iterations)
Excel Solver (RMS):	2.96 oz @ 354^{0}
Excel Solver (Mini-Max):	2.81 oz @ 359^{0}
Bently Balance:	2.96 oz @ 354^{0}

Test Case 2 – Vane-Axial Fan
Number of measurements: 6
Number of balance planes: 2
Note: W1cal left installed for the W2cal run

O(r)	O(θ)	O+C1 (r)	O+1C1 (θ)	O+C2 (r)	O+1C2 (θ)	W1cal (r)	Wcal1 (θ)	W2cal (r)	Wcal2 (θ)
0.060	292	0.09	295	0.12	297	25.0	270	22.0	337.5
0.070	248	0.12	250	0.15	254	22	337.5		
0.190	338	0.46	353	0.46	1				
0.180	318	0.29	300	0.19	310				
0.200	54	0.34	29	0.28	29				
0.190	145	0.4	155	0.41	162				

Python (PLS):	15.3 gram @ 86^{0} / 18.6 gram @ 1^{0}
Python (WLS):	16.6 gram @ 113^{0} / 24.3 gram @ 346^{0} (28 iterations)
Excel Solver (RMS):	13.4 gram @ 95^{0} / 30.7 gram @ 34^{0}
Excel Solver (Mini-Max):	14.3 gram @ 98^{0} / 26.3 gram @ 36^{0}
Bently Balance:	13.2 gram @ 95^{0} / 30.4 gram @ 31^{0}

Test Case 3 – Steam Turbine
Number of measurements: 12
Number of balance planes: 1 (Static-Pair)

O(r)	O(θ)	O+C(r)	O+C(θ)	Wcal(r)	Wcal(θ)
0.584	99	0.172	70	12.0	135
0.880	354	0.263	333		
0.855	239	0.451	186		
1.659	147	0.437	187		
0.694	293	0.899	278		
1.314	257	0.775	164		
2.868	19	1.955	49		
3.467	247	2.425	276		
1.497	32	1.056	96		
2.598	218	1.177	242		
1.221	151	1.105	177		
0.861	345	0.646	45		

Python (PLS):	23.1 oz @ 233°
Python (WLS):	76.8 oz @ 139° (73 iterations)
Excel Solver (RMS):	15.9 oz @ 161°
Excel Solver (Mini-Max):	14.8 oz @ 158°
Bently Balance:	15.9 oz @ 161°

NIAGARA FALLS, NEW YORK JULY 26 -28, 2023

CHARATERIZING OFF-LINE TO RUNNING MOVEMENT USING HIGH-RESOLUTION IMAGERY

John Piotrowski
Turvac, Inc.
6434 Coach Light Circle
Liberty Twp OH 45011-1198
(001) 513-759-2771
ContactUs@Turvac.com

Bill Marscher, CEO
And
Andy Lerche, Senior Staff Engineer
Mechanical Solutions, Inc.
11 Apollo Drive
Whippany, NJ 07981
(001) 973-326-9920
AHL@mechsol.com

Bill Marscher, P.E. founded MSI in 1996, which has grown to a consulting firm of 30. Bill has been an MFPT and VI participant since the 1980's, and is a long-standing member of the MFPT Board, and past chair. He has BS and MS degrees in Mechanical Engineering from Cornell University, and an MS in Mechanics from RPI. Bill previously worked at Dresser Pump and Pratt & Whitney. He is past president of the Society of STLE, and is on the Advisory Committee of the Texas A&M Pump Symposium. He is a member of the ISO TC108 Machinery Standards Committee, and is past Vice Chair of the Hydraulic Institute Vibration Standards Committee.

Mr. Piotrowski graduated from the University of Pittsburgh in 1974 with a degree in Mechanical Engineering. He has been working with rotating machinery in industry for over 44 years and is considered to be one of the leading authorities on alignment of rotating machinery. He has authored over fifty technical articles on the subject and is author of the Basic Shaft Alignment Workbook, the Shaft Alignment Self Study Guide, and the Shaft Alignment Handbook (now in its 3rd edition).

He was one of the contributing authors for the development of the ANSI/ASA Shaft Alignment Standards (ANSI/ASA S2.75 Parts 1, 2, and 3). He has also a contributing author to the 4th and 5th editions of the Shock and Vibration Handbook (ISBN 0-07-026920-3, Chapter 39. Part II: Vibration Induced by Shaft Misalignment). Mr. Piotrowski is an active member of the Vibration Institute.

Since founding Turvac Inc. in 1982, John has trained over 7500 people in shaft alignment principles and methodology, been a consultant to several major shaft alignment equipment manufacturers and participated in alignment service work in a wide variety of industries. He has designed training demonstrators for alignment and balancing, shaft alignment measurement systems, shaft alignment software programs, invented the Ball-Rod-Tubing Connector system used to measure off-line to running machinery movement, and performed several controlled alignment / vibration tests. But above everything else, John loves what he is doing and wants everyone who is involved with rotating machinery to understand this broad and fascinating topic.

Mr. Lerche has more than 20 years of professional experience in the areas of design, testing, analysis (fluid and structural), and troubleshooting of mechanical equipment and machinery such as pumps and turbomachinery. He is experienced in the areas of turbomachinery, high pressure fluid control equipment, and shipboard equipment. He has extensive experience with flow-induced vibration of turbomachinery blades, as well as shock and vibration testing and analysis of fixed mechanical and electrical equipment. Mr. Lerche has authored and presented several technical papers in the area of turbomachinery and holds three patents related to industrial compressors and valves. He holds a B.S. and M.S. in Mechanical Engineering from the University of Texas at Austin and the University of Texas at San Antonio, respectively.

ABSTRACT:
Adequate alignment of machinery is key to its reliable operation. Poorly aligned driver/driven machinery shafting can overload bearings, or cause internal rubs at close running clearances, and seal interfaces. Alignment can change in an unexpected manner between the at-rest "cold" condition of the machinery, and its warmed-up operating state. Conventional approaches can do a good job of assessing the degree and nature of resulting misalignment, but such approaches can be awkward and costly in many circumstances. Some new optical methods based on video motion magnification and/or video scene interpretation show promise in many such applications. Some case histories are provided.

KEY WORDS: Motion Magnification; Measurement Accuracy; Rotating Machinery; Machinery Diagnostics; Machinery Alignment

1. Introduction:

Turbomachinery, pump, and motor vibration is affected by various factors, some aerodynamic, and some mechanical. The most common mechanical causes are imbalance operating at 1x rpm, and driver/driven machine misalignment, which has been observed to act at 1x, 2x, or 3x rpm. When misalignment is suspected, and non-operating cold alignment has been verified to be as intended, various methods may be employed to determine the reason for alignment change between the off-line to the operating and warmed-up alignment condition. The conventional methods available to date often can do a good job investigating this, but sometimes at significant cost and/ or inconvenience. The advantages and disadvantages of these various methods are briefly presented, and a new video-based high-resolution motion interpretation technique is presented.

Some key issues that are discussed in this paper include:

- What is Off-Line To Running (OL2R) machinery movement?
- What causes this movement to occur?
- What types of rotating machinery will move enough from OL2R conditions to warrant measuring this movement?
- How to conduct an accurate OL2R survey.
- Four general categories of OL2R movement measurement.

2. Background:

Proper alignment of operating machinery is critical to its reliability [1], [2]. Improper alignment can overload bearings, and cause excess deflection:

=> internal rubbing, or
=> failure of mechanical seals

Evaluating driver/ driven machine alignment has typically been performed by:

⇒ The lowly straight-edge
⇒ Reverse Dial Indicator

In addition, "Hands Off" Optical methods are useful during warm-up and operation:

⇒ Lasers
⇒ High-resolution cameras.

What Is Off-Line to Running (OL2R) Machinery Movement?

Once the shafts are aligned within acceptable tolerances, as provided in [3], the coupling engaged, and the machinery started up, changes in the positions of the shafts may begin to occur that could alter the accuracy of the alignment when the equipment is running.

Virtually all rotating equipment will undergo a change in position during start up and while running that affects the alignment of the shafts. The driver machine may move a certain amount and in a certain direction as it goes from an off-line to a stable (we hope) operating condition. The driven element(s) of the system may move different amounts and in different directions.

In order for the shafts to run collinear under normal operating conditions, it is desirable to know the amount and direction of this movement to properly position the machinery during what is commonly called the "cold" (i.e. off-line or not running) alignment process to compensate for this change.

What Causes This Movement to Occur?
The most common cause is due to temperature changes in the machinery itself (as it compresses gases or heats the lubricant from friction in the bearings) and is therefore generally referred to as "thermal" or "cold" to "hot" movement. The temperature change in rotating machinery is rarely uniform throughout the casing which causes most equipment to "pitch" at some angle rather than grow (or shrink) straight up (or down).

For compressors, turbines, and pumps, thermal expansion (or contraction) of the attached piping may also cause the equipment to shift.

Other sources of movement in machinery can be caused by: loose foot bolts, varying weather conditions for equipment located outdoors, heating or cooling of concrete pedestals, changes in the operating condition of equipment from unloaded to loaded postures, and/or casing and support counter-reactions to the centrifugal force of rotors as they are spinning.

What Type of Machinery Is Likely to Change Its Position When Running?
Any rotating or reciprocating machinery with significant torque, speed, flexible supports, or temperature or pressure change is subject to potentially significant alignment change:

- Rotating machinery drive systems at or above 200 hp & speeds of 1200 rpm or greater.

- Machinery that undergoes a change in casing temperature. For example ...

    - electric motors and generators
    - steam turbines
    - gas turbines
    - internal combustion engines (diesels, etc.)

- Speed changers (e.g. gear boxes and fluid drives).

- Machinery that is pumping or compressing where the liquid or gas undergoes a change in temperature by 50 degrees or greater from intake to discharge (this could be either a rise or drop in temperature). For example:

    - centrifugal or reciprocating compressors
    - centrifugal pumps
    - HVAC air moving equipment
    - furnace fans

- Equipment with poorly supported piping attached to the machine where expansion or contraction of the piping induces nozzle loads for where fluid flows can cause a reactionary load in the piping

What Have the Authors Observed Over the Years?
Fewer than 10% of the people who are responsible for alignment of rotating machinery in industry have ever actually conducted off-line to running (OL2R) machinery movement surveys. Some people are aware of this phenomenon but only a handful of them have actually tried to measure it.

Only a small percentage of the rotating machinery in existence has ever been measured for off-line to running movement. (Estimated to be less than 5%). Since very few drive systems have been studied, much has yet to be learned about what happens to the alignment of machinery when it is running (and when it's off line). For example, many of the people who have conducted OL2R surveys or are aware of this phenomenon, believe that machinery only moves in the up and down direction disregarding the possibility/probability of sideways or axial movement.

In several cases where these data have been collected, and it was verified that the machinery was not in the right alignment position when off-line, there was a great reluctance to change the existing positions of the machinery to reflect the newly discovered information.

How to Conduct an OL2R Survey
Step 1: Gather information about the drive system such as:

- dimensional information on the location of the bearings
- how the equipment is attached to the frame, soleplates, or baseplate
- as found and final alignment measurements from the initial installation through each outage to present day
- vibration data (trends, spectral plots, orbits)
- evaluate the site and select an OL2R method (or maybe two for comparison purposes)
- set up the measurement process

Step 2: Ensure the drive system is in the final desired off line alignment position and record the final alignment readings at the couplings. Capture two sets of off line measurements to ensure the drive system is in a stable position. If the readings do not repeat (or nearly so i.e. within 1-2 mils), you should take a third (or fourth) set until you see repeatability in your measurements.

Note: If you observe that the machinery is still moving, you may want to recheck the shaft positions by taking measurements across the couplings ... that is, repeat the first item in this list.

Step 3: Start the drive system up and allow it to stabilize. Capture at least two sets of running measurements to ensure the drive system is in a stable position. If the readings do not repeat (or nearly so i.e. within 1-2 mils), you should take a third (or fourth) set until you see repeatability in your measurements.

Step 4: Capture a set of running measurements every two to four months with the drive system at normal operating conditions. It is a good idea to see perhaps, what the running alignment condition is in the winter, spring, summer, and fall.

Repeat this process for another 12 months to see if there is a pattern of movement or running position. If the drive system varies in load throughout the day, or through a week or month, you should consider measuring the running position at several of these various operating modes to establish a range of movement / position.

Advice: Be Prepared

Compared to the amount of time usually taken to align rotating machinery when off-line (which takes 3-6 hours on the average), off-line to running machinery movement surveys can take days, weeks, even months to complete. These types of surveys are frequently conducted on the more critical pieces of rotating machinery in a plant and these drive systems usually cannot be turned on and turned off at the convenience of the people performing the test. So be prepared to spend a lot more time planning, designing, fabricating, installing, measuring, and analyzing these data than you typically would for "off-line / inline" shaft alignment.

The measurements you take with the machinery off-line are often not going to be on the same day that you capture the measurements with the machinery running in a stable operating condition. This means one of three things:

1. The operator can keep all of the OL2R equipment set up and hope nothing happens to it during the "warm up" cycle.
2. The operator can keep some of the equipment set up and store the rest until your next measurement time, then set the stored equipment back up again.
3. The operator can set it all up, take it all off, and then set it all up again.

Four General Categories of OL2R Measurements

The reasons for OL2R alignment changes tend to fall into four categories, as described below.

Category 1: Movement of the centerlines of the machine cases with respect to their baseplate or frame:

- calculating machine case thermal expansion using the strain equation
- inside micrometer / tooling ball / angle measurement devices
- proximity probes with water cooled stands

Category 2: Movement of the centerlines of the machine cases with respect to some remote reference or observation point:

- Optical alignment equipment
- laser - detector measurement systems
- electronic theodolite system
- laser tracker systems

Category 3: Movement of one machine case with respect to another machine case:

- alignment bars or custom fixtures with proximity probes
- laser - detector systems with custom fabricated brackets or special mounting systems
- Ball-Rod-Tubing connector system

Category 4: Movement of one shaft with respect to another shaft:

- vernier - strobe method
- instrumented coupling system

Category Tools and Techniques
The particular alignment tools and techniques vary depending upon the category, as described below (in #/$, # means cases encountered, and more $ signs mean more cost):

Category 1: Movement of the Centerline of the Machine Cases with Respect to the Baseplate or Frame. Tools and techniques that fall into this category are:

- calculating machine case expansion using the thermal strain equation (3/$)
- inside micrometer / tooling ball / angle measurement devices (6/$$)
- proximity probes with water cooled stands (6/$$$)

Category 2: Movement of the Centerline of the Machine Cases with Respect to Some Remote Reference or Observation Point. Tools and techniques that fall into this category are:

- optical alignment equipment (9-10/$$$$)
- laser - detector measurement systems (4-5/$$$$$)
- electronic theodolite system (4-5/$$$$$)
- laser tracker systems (9-10/$$$$$$$$$$)

Category 3: Movement of Single Machine Cases with Respect to Another Machine Case. Tools and techniques that fall into this category are:

- alignment bars or custom fixtures with proximity probes (3-4/$$$)
- laser - detector systems with custom fabricated brackets or special mounting systems (3-4/$$$$$)
- Ball-Rod-Tubing connector system (3-4/$$$)

Category 4: Movement of One Shaft with Respect to Another Shaft. Tools and techniques that fall into this category are:

- vernier - strobe method (9-10/$$)
- instrumented coupling system (9-10/$$$$)

The Option of Taking "Hot" Alignment Measurements Immediately After Shutdown
One popular, but relatively inaccurate method people have done is to take a quick "hot" alignment reading with their shaft alignment measurement system immediately after a unit has been shut down. This method is not recommended for the following reasons:

1. The machinery is no longer running.
2. There is a safety related issue here.
3. The shafts are probably moving while you are taking the alignment readings.

A New Option- Motion high-resolution Video Visualization and Documentation

The authors have developed a video-based method for tracking key points on a machinery train. The tracking uses a set of relatively inexpensive high-resolution cameras, the output of which is acquired and processed by specialty software. The method is automated, and relatively easy to set-up and calibrate. 3-D alignment data is able to be obtained, and developed as actionable information. Some case histories that resulted from application of the new alignment method will be presented in a later section. First, the basics of what the method consists of will be presented, including its relationship to previous optical/vision based techniques.

Classical Motion Visualization Methods:

Around 1980, some researchers developed a methodology for "seeing" motion in extensive mechanical systems. Their approach was to acquire data at many locations and directions for a structure's motion, and then allocating those motions to modify a series of photographs, or a complete video.

This range of efforts, and others, led scientists and engineers to an appreciation for the power of video to uncover useful information in dynamic scenes, assuming that appropriate algorithms, implemented by software, were able to be used to evaluate the video footage.

In the US, several universities, including MIT [4], [5], have been actively pursuing video magnification and scene interpretation. At MIT, the particular technology which became the core of their research was an Eulerian approach [6], which essentially tracks the variation of individual pixels over time, and then exaggerates those differences. For a computer, the tiniest per-pixel fluctuations (between white and slightly-off-white, say) are easy to detect and quantify. MIT originally developed their software approach, known as Motion Magnification, to measure the vital signs of neonatal babies without physical contact, but they realized that there were other, far-ranging applications. For biological applications, Eulerian VVA detects changes in skin color, as well as exaggerates movements.

Research by the Authors:

The authors' organization initiated its detailed research into engineering uses for video scene interpretation in 2003.

During this process, the authors began applying Eulerian algorithms to measure and amplify motions of individual pixels [7]. Different procedures were tried, and after several years a successful method was achieved. Transient motions as low as tenths of a mil (i.e. several microns) on machinery surfaces were demonstrated first in the laboratory, and soon after in on-site machinery troubleshooting [7].

Some of the authors' methods are based on the Eulerian approach [4], [5], [6], which observes the intensity variation of individual pixels, combining their effects statistically into a full scene. Alternative methods typically keep track of motion pixel-to-pixel, the so-called Lagrangian techniques. Variations include feature tracking, and an approach called "optical flow". These Lagrangian methods can work for amplifying motions that occur over multiple pixels, e.g. motions of more than 2mm (0.08 inches) for a 3 m (10 foot) field-of-view. These approaches are limited in how small a displacement they can detect for a given field of view and a given camera resolution. This can be a problem in vibration detection and display, in that interesting vibration levels can be well below such limits, particularly at higher frequencies of vibration. If higher camera resolutions are used to overcome this limitation, the computational overhead becomes excessive. Some applications mediate this by selecting only a greatly reduced number of pixels in a scene.

The benefits of video use for detecting and displaying position changes and gradual motion are that it becomes a powerful and intuitive procedure for displaying complex motions. The technique does not require contact, so that surfaces above 1000 F for instance, or that are radioactive, or that are (for example) 50 feet vertically above in a ceiling, can be observed without difficulty. The results are obtained quickly, and at modest cost and effort.

Case Histories:
Some results videos will be presented as videos during VIATC2023. Still photos will be provided in this paper to represent the material in the presentation.

Camera-Based Alignment Monitoring Example
A 30 year old 3600 rpm 60 MW steam turbine, shown in Fig. 1, operated at a power plant was exhibiting high vibration after an overhaul. High-Resolution video tracking was used to visualize and measure the thermal growth of the turbine. Motion Magnification Video, Experimental Modal Analysis, and Operating Deflection Shape (ODS) testing helped identify vibration issues. As was eventually discovered using optical techniques, the root cause of the problem was misalignment due to uneven thermal growth, and the vibration effects were indirect.

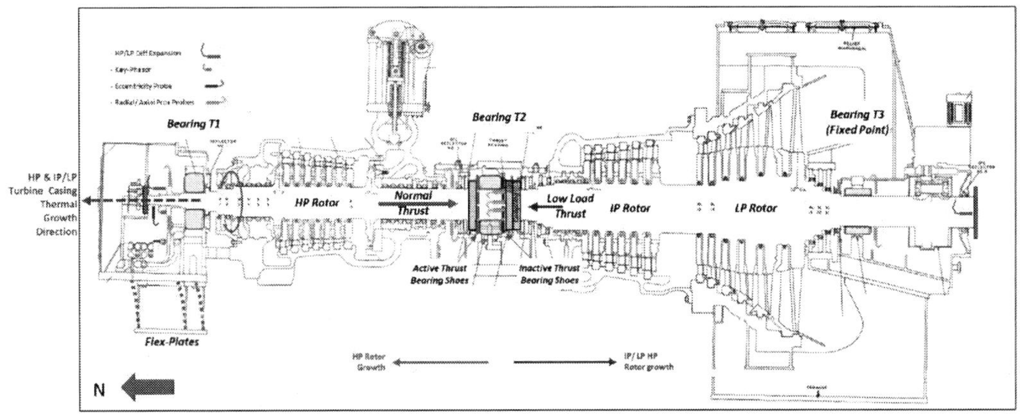

Figure 1: Steam Turbine General Info

Fig. 2 shows the distribution of COTS video cameras and enhanced lighting that was placed around the turbine to detect, visually display, and quantify/ document the motion at strategic locations that would indicate degree (and reason for) misalignment.

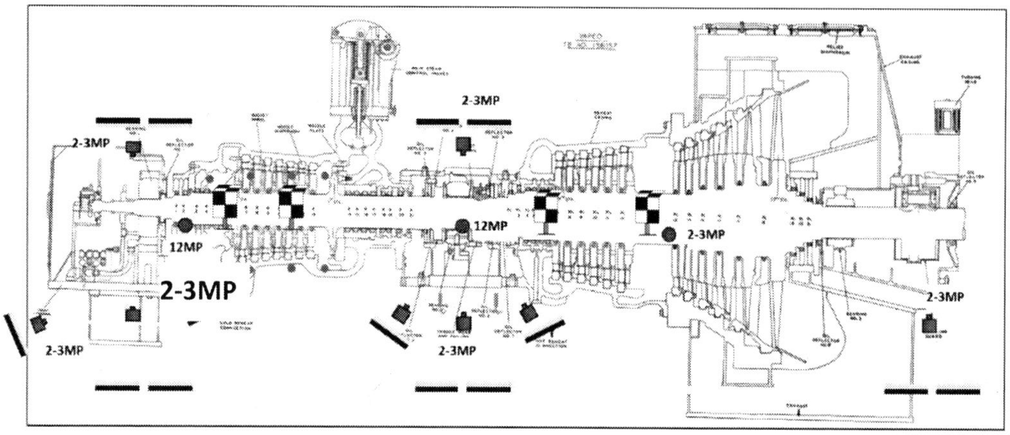

Figure 2: Test Plan Layout – Video Camera and Lighting

The background concerning the problem is described as follows:

- After 30 years, high vibration amplitude (5.2 mils pk-pk) was observed at the T1 bearing.
- The T1 vibration alarm was 4 mils pk-pk, and trip was 7 mils pk-pk, versus a diametral clearance of 8 mils.
- The dominant frequency of vibration was 1x rpm, and was highest at low load.
- The plant performed high-speed balance of the rotor, and then when vibration remained elevated, corrected the cold alignment.
- 1x rpm vibration problem persisted, and internal rubbing was discovered.
- The plant removed/reworked lower half of HP and IP casing to address the rubs.
- Upon further inspection, the plant found evidence of insufficient sliding of T2 "mid-standard" versus foundation, but it was expensive to address without additional evidence of its effect on the problem.
- The problem persisted, and at this point, and the plant suspected hot misalignment due to this, and possibly other contributing causes.

Fig. 3 shows two example views from video frames taken by the cameras. The example view is for the T1 bearing standard, from two separate directions. Fig. 4 documents the thermal growth detected by the cameras at strategic locations, and plotted/evaluated by the software. Fig. 5 shows a similar set of plots, cross-compared for the T2 standard, exhibiting a 53 mils differential growth axially, leading to a horizontal tilt.

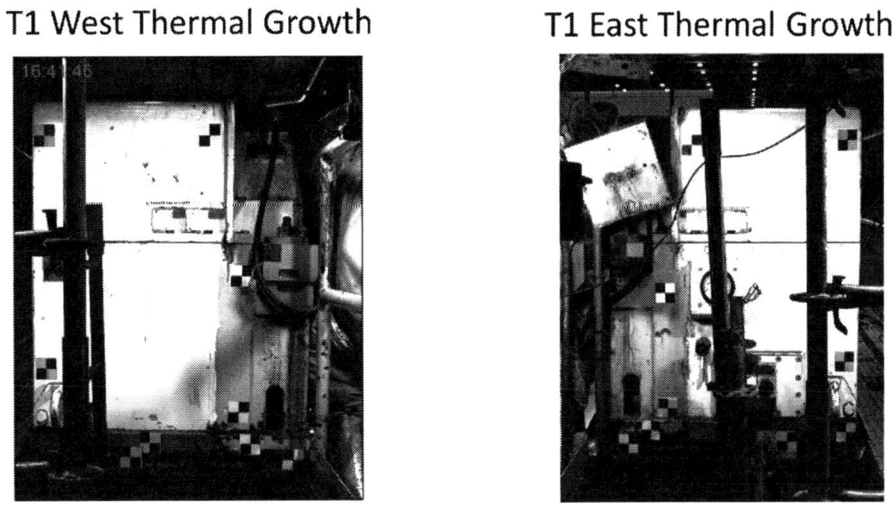

Figure 3: T1 East & West Thermal Growth – Time Lapse Video

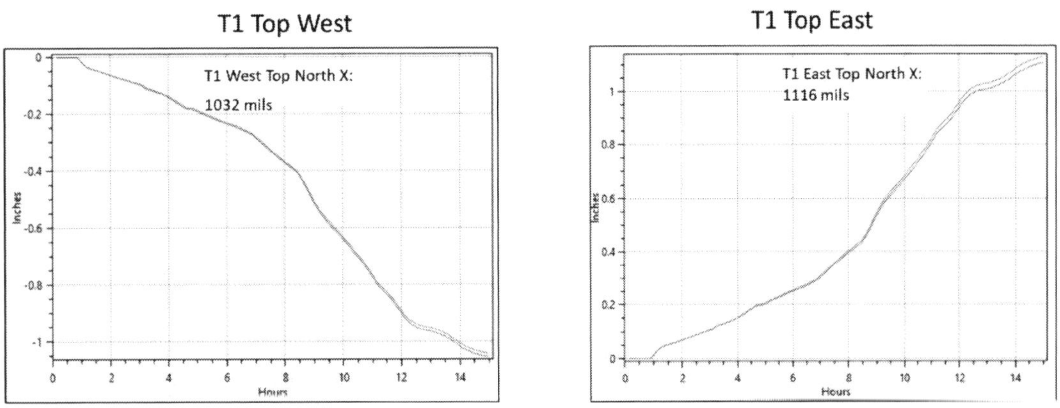

84 mils more displacement at T1 East than T1 West at the top

Figure 4: T1 Axial Growth

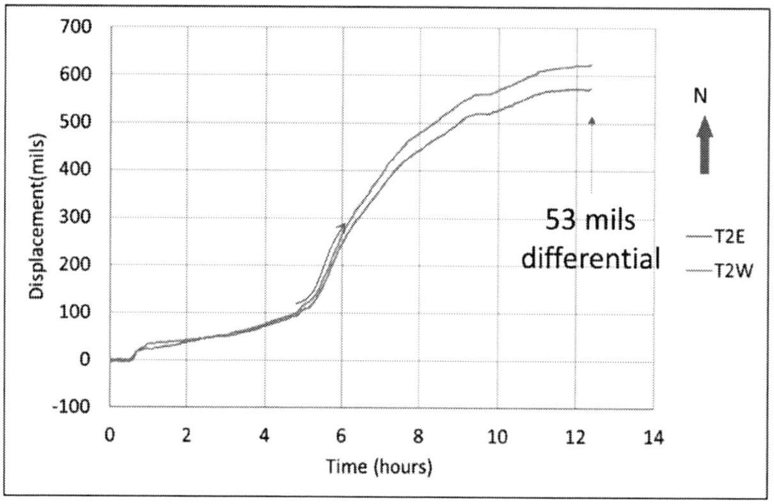

53 mils more displacement at **T2 West** than **T2 East** at the base

Figure 5: T2 Axial Absolute Displacement

Fig. 6 shows an Operating Deflection Shape (ODS) computer model, for which vibration data was gathered on site and impressed on the measurement locations, and then animated. These results agreed well with Motion Magnified Video (MMV) high speed camera footage, using the Eulerian technique to produce a roughly 1000x magnification of the vibrations at all locations. The advantage of the ODS is that it is fully 3-D, while the advantage of the MMV is that is can be implemented much more quickly and inexpensively. The detailed animations in each case exhibited a pedestal attachment/ soft foot issue that was enabled by the misalignment documented by the authors' video technique.

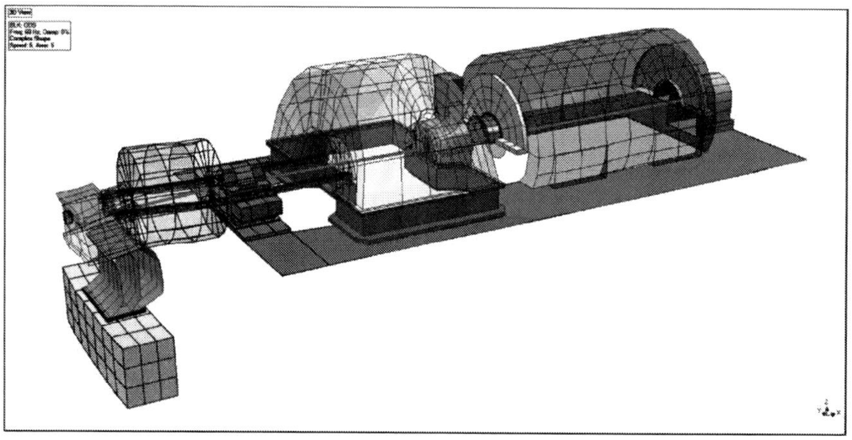

Figure 6: ODS Animation at 60 Hz (1x rpm), and Motion Magnified Video

The key findings of the combined vibration and alignment detection procedures were as follows:

1. Uneven axial thermal growth was occurring, causing hot misalignment.
    - T2: ~50 mils more thermal growth at the east side of the T2 base than the west side, where thrust bearing was located. T2 was restricting the HP casing growth.

- T1: ~80 mils more displacement occurred at T1 east than T1 west at the top.

2. Twisting motion of T1 and T2 directly caused subsequent vibration issues of the HP rotor at the T1 front standard, causing "cramping" of the HP rotor.
3. Uneven thermal growth also caused a twist and corner lifting of front standard.
4. This shifted the "axial swaying" axial structural natural frequency down to 60 Hz.
5. Axial structural resonance amplified axial vibration about 2x to about 0.30 in/s rms.

As a final remediation, all keys & grease pads were cleaned & re-lubricated during the next major outage. This resulted in the following satisfactory operation:

1. The T2 mid-standard no longer showed twisting behavior
2. The T1 front standard grew over 100 mils more than previously, with no more tilt
3. The T1 structural natural frequency increased to 70 Hz from the previous 60 Hz, alleviating the 1x rpm vibration resonance.
4. Overall vibration of the bearing standards decreased to under 0.20 in/s rms
5. The HP rotor vibration decreased to 2.1 mils pk-pk, vs. the original 5.2 mils pk-pk

Case History 2: Charge Pump Motor Alignment
The situation for the second case history can be described as follows:

- Six pumps were driven at 6036 rpm by 600 HP constant speed induction motors.
- The motors operated at 1780 rpm, through a gear increasing double-helical gear set.
- One motor stator was replaced by the OEM after successful test at the factory. The vibration in the factory test was low (0.01 in/s pk overall) at all frequencies.
- The motor was installed using a questionable alignment on an old baseplate skid.
- The pump ran 14 hours at a high 2xrpm vibration, with oil blackening & an IB bearing wipe.
- The motor was repaired, and re-aligned, but the motor vibration was still elevated at 2xrpm. However, pump bearing temperature and oil quality were now acceptable.

Fig. 7 shows the author's instrumentation, used for the primary vibration detection. It was later discovered that position and dimensional changes due to OL2R was a probable issue as well, and additional video instrumentation was implemented for a video alignment investigation, see Fig. 11 as discussed later.

Fig. 8 shows a typical vibration spectrum obtained from an impact modal test in this example as performed on the motor in the horizontal direction. Several natural frequencies are clearly evident, with probably resonances with 2x motor speed (2x is often driven by misalignment), and 2x motor line frequency, a common excitation is electric motors. The MMV video animations, a frame of which is provided in Fig. 9, as well as ODS test videos, a frame of which is exhibited in Fig. 10, verified that these natural frequencies were being excited to a mild resonance (mild because there was still some percentage of separation margin).

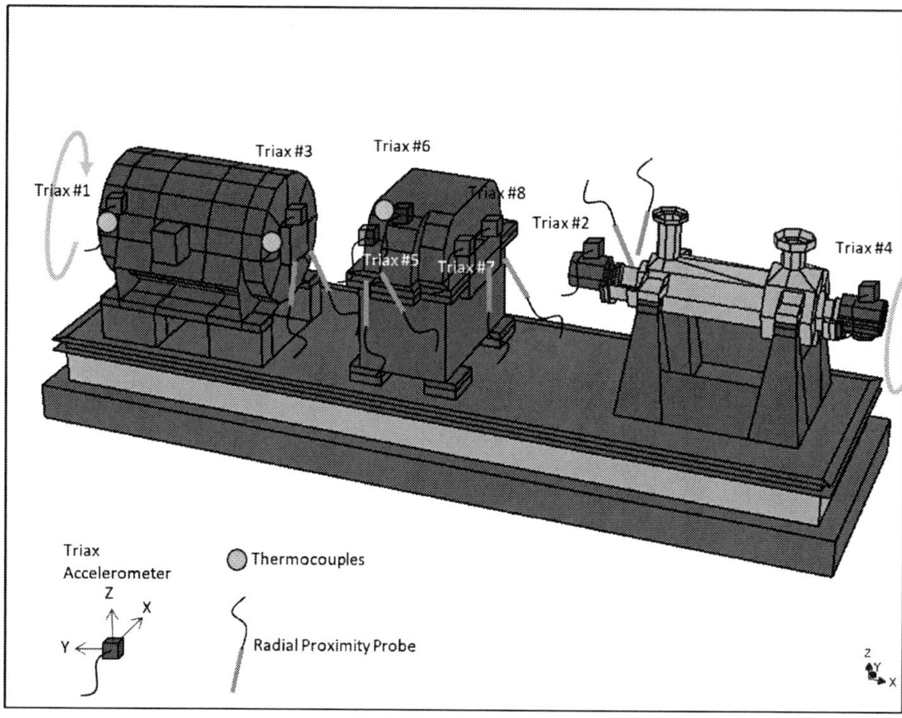

Figure 7: Authors' Vibration Instrumentation

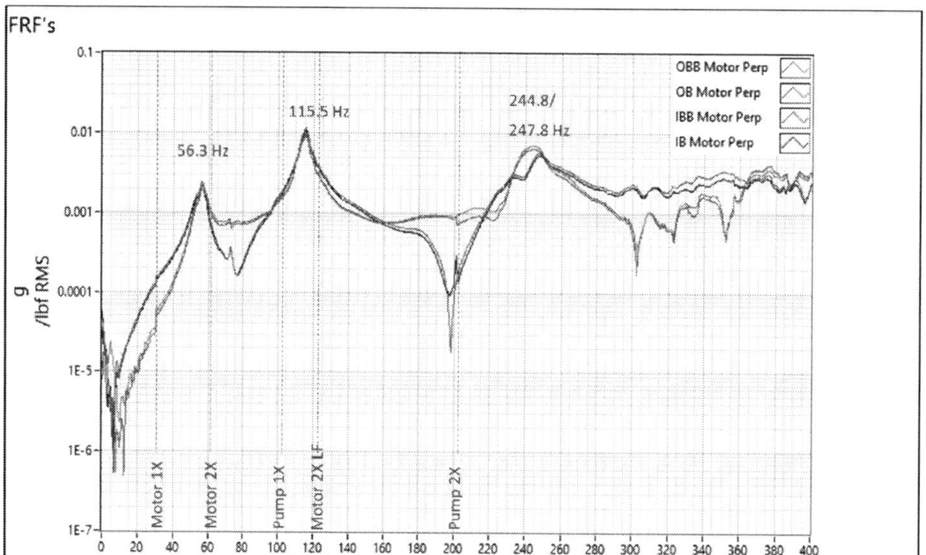

Figure 8: Motor Modal Test Results

Video at 1x rpm Motor Shaft (30 Hz) Video at 2x rpm Motor Shaft (60 Hz)

Gearbox Gearbox

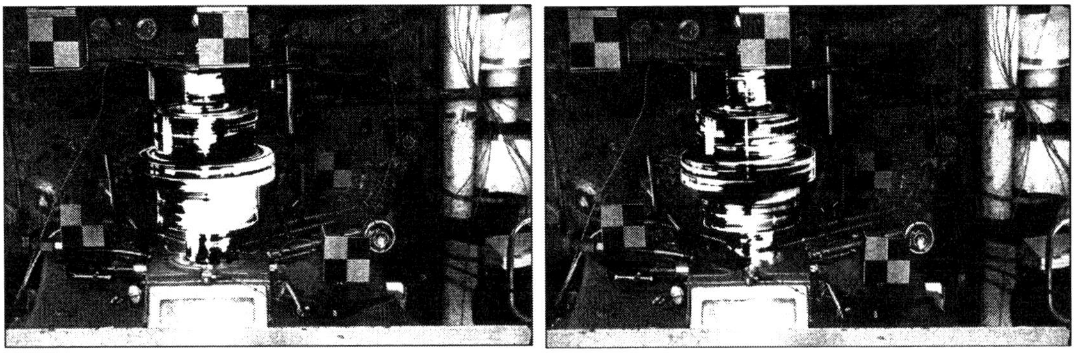

Motor Motor

Figure 9: Motor IB to Gearbox Input Using Motion Magnification

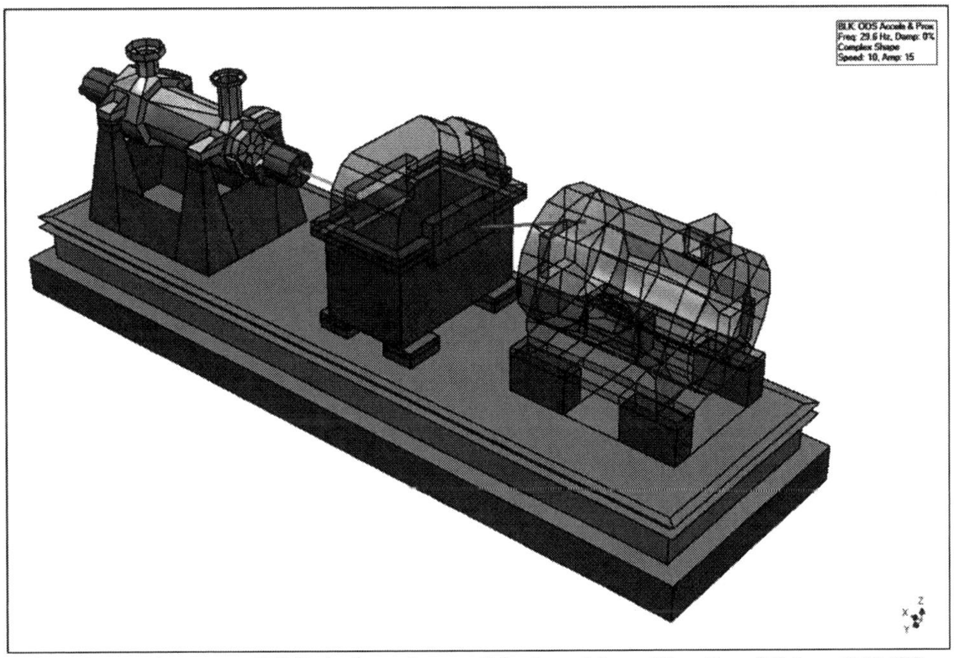

Figure 10: Motor 1x rpm ODS Animations. Motion Magnification Video (MMV) and the Operating Deflection Shape (ODS) testing confirmed the side-to-side motion and the excessive flexibility of the motor support structure, at 2x rpm.

The Optical MMV and ODS vibration tests detected the following vibration issues:

1. There was a verified structural resonance of the motor frame at 56.3 Hz, which was excited to resonance by the 2nd harmonic of the running speed of the motor at 59.3 Hz.
2. This offending structural natural frequency has only 5% of separation margin from the 2x rpm excitation source. The minimum separation margin required as measured in the field is 10%.

3. This offending structural natural frequency exhibits lateral side-to-side rocking motion of the motor frame on its pedestal.
4. This structural mode was originally higher in frequency, but gradually shifted downward due to the following issues:
    a. Soft-foot between the motor feet and the pedestals.
    b. Voids caused by delamination of the internal grout within the baseplate underneath the motor pedestals (aging effect).
    c. Separation between the baseplate and the concrete foundation due to simple looseness of the anchor bolts in combination with deterioration at the connection between the anchor bolt and the foundation.

The vibration issues were found to be significant, but not enough on their own to explain all aspects of the problem. Furthermore, cold vs. hot modal tests suggested that the soft foot appeared to be sourced in a hot alignment problem that lifted the foot, and therefore the resonance was affected by alignment as well.

Therefore, a detailed alignment quantification test was performed, using the authors' high-resolution time-lapse video technique. Fig 11 documents the placement of the vertical and horizontal video cameras and auxiliary lighting.

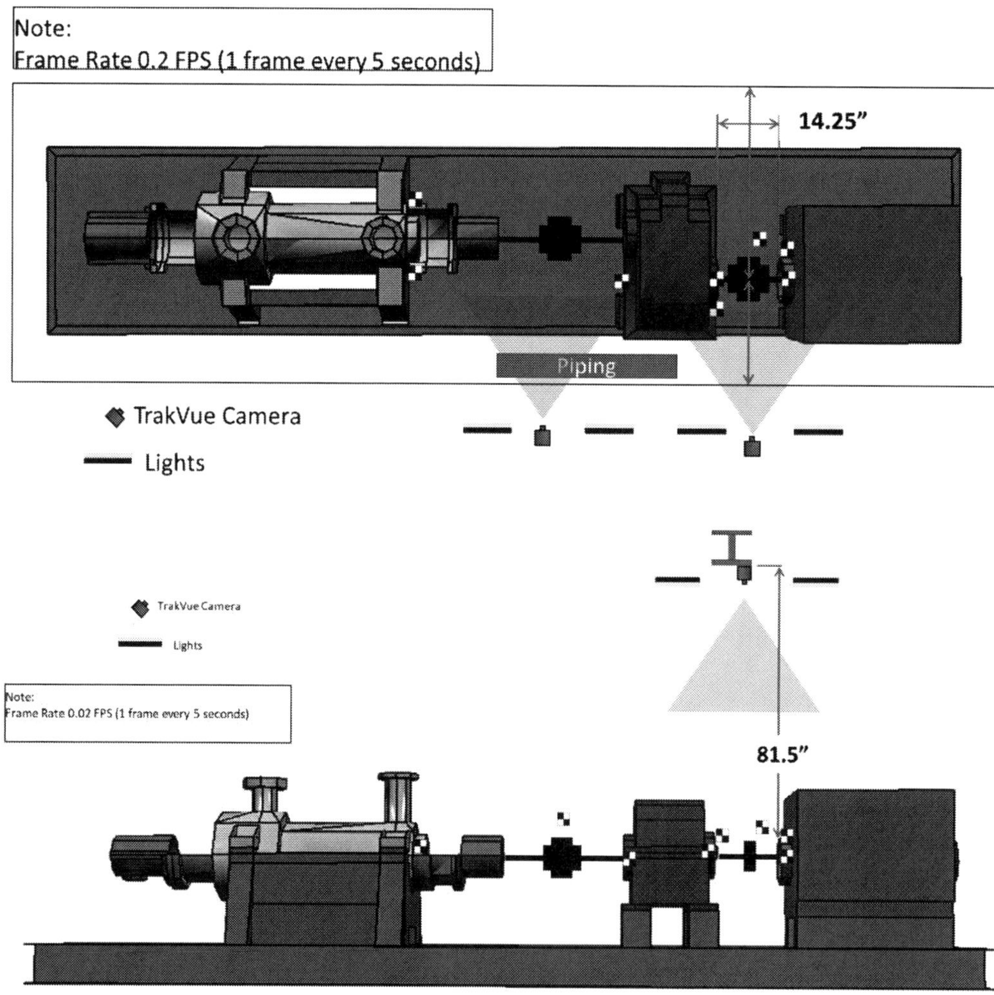

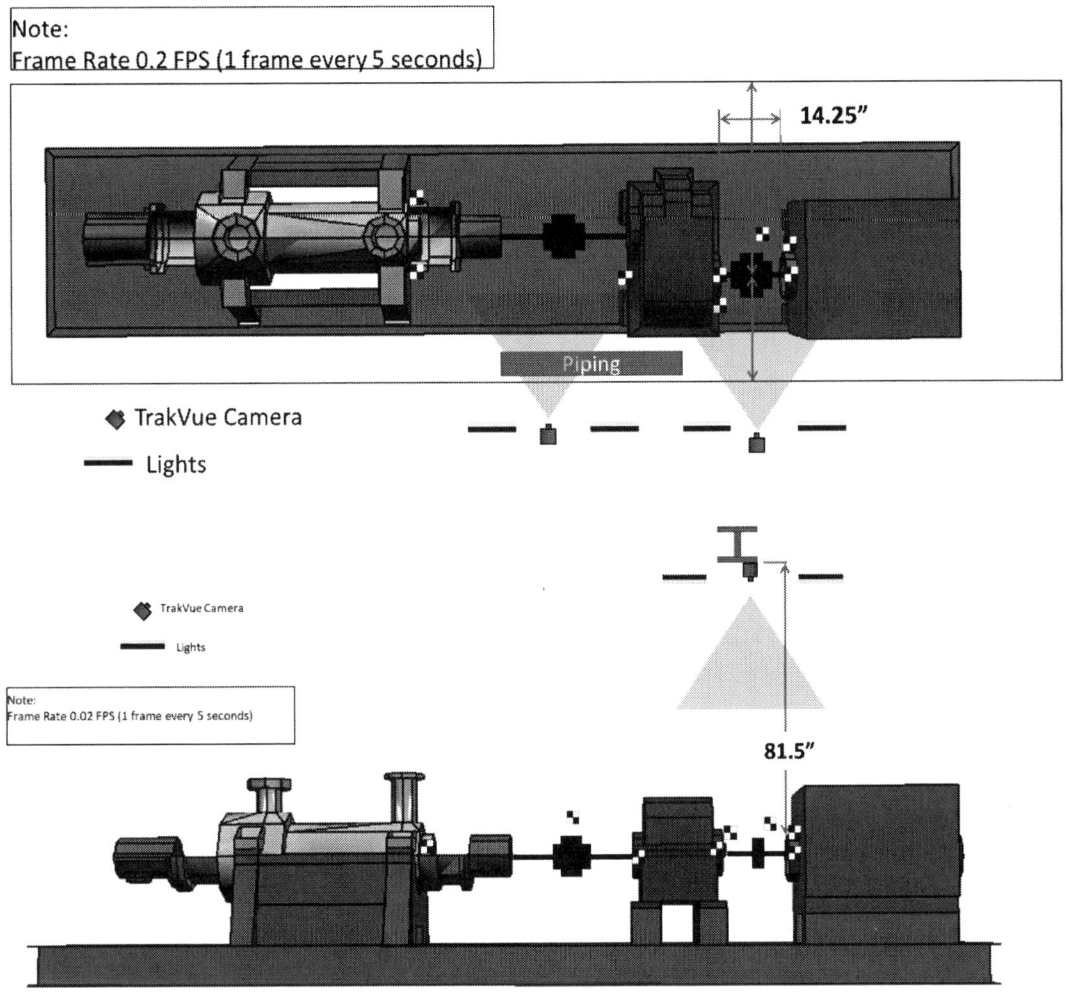

Figure 11: The Authors' Optical Alignment High-Resolution Video Instrumentation

Fig. 12 shows a representative screen for video target location, motion calibration, and tracking.

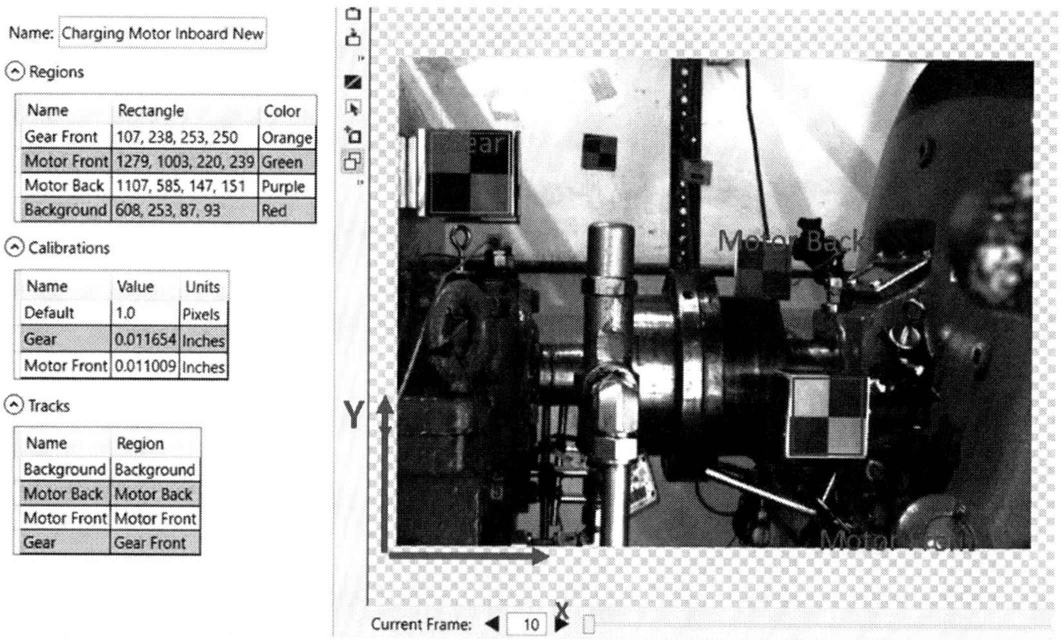

Figure 12: Dedicated software set-up and tracking of movement of particular points

Fig. 13 shows a typical set of plots that document the 2-D motion of the tracked strategic locations, in this case on the motor front. Both ends of the motor, gear, and pump were similarly tracked, in terms of vertical growth a tilt, and horizonal alignment. Fig. 14 shows a summary table of the alignment tracking results, providing clear evidence of a hot alignment problem, and the detailed motions associated with it. This allowed the physical root causes to be identified.

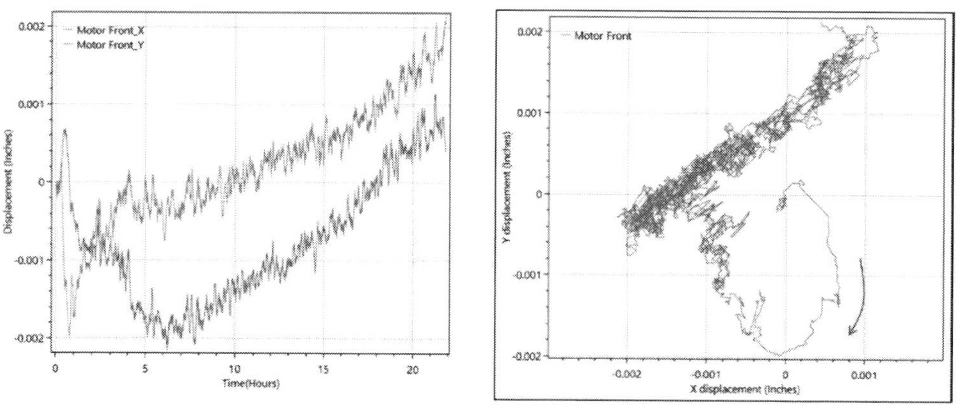

- X axis for the axial motion, Y axis for the vertical motion
- Well-calibrated with the background motion subtracted

Figure 13: Example: Tracking of the Front Motor Target

Camera	Target Location	Axis	Direction of Motion	Total Displacement (mils)	Observations
Side View Gear Left and Motor Right view	Gear LS Input Side view	X	Axial to Motor	3.00	Gearbox LS Input end moved 5 mils up and 3 mils towards the motor
		Y	Vertical	5.00	
	Motor IBB Front Target	X	Axial to Motor	0.50	Motor IBB moved 2 mils up and 0.5 mils towards the OB end
		Y	Vertical	2.00	
	Motor IBB Back Target	X	Axial to Motor	0.25	
		Y	Vertical	0.75	
Motor Overhead View Gear top and Motor bottom view	Motor Top view Right	X	Horizontal to Right	3.00	Motor IBB moved 3 mils towards right (viewed from OB end) and 1 mil towards motor OB end
		Y	Axial to Motor OB	-1.00	
	Motor Top view	X	N/A	N/A	
		Y	N/A	N/A	
	Gear LS Input Right	X	Horizontal to Left	-2.00	Gearbox LS Input end moved 9.5 mils to the left (viewed from OB motor) and 2.5 to 3.0 mils axially toward Pump (twist motion)
		Y	Axial to Pump	2.50	
	Gear LS Input Left	X	Horizontal to Left	-9.50	
		Y	Axial to Pump	7.00	
Side View Pump Left and Gear Right view	Pump IBB	X	Axial to Motor	0.00	Pump IBB moved down 3.5 mils
		Y	Vertical	-3.50	
	Gear HS Output DE	X	Axial to Motor	0.00	Gear HS DE output DE moved down 3 mils
		Y	Vertical	-3.00	

Figure 14: Motor-Gear-Pump Video Tracking Summary of Results

The net alignment results in this case, as determined by the video motion method, are as follows:

- The gearbox LS input shaft end bearing shifted 5 mils upwards, while the motor grew 2 mils in the same direction.
- This would cause the motor shaft to operate in the lower half of the bearing bore, as seen in orbit plots.
- The gearbox LS input end moved 9.5 mils to the left (viewed from OB motor) and 2.5 to 3.0 mils axially toward pump (twist motion), while the motor IB end moved 3 mils towards right, in the opposite direction, (viewed from OB end) and 1 mil away from the gearbox.
- This would cause the motor shaft to operate against the 3 O'clock position of the bearing, as seen in orbit plots.

By performing a cold alignment equal and opposite to these motions, with the initial cold alignment setting subtracted, an excellent alignment was achieved. The soft foot that was being enabled by the misalignment was relieved, and the 2x rpm motor resonance was detuned, with vibration settling to acceptable levels.

3. FUTURE RESEARCH:

The following summarizes the research currently underway at the authors' organization:

1. Improving video motion monitoring capability using low-cost COTS cameras
2. Continuous improvement of displacement sensing accuracy
3. Processing and displaying 3-D motions

4. CONCLUSIONS:

Off-line to running alignment determination is important to resolve certain chronic machinery reliability problems, particularly if improving balance is found to have little benefit. A number of approaches were discussed in this paper.

As presented, the new video-based alignment quantification, documentation, and display procedures and equipment now on the market can be very useful for understanding and compensating for position change and alignment of rotating machinery, and associated piping and systems, when the equipment progresses from off-line to running. The new method has advantages over the traditional optical measurement methods (such as lasers) in many instances. Video-based methods also typically take much less time and "logistics" than conventional methods for detection and documentation of complex motions, such as dial indicator or proximity probe based methods, to implement.

Video-based motion detection benefits are:

- ✓ It is a powerful and intuitive diagnostic tool
  - Realistically determines and displays motion relevant to alignment
  - Easy for non-experts to understand (management, etc.)
  - Able to observe, and display, the entire system of interest
- ✓ It does not require "contact" – perfect for restricted areas
  - Avoids heat, radiation, accessibility/scaffolding issues
- ✓ It is fast and cost-effective
  - Millions of data points ready to evaluate in minutes, not days

Video-based motion detection and alignment determination methods are not always superior to existing conventional methods, but should be considered one more important tool is the machinery alignment toolbox.

REFERENCES:

[1] Piotrowski, John, The Shaft Alignment Handbook, 3rd Edition, CRC Press, c. 2006
[2] The Importance of Motor Shaft Alignment, US Dept. of Energy, Motor Systems Tip Sheet #4, No. 2012
[3] ANSI/ASA S2.75-217, Shaft Alignment Methodology
[4] Timoner, SJ, and Freeman, DM (2001), "Multi-image gradient-based algorithms for motion estimation", Optical Engineering 40.9, pp. 2003-2016
[5] Liu et al (2005), "Motion magnification", ACM Transactions of Graphics (TOG) – Proc. of ACM SIGGRAPH 2005, TOG Homepage, v. 24, Issue 3, July
[6] Wu, H-Y et al (2012), "Eulerian video magnification for revealing subtle changes in the world", ACM Trans. of Graphics (Proc. SIGGRAPH 2012), v. 31, August
[7] Marscher, W.D., and Onari, M.M. (2018), Video Motion Magnification vs. Operating Deflection Shapes for Machinery Diagnosis, MFPT2018, Virginia Beach, May

VIATC TRAINING CONFERENCE & EXPO
NIAGARA FALLS, NEW YORK JULY 26 -28, 2023

HOW AI TECHNIQUES CAN AUGMENT THE WORK OF AN EXPERIENCED VIBRATION ANALYST

Ed Spence, The Machine Instrumentation Group, ed.spence@machineinstrumentation.com
, Dr. David Siegel, Predictronics Corp.,
Matthew Moore V-CAT IV

Ed Spence is the Founder and Managing Director of The Machine Instrumentation Group, a consultant representing a network of consulting and contract engineering service providers with expertise in Condition Monitoring, sensor design, signal processing and data engineering.

Previously, Ed was the Marketing Manager for Analog Devices MEMS Sensor Technology Group, where he defined the MEMS accelerometer roadmap for vibration based Condition Monitoring, winning an Innovation Award at Sensors Expo 2017 for the flagship ADXL1002 MEMS accelerometer. Ed has been engaged in the development of Condition Monitoring sensor solutions since 2008. Ed regularly publishes or presents on the subject of sensors for Condition Monitoring.

Ed serves on the Board of Directors for The Society for Machinery Failure Prevention Technology, and is a member of the Vibration Institute, The Society of Maintenance and Reliability Professionals and The Society for Prognostics, Health and Maintenance. Ed also volunteers for non-profit IBEC Ventures, providing business consulting to small business start-ups overseas. Ed holds a B.S.E.E. from University of Massachusetts, and a certificate in Strategy and Innovation from MIT Sloan Business School.

Matthew Moore BSc (Hons) is a Global Subject Matter Expert in Condition Monitoring with over 25 years of experience in the field of vibration and condition monitoring within the Oil & Gas industry.

He is a certified Mobius Institute ISO 18436-2 Category IV vibration analyst and training instructor. He is also the United Kingdom Community Leader for Mobius Connect, an Editorial Board member for CBM Connect and Reliability Connect as well as a member of the Mobius Institute Board of Certification Vibration Analysis Technical Committee.

Dr. David Siegel is the chief technology officer for Predictronics Corp. His current role includes developing the technology roadmap for the company's predictive monitoring software and service solutions, developing new algorithms and methodologies, as well as leading the data science team to carry out the customization and deployment of various predictive monitoring solutions.

Prior to joining Predictronics, Dr. Siegel was a research assistant at the Center for Intelligent Maintenance Systems at the University of Cincinnati. During his time there, he led numerous research efforts on customized diagnostic and prognostic software development for a variety of industrial customers and applications. Examples of these research efforts include advanced diagnostic methods for industrial robots, health monitoring systems for railway applications, failure prediction tools for machine tool bearings, and intelligent maintenance systems for military ground vehicles.

Dr. Siegel is also a two-time winner of the Prognostics and Health Management Data Challenge and has won several best paper awards at various conferences focused on predictive monitoring and data analytics.

ABSTRACT:

With the expanded volume and availability of asset condition monitoring data emerging from the deployment of IoT systems, there is great interest in the potential for data driven (AI) approaches to automate identification, classification and diagnosis of asset anomalies. In this paper, a data engineer and an experienced vibration analyst collaborate to discuss and compare analysis for their respective disciplines. Moderated by a wireless sensor technologist with long experience in vibration sensing technology, we explore how AI approaches, when coupled with this increase in available data, might impact the practice of Condition Monitoring and Predictive Maintenance.

KEY WORDS: AI, Artificial Intelligence, Machine Learning, ML, IoT, IIoT, Vibration, Analysis, Vibration Signal Processing, Condition Monitoring, Predictive Maintenance

HUMAN VS ARTIFICIAL INTELLIGENCE TRAINING IN VIBRATION ANALYSIS

TRAINING TO BECOME A VIBRATION ANALYST

Many fine programs exist for the training of anyone interested in the field of vibration analysis as a means of monitoring the health of industrial machinery and diagnosing emerging faults. The industry has existed for decades with standards such as ISO 18436-2 developed to specifically address the training and competency requirements for vibration analysts. Useful tools for data collection and techniques for signal processing continue to improve each year. Field tools to aid diagnostics, such as FFT spectra wall charts and other references, are invaluable resources to the practitioner.

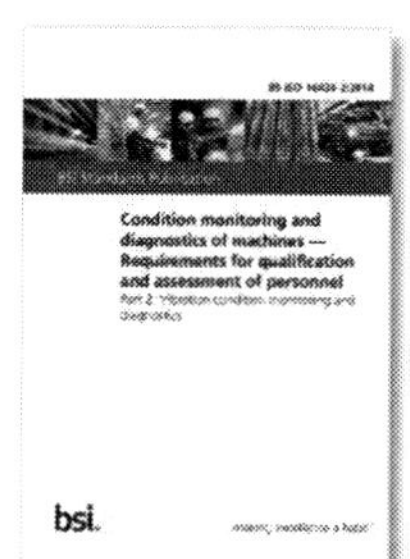

All that said, the most valuable resource may be a good mentor. On site exposure monitoring and diagnosing a variety of equipment in different facilities provides the breadth and depth of experience to help an analyst become truly effective.

Good training coupled with experience can help the individual analyst achieve not only competency but consistency on the job. That said, this does not necessarily apply to every analyst. Variations in experience, development of individual philosophies and personal tool preferences can lead to variations in results as well. This may be why the exchange of ideas at training venues are popular with practitioners.

TRAINING AI IS AN OVERALL PROCESS

It is important to note that application of AI techniques is part of an overall process that includes data selection and clean up, signal pre-processing, feature extraction and selection. Typically, machine learning model training and evaluation aspects are only considered after the feature selection step. This emphasizes that the accuracy of the health index and diagnosis information, output by the machine learning algorithm, is dependent upon the pre-processing and feature extraction steps.

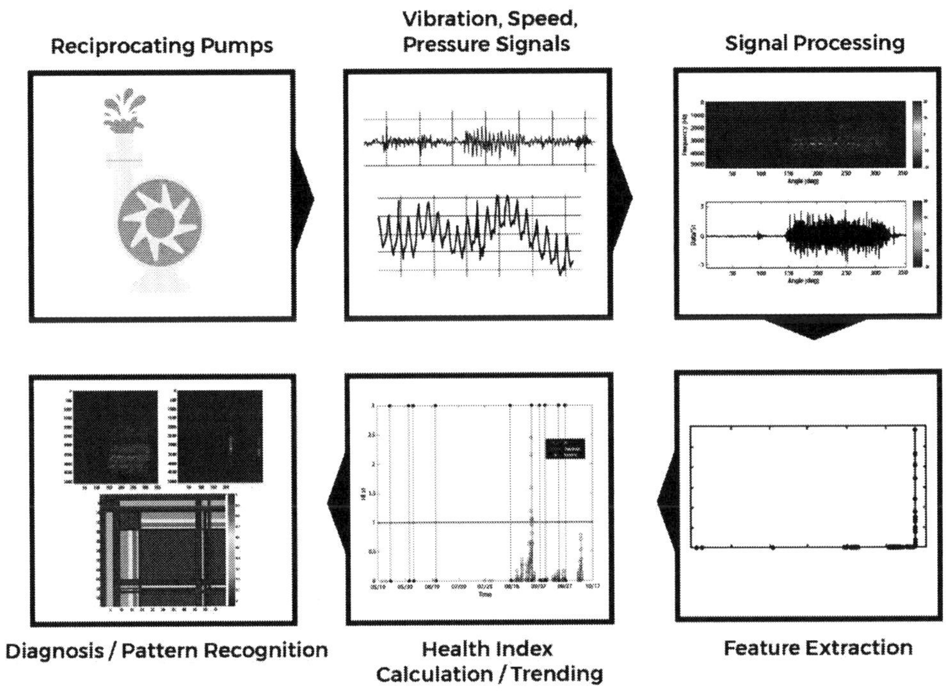

Figure 1. AI for asset fault diagnosis. Implementation of AI techniques to asset fault diagnosis is a process that involves several discrete steps. Source: Predictronics Corp.

DATA USED FOR TRAINING AI

AI has the ability to efficiently sift through large quantities of data with greater ease than a human analyst could alone, as well as consider inputs from multiple signals and locations at a time.

Training machine learning algorithms requires baseline data, but as the chart in Figure 2 highlights, there are many options to consider. Ideally, the baseline data should capture all operating modes for the monitored asset, as well as normal variation due to process or loading changes. This requires the selection and use of several data sets, over a variety of conditions, as well as data from similar machines. In our experience, 2-4 weeks of operation is generally needed to obtain the baseline data for most assets (without considering seasonality).

For automated equipment health diagnosis, the number of examples from each failure mode needed for training the model is not always clear. From our experience, 10 or more training examples are recommended for diagnostic purposes. For both detection and diagnosis, pre-processing steps or a visual review of the data could be needed to ensure that the machine learning models are trained with data that is representative of the health conditions and of good quality.

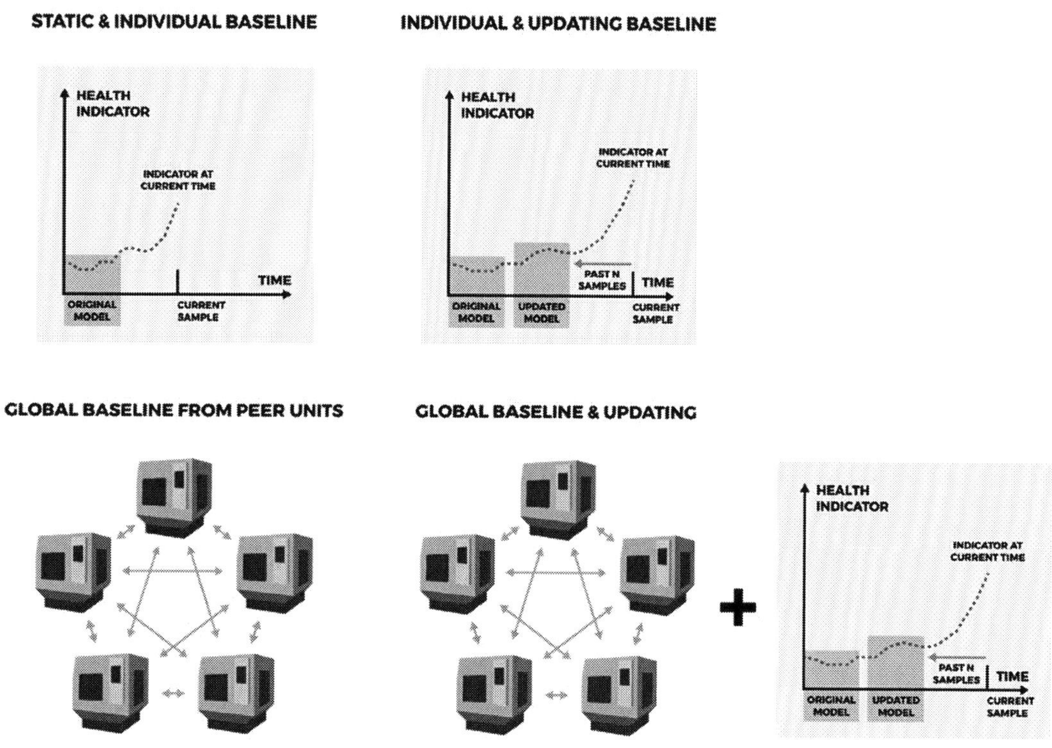

Figure 2. Algorithm training. Training requires multiple sets of good quality, or 'clean,' data, including both normal process variations, as well as faults to be detected. Source: Predictronics Corp.

Results often vary from one AI model to the next, highlighting the importance of algorithm selection and validation. However, once trained, the AI model will be consistent over time.

Validation of the AI machine health solution is then conducted by experienced data engineers. Patterns may be discovered in the data, but the physical meaning of these results and validation of the fault must come from human expertise.

CHOICE OF ALGORITHM IS A CRITICAL STEP

Detection, diagnosis, and prognosis are three different machine health monitoring objectives, each requiring a different set of tools, as well as approaches for model training. The table in Figure 3 highlights the different machine learning algorithms and methods one would consider for detection, diagnosis, and prognosis, respectively. For early detection, baseline data is used to learn the normal behavior of the machine and then train anomaly detection algorithms to automatically detect departures from this baseline data. For diagnosis, machine learning algorithms must be trained with data from different failure modes. Classification-based and rules-based models are the most common methods under the diagnosis approach. For prognosis, past failure trends are needed to train similarity-based prediction methods along with machine learning regression models. Alternatively, for prognosis, if past failure trends are limited, time-series prediction models can be considered. In general, AI for condition monitoring, i.e., fault detection, of many common assets is rapidly improving, whereas diagnosis and prognosis are not as mature. The lack of maturity for diagnosis and prognosis is likely due to the amount of data required for training those models, as only baseline data is needed for detection, but past failure instances and failure trends are needed for diagnosis and prognosis.

Condition Monitoring Task	Type of Machine Learning Model
Automated early detection of equipment or asset health issues	Anomaly detection algorithms: • Distance-from-normal methods • Regression based methods
Automated diagnosis of failure mode occurring for equipment or asset	• Classification based methods • Rules-based /expert system
Prediction of future health	• Time series prediction • Similarity-based prediction • Machine learning regression

Figure 3. AI Toolsets. Toolsets typically applied to the development of detection, diagnosis, and prognosis of a machine fault. Source: Predictronics Corp.

There are various ways to visualize the outputs from machine learning models; however, a common way of displaying this information is provided in Figure 4. For detection, the output from machine learning algorithms is displayed as a health index, which is commonly viewed with a corresponding warning and alarm threshold. The health index could be a normalized version of output from anomaly detection algorithms, a probability of fault value from a supervised machine learning algorithm, or a distance-from-normal metric. Regardless of how it is calculated, the health index is customarily plotted over time in a way that is similar to what is plotted in Figure 4.

Diagnosis information can be presented as a bar graph to indicate which failure mode is most likely to occur while an unhealthy condition is developing. In Figure 4, the bar graph shown does not display a diagnosis model, but rather highlights which variable is contributing to the anomalous health condition. Diagnosis information would typically be displayed in a similar manner. Although it is not shown in Figure 4, a prognosis time-series prediction could also be overlayed on the same dashboard to indicate the future health trend prediction and the remaining useful life estimation.

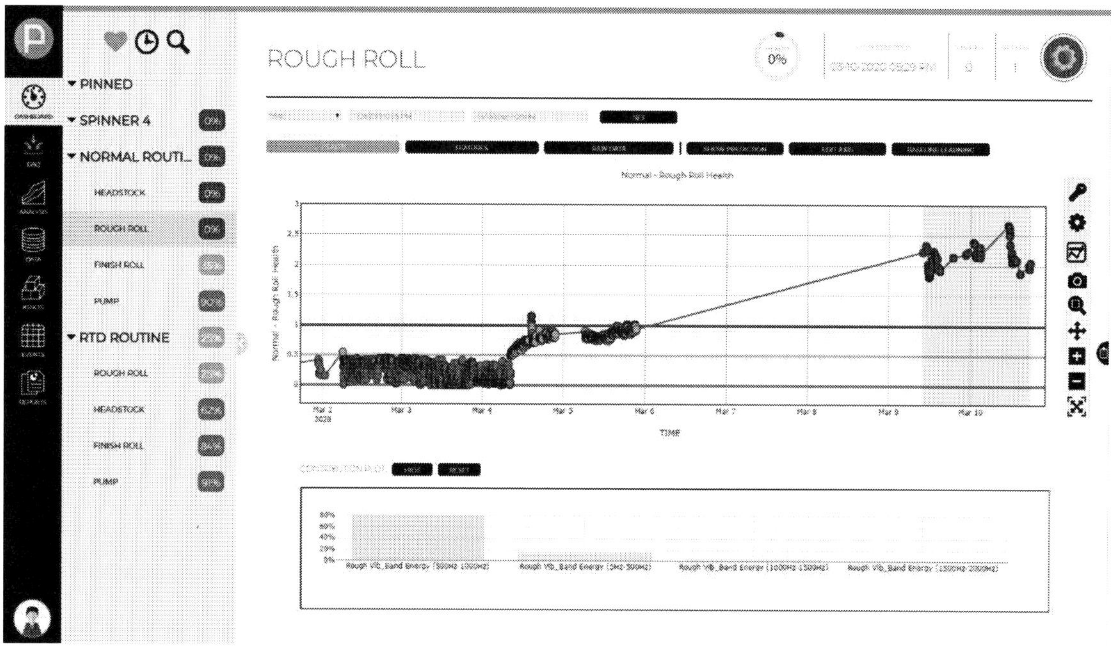

Figure 4. Health Index for Flow Forming Machine. A lubrication issue for the rough roll is identified from vibration in a certain frequency band. Note that the Health Index reflects a process composite of several data inputs. Source: Predictronics Corp.

HUMAN VS ARTIFICIAL INTELLIGENCE REGARDING VIBRATION DIAGNOSTICS

GATHERING CONTEXTUAL INFORMATION A CRITICAL STEP

Much of classical vibration signal analysis and diagnostics hinges on knowledge of the machine details. Operating parameters and physical characteristics enable more accurate (and faster) diagnosis.

Parameters such as RPM, Component Type, Component Details (e.g. vanes, blades), Orientation, Mounting / Structure, Bearings, Duty / Standby / Similar Equipment, Maintenance History and Operating Context / Process Conditions are typical inputs to the human diagnostic process.

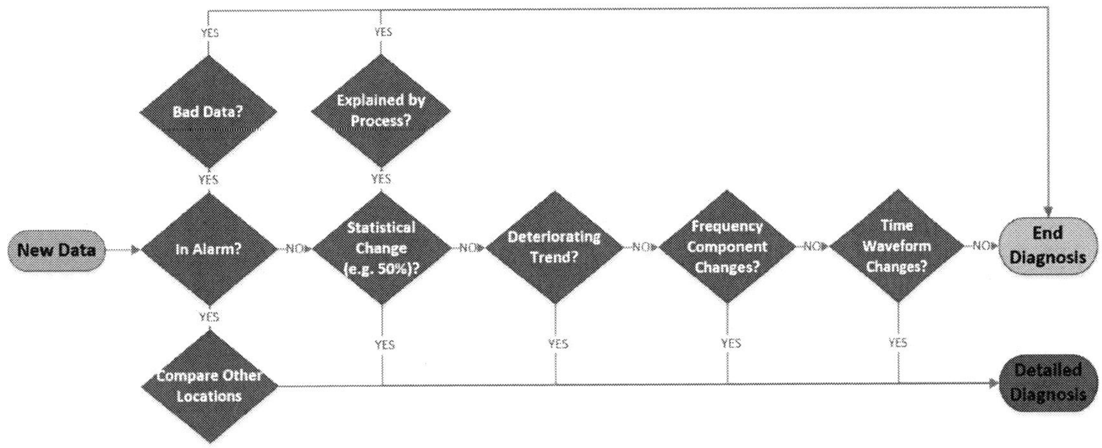

Figure 5. Vibration condition monitoring diagnostics example workflow. Source Matthew Moore.

For humans, diagnosis of vibration signals is primarily a visual analysis, using graphics such as FFT spectra plots. Although tools such as frequency markers can help identify key features of the spectrum to aid anomaly detection, uploading and graphing data can still be a labor intensive, time consuming process, particularly if large quantities of data need to be sorted through for various scenarios.

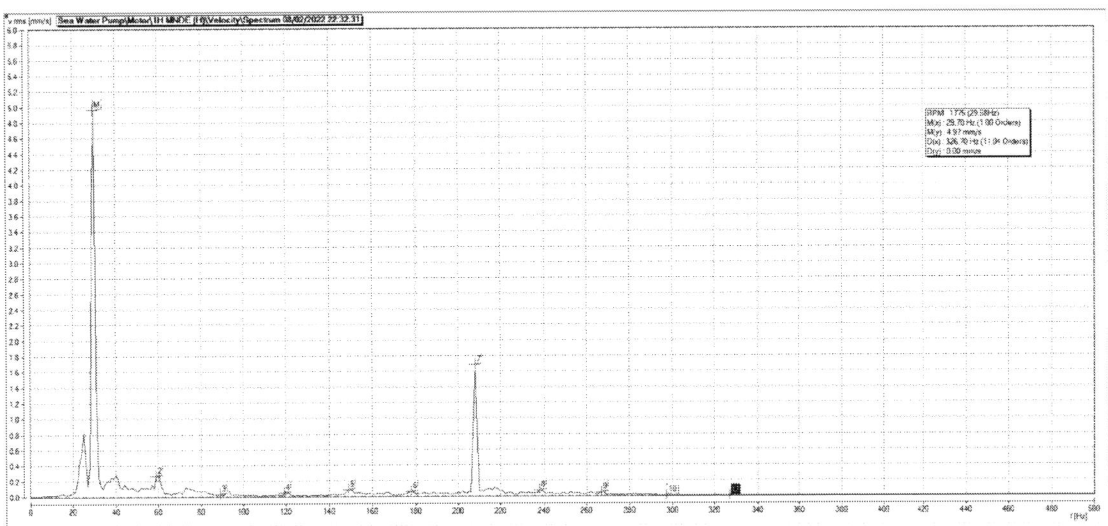

Figure 6. Vibration velocity spectrum with harmonic cursors. Source Matthew Moore.

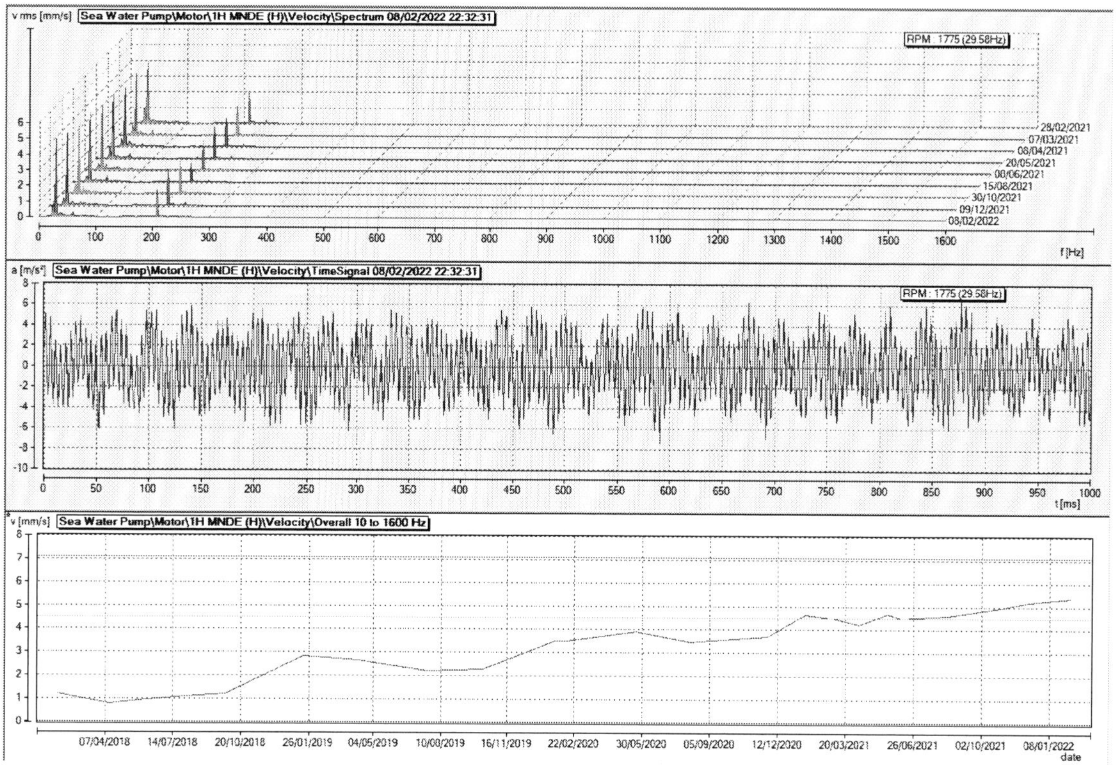

Figure 7. Multi plot. **Waterfall plot, time waveform and vibration velocity rms trend (with alarm thresholds) displayed simultaneously for ease of comparison.** Source Matthew Moore.

Much as the human eye is quicker to detect movement, visual aids can help the analyst see changes in a visual graph. The analyst is constantly asking – has anything changed? How does this data compare with that taken previously? Many diagnostic tools developed for the analyst are graphical ways of portraying various machine behaviors, but the task can be summed in the words 'identify, visualize and compare'.

At the same time, the human analyst is aware of broader context such as the machine type, location, maintenance history and other information.

Finally, the human analyst has access to data that for the most part eludes the AI algorithms – through their human senses!

AI PERCEPTION OF VIBRATION DATA

Although AI techniques can operate on images, scalar or numerical values are more typically used. For vibration analysis, these are typically metrics calculated from the raw data, such as the magnitude of peaks in the spectrum, RMS levels in certain frequency bands, etc.

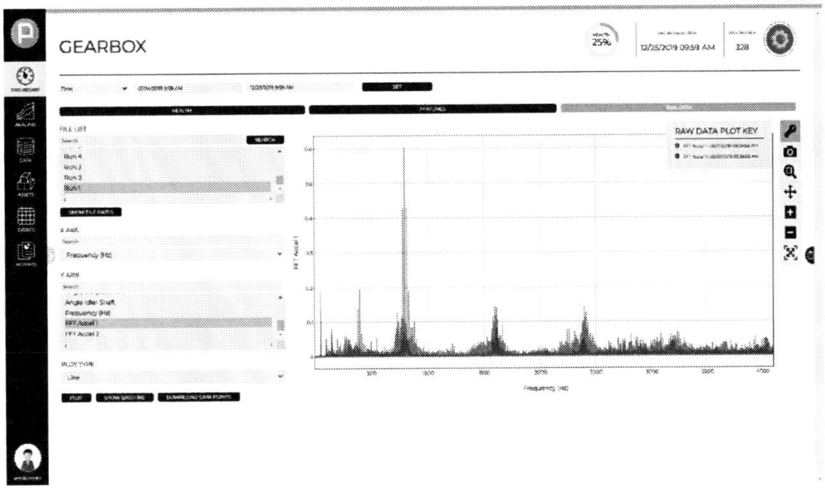

Figure 8. ***Vibration Spectrum.*** Spectrum of vibration data. AI techniques to analyze this raw data from a gearbox include extracting magnitude in the frequency domain of the sideband magnitudes, synchronous averaging related methods to consider typical gear meshing patterns and residuals from that pattern.

Unlike humans, AI software has the ability to consider large amounts of data including multiple vibration signals, hence fault detection and analysis is easier to scale. Data inputs from multiple sensor signals/locations can be difficult for humans to review efficiently, at least without augmentation from some kind of data analysis software. Much like a human, the output of an AI algorithm will be a probability score for each failure mode.

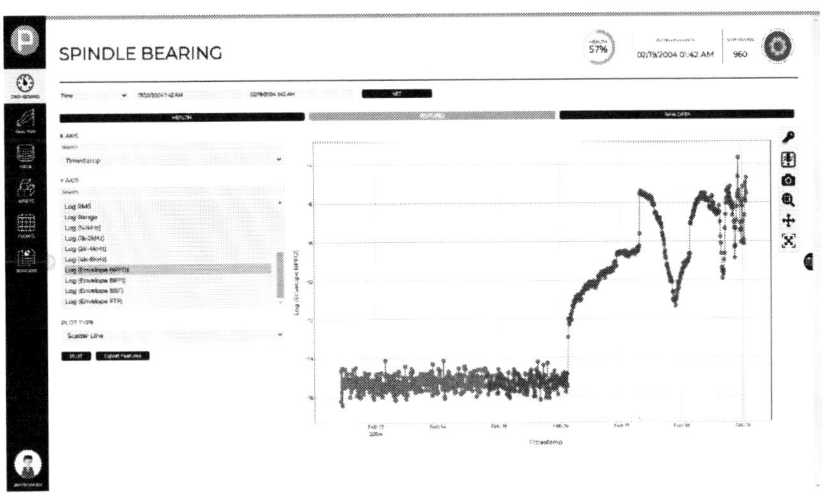

Figure 9. Spindle fault detection. This spindle fault detection index has been trained with data from multiple components at multiple locations, in which the earliest indication was related to the outer race frequency magnitude as shown here

IMPACT OF WIRELESS VIBRATION SENSORS (IOT)

A discussion of AI would be incomplete without reviewing the impact of IIoT. Permanently installed sensor networks will greatly increase the amount of data collected by offering more frequent measurement intervals.

With data available to a network rather than stored on a handheld data collector, the amount of data can not only be scaled, but accessed virtually anywhere in the world. Once digital data is available for analysis anywhere, anytime, the application of AI approaches to anomaly detection is a natural next step.

Some wireless sensors are designed to generate alerts regarding changes in machine operation, in order to call for potential review by an experienced analyst. This approach can monitor machines more closely, helping the analyst to use their time more efficiently, and is the first and most natural application of AI techniques.

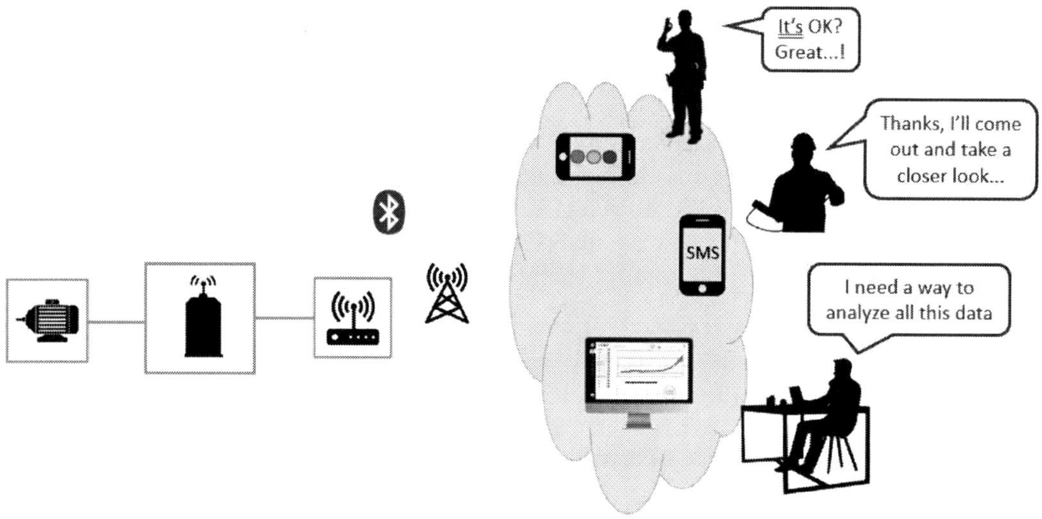

Figure 10. IOT means more data. Wireless systems for condition monitoring come with varying levels of vibration measurement fidelity and a broad range of features. Source: The Machine Instrumentation Group.

LOOKING AHEAD - WORKING TOGETHER AS TECHNOLOGY ADVANCES

Like many new technologies, we can hope that AI approaches to fault detection in vibration data will automate repetitive tasks, freeing up human faculties for tasks requiring more contextual inferences, creativity, experience, sensory perception etc and therefore improving the efficiency of the trained analyst and data collector alike. With relentless pressure on maintenance to do more with less, implementation AI offers hope for some relief.

AI is not likely to replace human processing anytime soon (if ever), but we can hope it will free humans up to apply their natural creativity and superior ability to draw inferences from sources entirely outside of the context near at hand.

One of our concerns for the future of human vibration analysts is that some analysts will be deprived of site experience due to IoT collecting the data. However, there are advantages of more closely monitoring machines as emerging faults are detected, with AI based analysis as more of an advisory tool to help humans make decisions regarding disposition and required actions.

Further, contrary to much of the hype surrounding AI, the most effective case studies we know of resulted from collaboration between the data engineer and the subject matter expert.

Recent experience of AI approaches to assets and processes has resulted in quality improvements as well. In the longer term, we can hope that the additional information available from wireless data collection and AI driven data analysis will enable more robust design, installation and maintenance of assets as well.

As with any new technology that is difficult for the layman to understand, the hype can induce a considerable amount of fear, uncertainty and doubt. But as with many innovations, familiarity can open up opportunities to make us more efficient and effective on the job.

BIBLIOGRAPHY:

BS ISO 18436-2:2014 – Condition Monitoring and Diagnostics of Machines – Requirements for Qualification and Assessment of Personnel – Part 2: Vibration Condition Monitoring and Diagnostics

"Rebooting AI: Building Artificial Intelligence We Can Trust", Gary Marcus and Ernest Davis. Vintage 2019.

VIATC TRAINING CONFERENCE & EXPO

NIAGARA FALLS, NEW YORK JULY 26 -28, 2023

TWO CASE STUDIES: VIBRATION FATIGUE CRACKING IN RECIPROCATING COMPRESSOR SUCTION AND DISCHARGE BOTTLE NOZZLES

Michael Bifano, Ph.D., PE and Thomas Calko, PE
E^2G | The Equity Engineering Group, Inc
Shaker Heights, Ohio

Thomas Calko, PE, is a consultant at E^2G in the REVD Team primarily working in the chemical and petrochemical industries. He performs advanced vibration data analysis and root cause evaluations to solve structural dynamics problems on structures and rotating equipment. He is a Category IV vibration analyst and a certified welding inspector. Prior to working at E^2G, he worked in maintenance engineering at nuclear power plant as a welding engineer and vibration engineer. He holds engineering degrees from Youngstown State University (BSME) and the University of Pittsburgh (MSME).

Michael Bifano, Ph.D., PE, is a consultant at E^2G and the Team Leader of the Rotating Equipment, Vibration, & Dynamics (REVD) Team. He performs fitness-for-service (FFS) evaluations of in-service equipment using the rules of API 579-1/ASME FFS-1. He is the E^2G technical authority on vibration measurements, diagnostics, and piping and structural vibration fatigue assessments. He is a voting member of the API 579 FFS Committee and one of the lead authors of the upcoming API 579 Part 15 methodologies on the assessment of Piping Vibration. Additionally, he specializes in transient thermal-mechanical fatigue analysis, computational fluid dynamics, elevated temperature creep and creep-fatigue simulations, and fracture mechanics using analytical and computational techniques. He holds a bachelor's degree in physics from Lebanon Valley College and a PhD in Mechanical Engineering from Case Western Reserve University. He is also a Category IV vibration analyst.

ABSTRACT:
Two separate case studies are presented involving large reciprocating compressors with elevated vibration levels that resulted in cracking in the suction and discharge bottle nozzles. Both sets of failures required expensive repairs and extensive production downtime. Vibration testing methods were utilized to show that the suction/discharge bottles were resonant with a running speed harmonic. Motion Amplification Videography and Modal Testing results are presented to show the mode shapes of the bottles. Vibration levels were reduced by utilizing damping, mass, and stiffness. When troubleshooting reciprocating compressors with high suction/discharge bottle vibration levels, the discharge bottle supports are an important location to include in the root cause analysis.

KEY WORDS: Reciprocating compressor, resonance, suction, discharge bottle, chock, damping, stiffness, piezoelectric strain gage, fatigue

General Compressor Layout and Support Background

Reciprocating compressors are utilized extensively in the petroleum and petrochemical processing industries; a typical configuration is shown in Figure 1. The suction bottles are often a cantilever type arrangement and discharge bottles have a similar support structure and also typically include a vertical support.

Figure 1: Reciprocating compressor with major components

Due to the positive displacement nature of reciprocating compressors, significant dynamic energy can be present to cause vibration and fatigue problems. These dynamic forces can be mechanical, acoustic, or both. Generally, the suction and discharge bottles do not have many support points, therefore when these machines develop high vibration problems at machine running speeds (i.e. resonance) the suction and discharge bottles often become locations for cracking problems. Figure 2 and Figure 3 show other suction bottle arrangements.

Figure 2: Suction bottle on single throw compressor

Figure 3: Suction bottle for two compressor cylinders

Discharge bottle supports are important locations to check when trouble-shooting high vibration problems. They come in many different shapes and sizes, with and without anti-vibration liners as shown in Figure 4 to Figure 6.

Figure 4: Discharge bottle and support for single throw compressor

Figure 5: Discharge bottle and support two compressor cylinders

Figure 6: Discharge bottle and support for three compressor cylinders

Acceptance Criteria ISO 20816-8

ISO standard 20816-8 is titled Mechanical Vibration – Measurement and Evaluation of Machine Vibration for Reciprocating Compressor Systems. It provides guidelines for the location of vibration measurements and the classification of vibration amplitudes based on avoidance of fatigue problems. The reciprocating compressor components covered in this standard include the compressor frame, cylinder, suction and discharge bottles, connected process piping, and small bore connections. ISO 20816-8 is not used for detecting problems on internal machine components such as pistons and valves and it is not generally utilized for condition monitoring purposes. However, the guidelines provided in ISO 20816-8 combined with condition monitoring trend data can be a useful tool when diagnosing vibration related problems with reciprocating compressors.

Case Study 1 – Overview

This case study presents the resulting dynamic response to two different vibration control solutions: increased damping and added mass.

Machine Layout and Problem Description

A natural gas transmission compressor station had five engine driven reciprocating compressor units. The compressors were engine driven, 6-throw, double-acting, single stage machines with two suction and two discharge bottles each (Figure 7). The operating speed range of the compressor was 750-1000 rpm. The compressor station was built in 2001, and after about 10 years in operation many of the suction and discharge bottles developed cracks at the nozzle to bottle connection as shown in Figure 8. In 2011, all the suction and discharge bottles were removed and the reinforced pad design was upgraded to a reinforced saddle design with the intent of reducing the stresses and preventing the fatigue cracking. However after 6 years of operation, cracking started to occur again in several suction and discharge bottles.

After the cracking problem recurred at the compressor station, vibration testing was performed to identify the cause of the high vibration levels and vibration reduction strategies were developed to decrease the alternating stresses and prevent the cracking problems.

Figure 7: View of six throw compressor with suction (top) and discharge (bottom) bottles

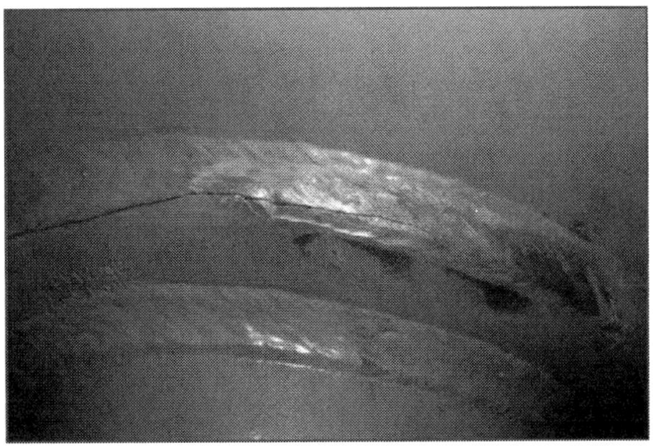

Figure 8: Cracking that developed on bottles at nozzle to bottle weld

Data and Analysis for Root Cause

The compressor operating speed range is 750-1000 rpm, so variable speed vibration testing was performed to identify the highest vibration levels over the operating speed range. A tachometer was setup and order tacking was utilized. The compressor frame, cylinder heads, and suction bottles were instrumented with accelerometers. A typical plot of the vibration levels vs engine speed is shown in Figure 9. The order tracking data clearly shows an increase in vibration levels around 860 rpm – this data indicates a suction bottle resonance condition.

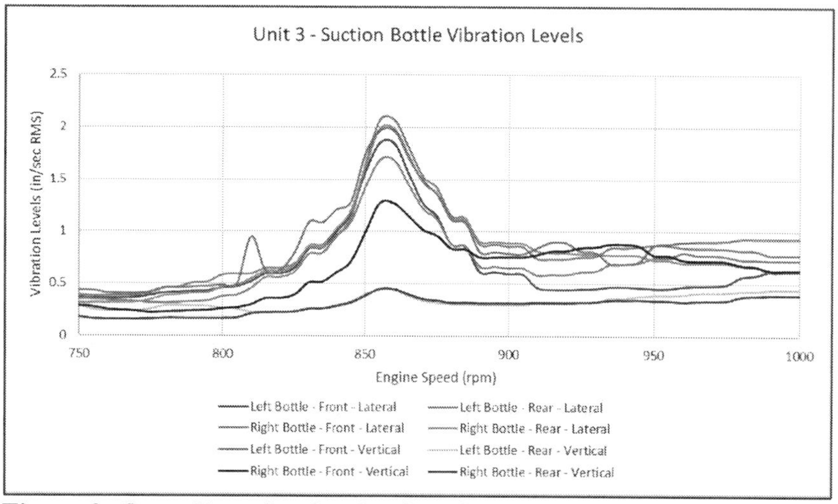

Figure 9: Overall suction bottle vibration levels vs engine speed

During the variable speed sweep testing, amplified motion videography (AMV) was utilized to capture the operating deflection shape. At the engine speeds with the highest vibration levels, AMV showed the in-phase bending motion of all cylinder heads on one side of the compressor and the resulting suction bottle rocking vibrations. This motion was identified on each of the compressor units and the motion of concern occurred between 38 Hz and 44 Hz.

A modal test was performed on the suction bottle and nozzle to verify the natural frequency and mode shape at frequency range of concern. The mode shape shown in Figure 10 was identified at 44 Hz, which shows a (mostly) vertical translating motion of the compressor cylinder head and nozzle and a strong rocking motion of the suction bottle. This motion appears to generate high alternating tensile stresses at the bottle-nozzle welds.

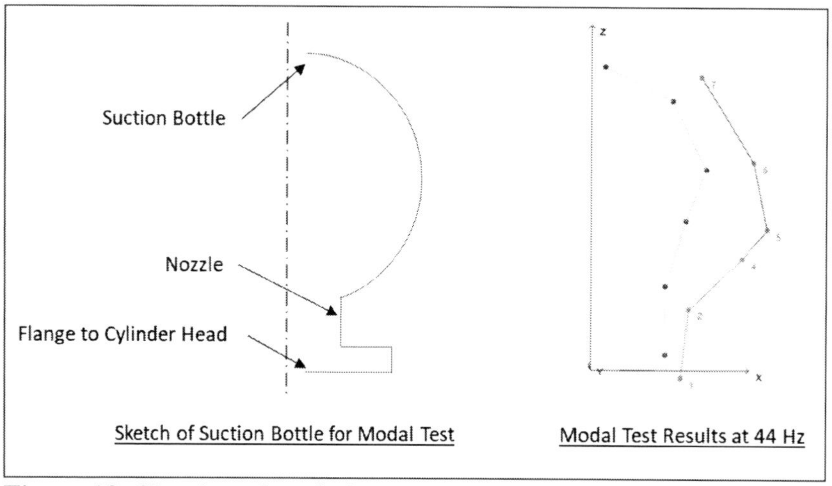

Figure 10: Sketch and typical modal test results for suction bottle

Finally, strain gages were also installed to confirm the natural frequency and mode shape that resulted in the highest strains and calculated stresses. Strain gages were not installed at the weld toe. Instead, two strain gages were installed at different distances from the weld toe (0.6" and 2.1") and the strain at the weld toe was extrapolated from these measured locations. This method is referred to the "hot-spot" method and the procedure was followed from BS-7608, Guide to Fatigue Design and Assessment of Steel Products.

The basis for this method of stress measurement is that at welded joints, alternating tensile membrane and bending stresses are the primary crack driving force. Membrane and bending stresses are defined in ASME Section VIII Division 2 as stresses that act across the section thickness of the welded joint. The peak stresses associated with changes in geometry are localized and may result in crack initiation or local yielding, but these stresses do not contribute significantly to the growth of fatigue cracks in welded joints.

By placing the strain gages the specified distance away, the membrane and bending stresses are measured at two locations, and the membrane and bending stress at the weld toe can then be determined by linear extrapolation. Using this method during the variable speed testing, the maximum strains were measured at 44 Hz (Figure 11), which corresponds with the measured mode shape and natural frequency of concern. This data provided further confirmation that the problematic mode shape and natural frequency is between 38-44 Hz for the compressor units and corresponds with the translating and rocking motion of the cylinder head and suction bottles.

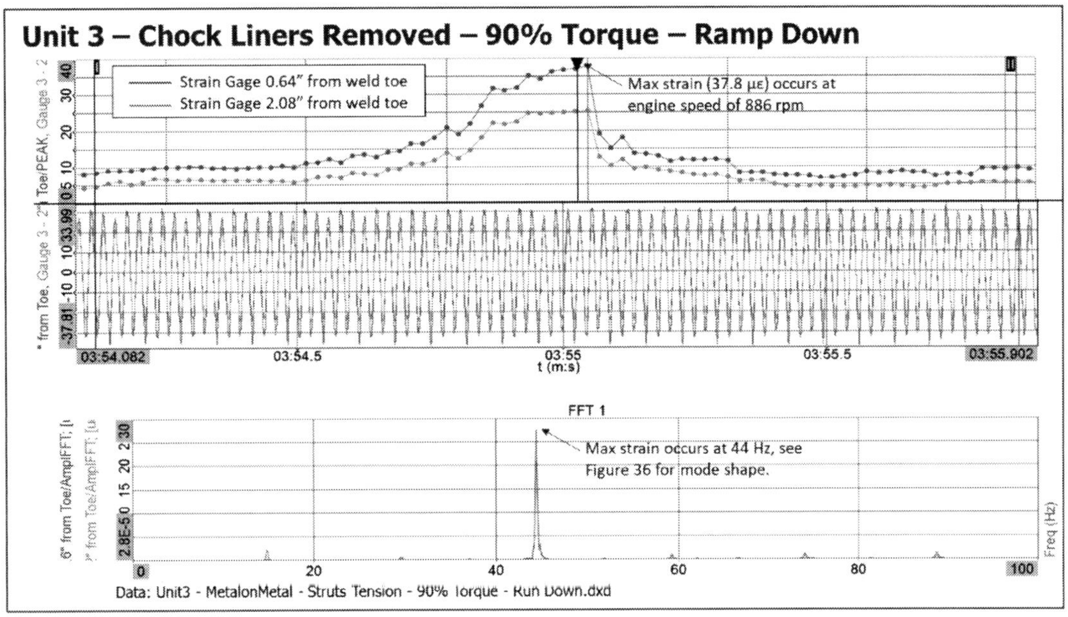

Figure 11: Measured dynamic strain at suction bottle nozzle near bottle-nozzle weld

The data acquired from the vibration testing shows that the root cause of the bottle-nozzle cracking is the translating/rocking mode of the cylinder heads and bottles at approximately 40 Hz. The variable speed testing, AMV ODS, modal testing, and strain gage data all support this conclusion.

The driving dynamic frequency exciting this mode was determined to be the 3rd order running speed of the compressor. Note that the 3rd order coincides with the fact that each suction bottle is connected to three cylinders, so the 1st order component for each cylinder acts on the suction bottle and is considered the driving dynamic force exciting the problematic mode.

Solution #1 – Damped Braces to Reduce Vibration Levels

With the root cause of the high vibration levels known, two methods were implemented to determine the effectiveness of reducing vibration levels: (1) increasing damping and (2) increasing mass.

A damped bracing arrangement was designed and installed as shown below. The arrangements consisted of heavy pipe sway struts pinned to heavy tube steel section that was bolted to the compressor cylinders. On the other end of the sway struts was a pipe clamp sized to the suction bottle lined with damping material. The low angle of the damped braces was established to maximize the damping of the problematic translating/ rocking mode of the suction bottle. Figure 12 shows the damped brace arrangement installed on a compressor.

Figure 12: View of compressor with four damped braces installed

Resulting Vibration Levels

The effectiveness of the damped braces was quantified in two manners: (1) modal damping values (2) operating variable speed compressor tests.

Modal testing was performed on Unit 1 to compare the modal damping levels for the problematic mode before and after the damped braces were installed. Table 1 shows the damping ratio improved by a factor of more than 2.2x. The Nyquist circle-fit method was utilized to calculate the damping ratio as a percent of critical damping.

Table 1: Summary of Improved Damping Ratio at Problematic Mode

Unit 1 Suction Bottle	Unbraced Damping Ratio	Braced Damping Ratio	Increase in Damping
West	0.6%	1.3%	2.2x
East	0.5%	1.2%	2.4x

The damped braces were installed and tested on four of the five units for operating vibration variable speed testing. A decreasing relationship was observed between the maximum operating vibration levels and the damping ratio. This data clearly shows that damping ratio can be effectively utilized to decrease vibration levels.

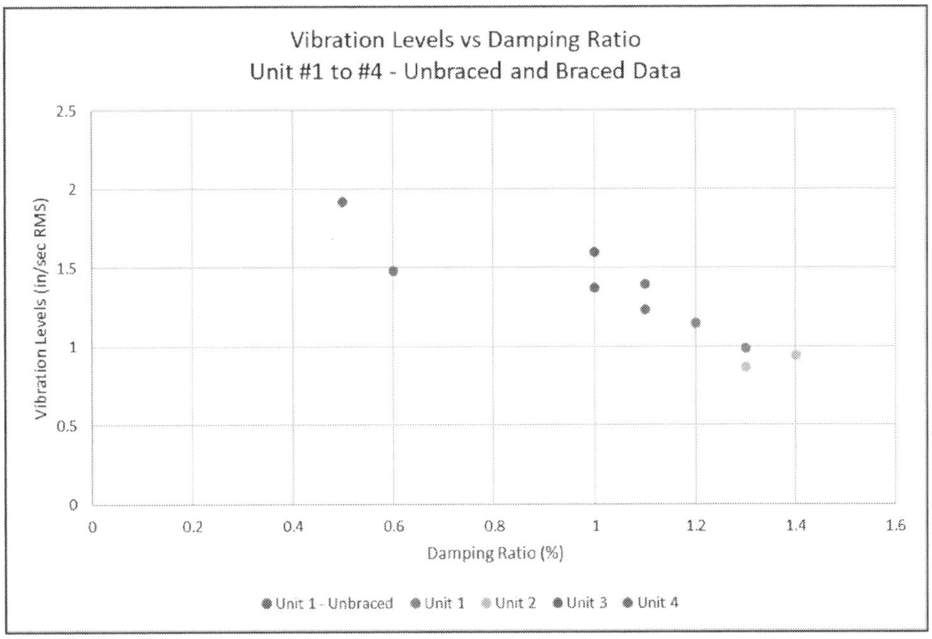

Figure 13: Comparison of vibration levels and damping ration with and without damped braces

<u>Discharge Bottle Chock Settings</u>
Since the problematic mode shape includes modal participation of the suction and discharge bottles, the discharge bottle supports were inspected and adjusted to ensure optimum structural dynamic performance of the compressor arrangement.

For this compressor unit, each discharge bottle is supported by three sets of chocks shown below. A cross section schematic is shown as well. The chock is set by adjusting two studs that bring the chocks together and lift the bottle or further apart and reduce the lift of the bottle.

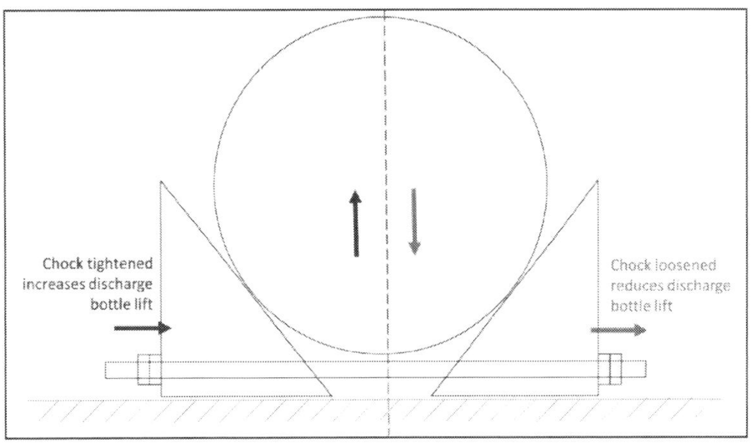

Figure 14: Discharge bottle chock arrangement

As part of the damped brace installation, the following checks and adjustments were made:

- Squareness of all chocks to discharge bottle.
- Setting discharge bottle lift per compressor manufacturer requirements vs standard plant installation procedure.
- Use of damping liner material between chock and discharge bottle.

In some cases, the discharge bottle supports can be critical to the dynamic response of the machine. However for these machines, the adjustments that were made either had a small impact on vibration levels or were considered inconclusive regarding the resulting vibration levels.

Solution #2 – Increased Mass Solution

Since stiffness was not provided to the compressor arrangement, the damped brace solution did not shift the natural frequency for the mode shape of concern. Another solution that was utilized to reduce vibration levels was to add mass to the suction bottle and decrease the natural frequency of concern to a frequency that corresponds with a running speed less than the operating speed. The compressor suction bottle with installed mass is shown in Figure 15

Figure 15: 600-lb weight secured to the compressor suction bottle.

A simplified FEA model of the compressor suction bottle indicated that 1134 lb of mass added to the top of the suction bottle should provide sufficient modal mass to decrease the natural frequency 6 Hz from 42 Hz to 36 Hz. The engine speed that excites the 42 Hz mode is 840 rpm (840 rpm / 60 sec/min x 3rd order = 42 Hz). Shifting this mode shape to 36 Hz results in a running speed excitation of 720 rpm (720 rpm / 60 sec/min x 3rd order). This speed is below the operating speed range of 750 to 1000 rpm, so vibration levels from this problematic mode are expected to decrease. Additionally, the alternating stress amplitudes at the bottle-nozzle welds that resulted in the nozzle cracks are also expected to decrease.

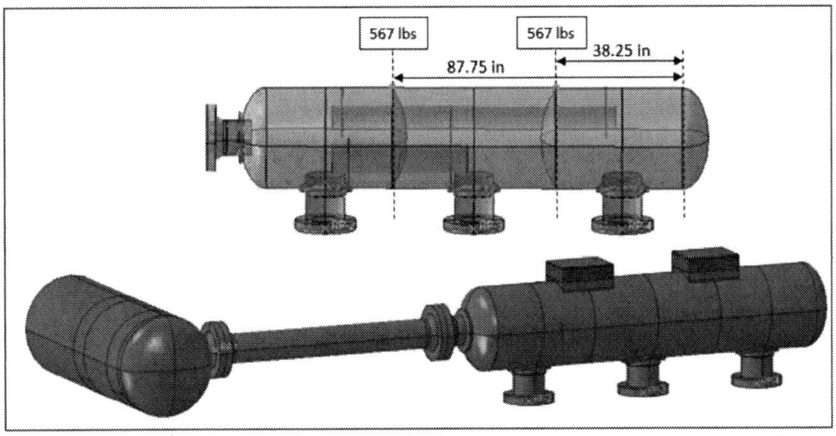

Figure 16: View of FEA model with increased mass solution to decrease the natural frequency

Resulting Vibration Levels

Proof of concept testing was performed on a single compressor unit by installing approximately 600 lb of salt bags to the top of the left suction. This solution resulted in a reduction of compressor suction bottle vibration levels. Figure 17 summarizes the results from the variable speed testing with and without mass added to the suction bottle. The vibration levels were reduced from 2.2 in/sec to 1.3 in/sec, which is a 60% reduction.

The order tracking data also shows that without mass installed, the peak vibration occurs at 880 rpm which corresponds to a natural frequency of 44 Hz (880 rpm / 60 sec/min x 3 = 44 Hz). With the mass installed, the frequency of peak vibration occurred at 855 rpm or 42.75 Hz, which is a decrease in natural frequency of 1.25 Hz. Based on the FEA results and assuming the effect of the modal mass is linear, increasing mass by 600 lb should have resulted in a decrease in natural frequency of about 3 Hz, but the test results indicated a natural frequency shift of 1.25 Hz. This indicates that the structure is stiffer than the FEA model assumptions.

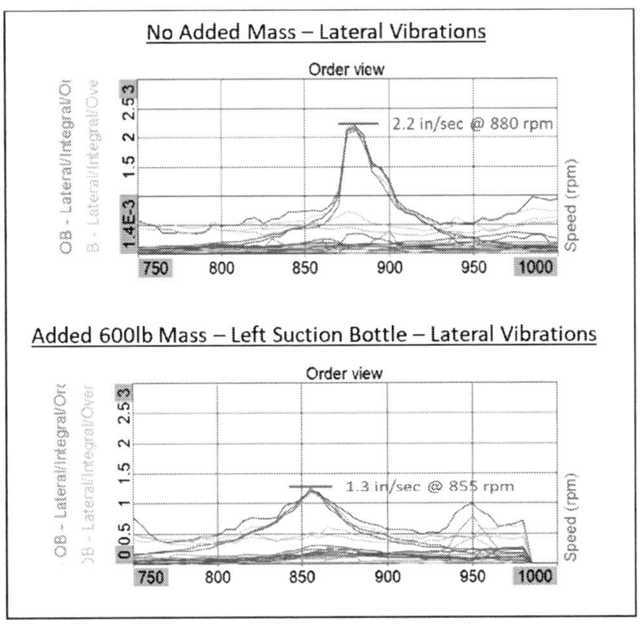

Figure 17: Comparison of overall suction bottle vibration data with and without added mass

Case Study 2 - Overview
This case study focuses on the importance of discharge bottle support stiffness. Improper setting of the discharge bottle supports was the root cause of the high vibration levels in this case study.

Machine Layout and Problem Description
A petrochemical plant had two ethylene compressors units in operation. The compressors were synchronous motor drive, 2-throw, double-acting, two stage machines with a single suction and discharge bottle on each compressor throw. Figure 18 shows one throw and the suction bottle on the compressor. The operating speed of the compressor was 327 rpm and the compressor operated at three different load steps.

Figure 18: View of one cylinder of a 2 throw compressor

The compressors had been in operation for many years and the B compressor was recently overhauled, which included a complete foundation replacement. Following approximately 1 month of operation after startup, the online continuous monitoring system showed elevated compressor rod vibration levels on the B compressor – the A compressor rod vibration levels did not change. The high rod vibration levels was sporadic and occurred multiple times over the course of a week.

Then a plant power issue occurred, which resulted in a pressure safety valve lift event on the B compressor. During the unscheduled outage to repair the power issue, the following issues were identified and repaired on the B compressor: (1) compressor cylinder support bolts sheared (2) three 1st stage cylinder suction line circumferential welds were cracked (3) several 1st stage cooling water lines cracked.

After these repairs were made, compressor rod vibration levels remained elevated. Vibration testing was performed to identify the cause of the high compressor rod vibration levels and failures that occurred on the 1st stage suction side of the compressor. Figure 19 summarizes the trend parameters including the increased compressor rod vertical vibration levels.

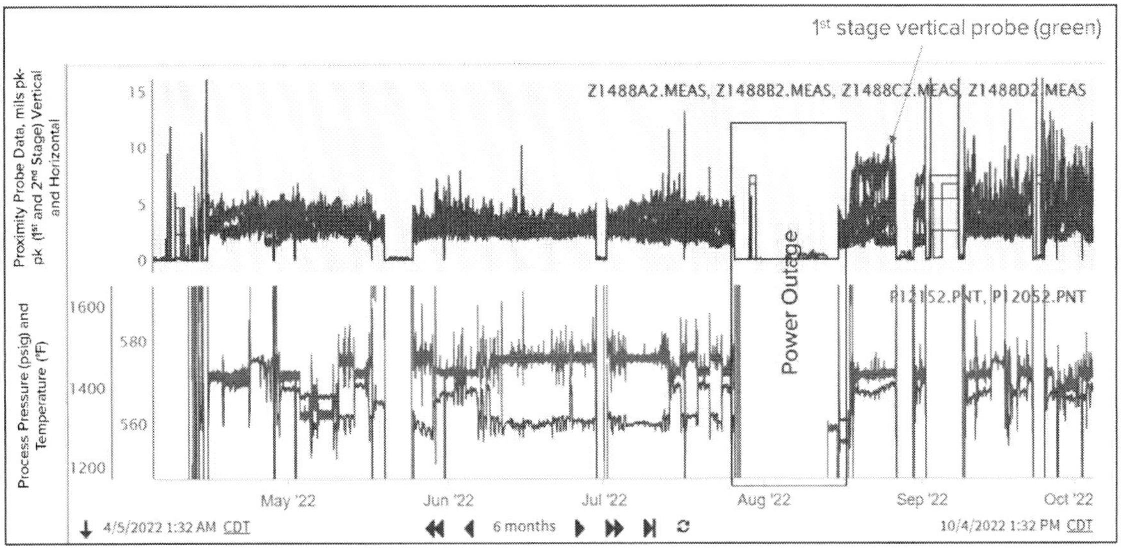

Figure 19: Plot of process temperature and vibration data over time including the vertical compressor rod vibrations

Suction and Discharge Bottle Supports

The 1st stage suction bottle on the B compressor had a pipe support installed to increase stiffness; a similar support was installed on the A compressor. During vibration testing efforts, an additional temporary support was added as shown in Figure 20.

Figure 20: 1st stage suction bottle

The as-found condition of the A and B compressor discharge bottle supports was also noted to be different. The B compressor support had much less stiffness as in Figure 21. The amount of spring compression on the B 1st stage discharge bottle support was much less than A 1st stage discharge bottle support. This means there is less torque on the studs and therefore less lift on the discharge bottles. Additionally, on the B compressor, a 1/2" thick rubber liner was installed between the support and bottle. This provided a significantly less stiff support arrangement compared to the A compressor.

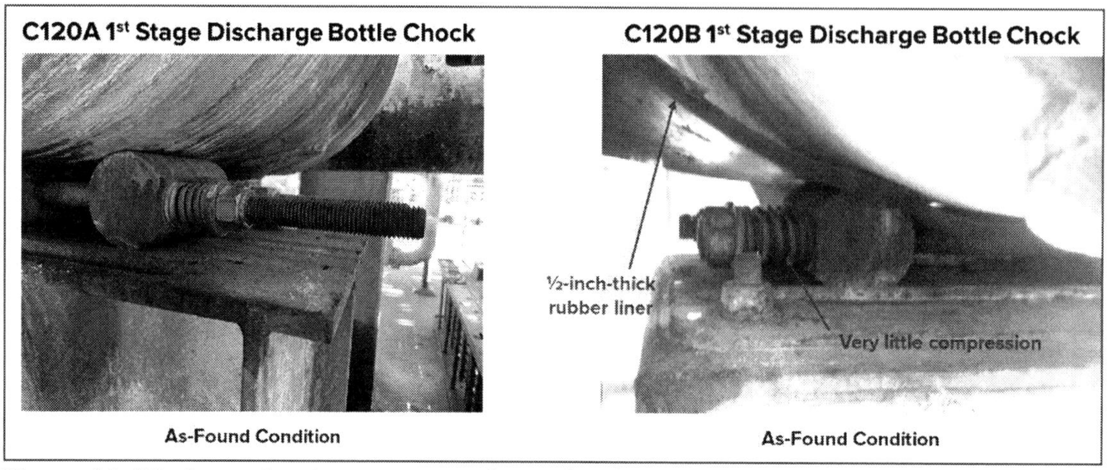

Figure 21: Discharge bottle support configuration for A and B compressors

Vibration Testing

A 32 channel data acquisition system was setup to collect vibration data on the compressor frame, feet, distance piece cylinder head, cylinder head support, suction bottles, suction piping, and discharge bottles. Four dynamic pressure transducers were installed in the piping system to determine if the vibration on the compressor correlated with pressure pulsations in the piping. High resolution spectral data was continuously collected over 26 hours to identify mechanical natural frequencies in the low amplitude range of the spectrum (using log scale). Figure 22 shows an increased noise floor around the 6x and 7x rpm frequencies. This raised noise floor indicates the locations of the mechanical natural frequencies where the natural frequency is closest to the running speed multiples and the response of the running speed is amplified. During the 26 hour timeframe data was collected, the peak of the raised noise floor shifted from near the 7x rpm to 6x rpm and back to the 7x rpm – the structural natural frequency of the system was moving over time while the excitation frequencies from the compressor operating speed remain fixed. This same trend was not noted on the A compressor, which indicated this may be part of the high vibration problem.

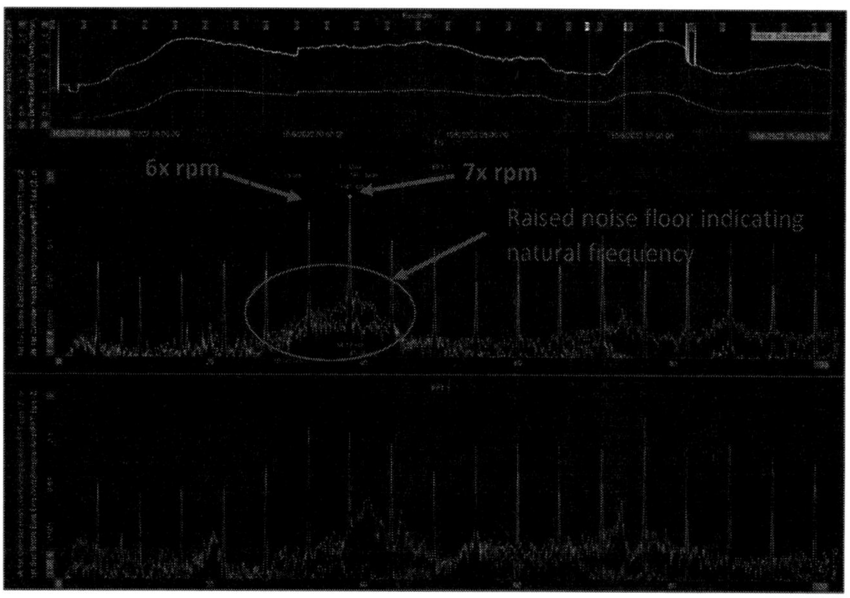

Figure 22: Vibration data showing raised noise floor indicating a natural frequency near the 6x rpm and 7x rpm frequencies.

The following vibration tests were completed to assist in identifying the cause of the change in natural frequency over time. Vibration data was acquired while making adjustments to the system to easily quantify the effects of the adjustment.

- The original suction bottle support on the head-end of the 1st stage suction bottle (Figure 20) was gradually tightened to reduce relative motion between the bottle and pipe clamp to determine if too much rubber liner was installed between the bottle and clamp.
    - This did not have an observable effect on the vibration levels.
- The temporary jack stand (Figure 20) was installed below the crank-end of the 1st stage suction bottle and gradually tightened to increase the vertical stiffness and evaluate the suction bottle vibration response. Additionally, the 1st stage discharge bottle supports (Figure 21) were tightened to achieve identical spring compression as found on the sister compressor.
    - The discharge bottle drain line vibrations reduced, but the discharge bottle vibrations increase.
- The rubber liner beneath the 1st stage discharge bottle support (Figure 20) was removed with the machine running and the bottle as placed on the supports with metal to metal contact – similar to the sister compressor arrangement.
    - This resulted in a significant reduction in vibration levels.

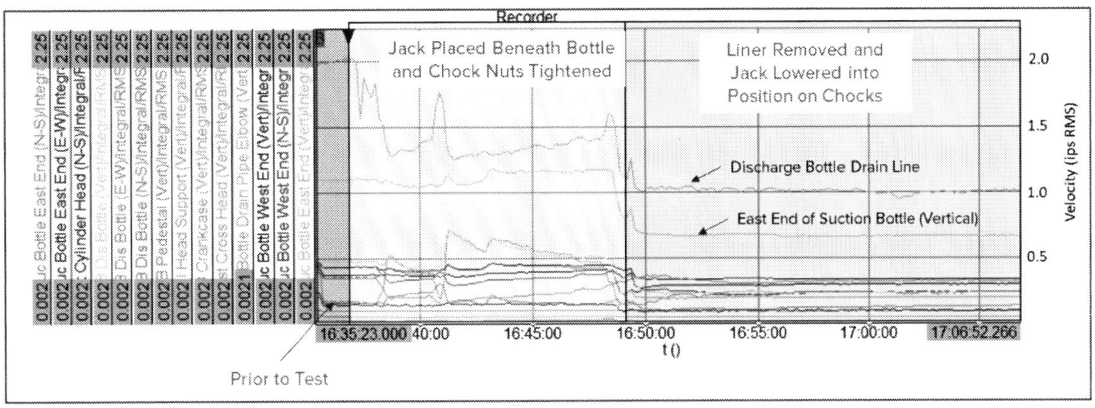

Figure 23: Vibration data trend during support adjustments

Root Cause and Solution

It was concluded that the increase in vibration levels was due to the resonance condition induced by the change in stiffness of the discharge bottle chocks. The only way for this to occur is for either the stiffness or the mass of the machine to change in time. It is reasonable to assume that it is the vertical stiffness of the machine changing in time, which must be occurring due to dimensional changes (expansion/contraction) between the cylinder head and discharge bottle chock support or stiffness of the ½-inch rubber bottle chock liner due to subtle changes in ambient/metal temperatures.

Overall Conclusions from Case Studies

- Suction and discharge bottles don't have lots of support points – discharge bottle supports are an important location to check when troubleshooting high compressor bottle vibration levels.
- Always test to determine mode shapes. The mode shape may be more complicated than assumed and the entire modal mass of the machine needs to be understood to implement to most effective vibration control solution.
- Don't discount damping solutions, they can be used to effectively reduce vibration levels.
- Strain and stress data can be great supplemental tools to vibration analysis for fatigue concerns.

MEASURING SHAFT CENTERLINE DISPLACEMENT WITH RELATIVE DISPLACEMENT SENSORS MOUNTED ON THE SIDES OF BEARING HOUSINGS

Eduardo Murphy Arteaga
Engineer, Murphy & Pérez
Emiliano Zapata, Morelos, México
emurphya@aol.com

Eduardo Murphy Arteaga mechanical-electrical engineer began static balancing of fans in 1978. In 1980 he had the opportunity to work at the Institute for Electrical Research at Cuernavaca, Morelos, México as a manager for the incipient Vibrations Laboratory. He attended his first Vibrations Institute Annual Meeting at Cherry Hill, New Jersey (1980) Since 1987, he, supported by his wife Teresa and sons, works on vibration analysis, balancing and alignment. He has developed computer programs for frequency analysis and multichannel polar plot generation. He is currently a Category IV analyst.

ABSTRACT:
Knowledge of shaft centerline displacement is required when analyzing bearing failures to determine the position of the shaft in the bearing. When the bearing has no displacement sensors permanently installed, common practice is to attach sensors to the bearing housing side. However, in some cases, the bearing housing itself will tilt and/or twist resulting in erroneous data. To obtain true displacement of the shaft, a second pair of sensors must be attached at the other side of the bearing. Simple linear equations provide true shaft displacement.

Two examples obtained from hydro generators are presented, both had the runners mounted on the generator shaft. In the first case, with a Francis type runner, no correction was made, however data was erroneous. On the second case, with a Pelton runner, correction was obtained using four sensors. Finally, a simple model is presented to demonstrate forward and backward orbits.

KEY WORDS: Attitude angle, orbits, relative displacement sensors, shaft centerline displacement.

HYDRO GENERATOR WITH FRANCIS RUNNER:

A repowering project was implemented in a Hydroelectric facility increasing power output from 14 to 16 MW. A new generator replaced the original unit. The new generator was smaller, and the bearing housings were independent. During installation, alignment was performed between the runner and the volute. Diametral clearance was only 1.5 mm (60 mils) so a quality alignment had to be done. Concern of some engineers regarding the change in vertical position due to hydrodynamic lift of the shaft in normal operation resulted in an increase of the upper gap in the

Figure 1

Figure 2

no run condition. Figure 1 shows the runner and volute with the front (discharge side) cover off. Note 15 blades and 16 wicket gates.
Figure 2 shows detail of the runner and back wear ring gap, about 1 mm (40 mils)

To maximize efficiency, most of the water must go through the runner, however, some amount will seep through the gaps. Pressure on the back side of the runner will impose an axial force which will be handled by the axial pads on the bearing with limited thrust capacity. A labyrinth seal is integrated in the runner and mating grooves on the back cover. Wear and axial displacement will allow more back pressure to build up, hence increasing thrust. A "compensation" valve is used to drain water and lower thrust forces.

The main bearing, located between the runner and generator, is partially mounted in a cantilever arrangement as shown in Figure 3.

Figure 3

Figure 4

Figure 4 shows a relative displacement probe mounted on the turbine side of the main bearing, vertical direction. Gap voltage in the no run state was measured as -15 V. At 14 MW, the reading was -12 V, resulting in a 380 μm (15 mils) change. Of course, only 50 or 100 μm (2 or 4 mils) was expected. A manometer located on the back plate indicated 2 kgf / cm^2 (196 kPa or 28.43 lbf / in^2) and the exposed runner area was calculated in 1.067 m^2 (1654 in^2) resulting in a force equal to 209 kN (47,000 lbf) Rated axial load for the bearing is 300 kN. Figure 5 shows a simplified drawing of the arrangement.

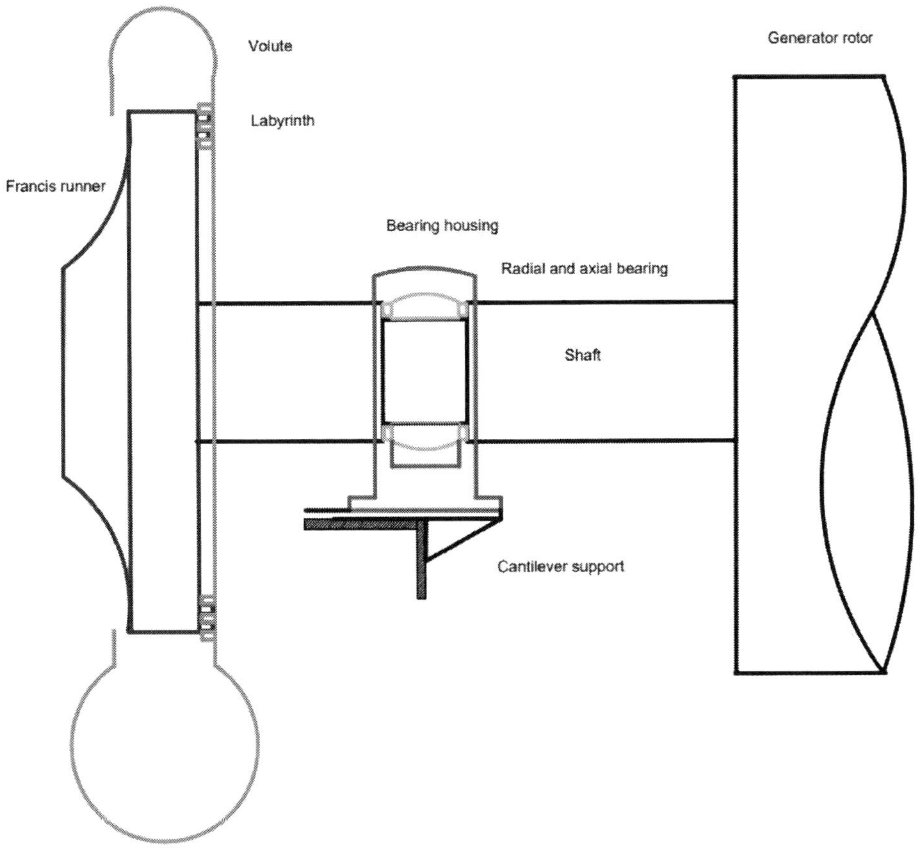

Figure 5

The bearing housing has a spherical seat for the bearing itself, so it was concluded that the axial force was tilting the housing towards the turbine, therefore closing the gap between the probe and shaft at the turbine side and increasing it at the generator side. At the moment, only two relative displacement probes were available: one mounted in the vertical direction and the other one, in the horizontal direction. A dial indicator was used to measure shaft axial displacement. Total gap between axial pads is 0.5 mm (20 mils) The indicator was placed at the exciter end to measure the shaft movement from 0 to 14 MW. A negative displacement of 1 mm (40 mils) was measured, so even if the initial position of the shaft was towards the exciter end, there is -at least- a 0.5 mm difference. Indeed, bearing

housing was tilting towards the volute and the runner towards the discharge side. This can result in a rub between the runner and discharge side cover, and hence, high vibration.

HYDRO GENERATOR WITH PELTON RUNNER:
As in the above case, this hydro generator was subjected to a repowering process from 16 to 18.5 MW. Runner geometry was unchanged, only nozzle area was increased. The runner is mounted in a cantilever fashion on the generator shaft. The arrangement is horizontal. There are two nozzles separated by almost 90° (Figure 6) and the resultant force is nearly horizontal. Total force acting on the main bearing (runner end) involves the calculation of horizontal and vertical moments with respect to one of the bearings and the determination of the resultant horizontal and vertical forces, including runner and generator rotor weight load proportion. Because of the large horizontal force involved, bearing design involves a clockwise rotation (turbine side view) of the lower and upper halves to move the support area further from the vertical. Rotation on the exciter end halves is counterclockwise. Figure 7 shows a simplified sketch of the horizontal forces involved.

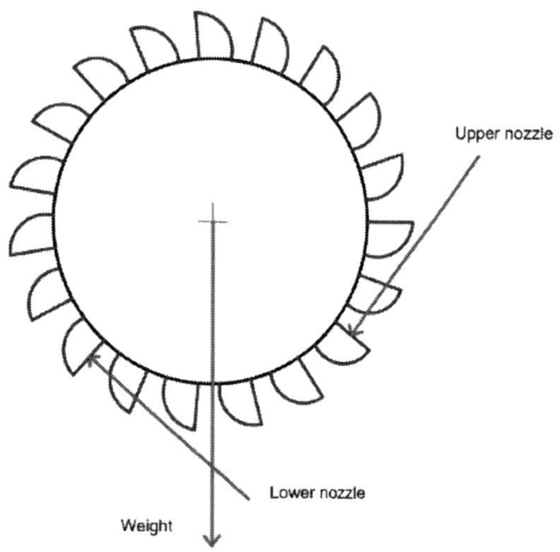

Figure 6

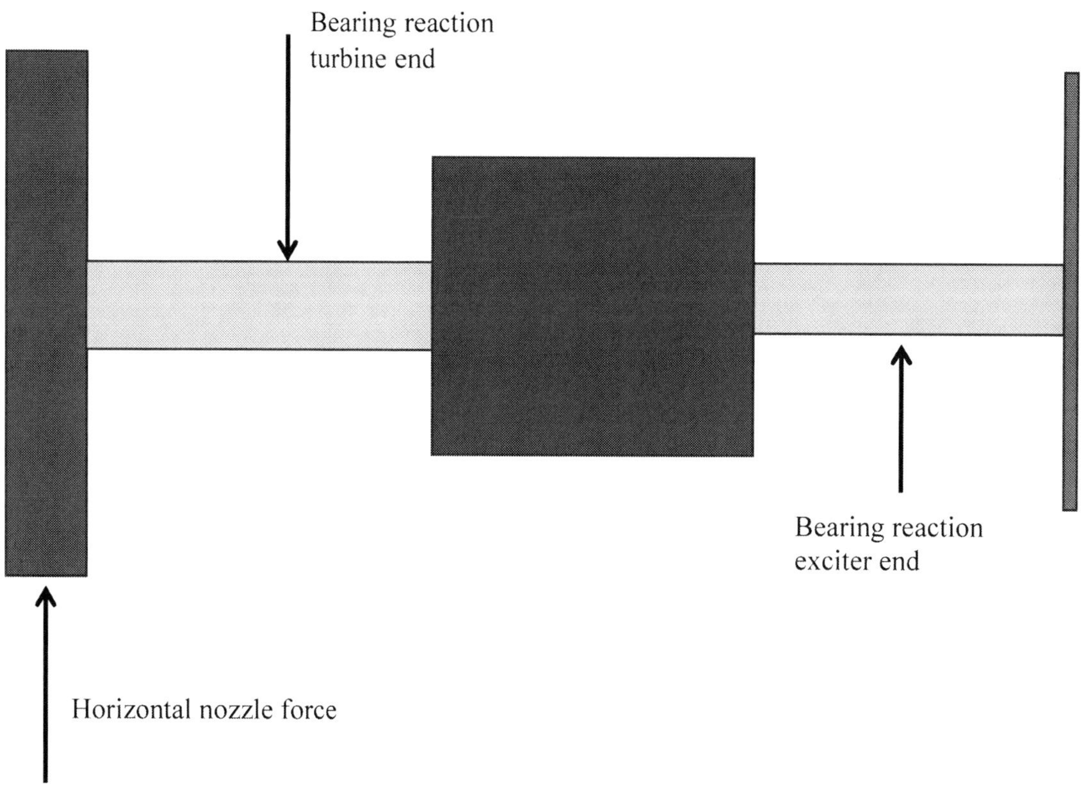

Figure 7

Nozzle forces were calculated each as 165,786 N (37,256 lbf) The resultant force magnitude is 252,129 N (56,661 lbf) at 190°. In rectangular components, -247,907 N horizontal and -45,947 N vertical (-55,712 and -10,326 lbf respectively) As can be seen in Figure 7, the horizontal force acting on the main bearing is larger due to the reaction at the exciter end bearing. Total horizontal force acting on the main bearing was calculated as -344,725 N and in the vertical direction as -226,073 N In polar coordinates, 412,243 N (92,643 lbf) at an angle of 213°

Operation at 18.5 MW resulted in main bearing failures approximately every 6 months. Visual inspection indicated fatigue as the origin. However, concern about the location of the working area on the bearing required a "shaft centerline" analysis. It was proposed to mount two displacement probes on the upper half of the bearing in true vertical and horizontal directions (recall that the halves are rotated) Several consultants opposed this arrangement and suggested to mount the probes on the outside faces of the bearing housing as it is commonly practiced. Because of our previous experience with tilting and or twisting housings, we agreed to perform the study except by using four probes: two on the runner side and two on the generator side.

A test was performed with 1 MW increments up to 18 MW and one-half increment to 18.5 MW. Gap voltage was measured using a multimeter directly on the oscillator-demodulators. Displacement was obtained by dividing the measured voltage by the system sensitivity: 7.874 mV/µm (200 mV / mil) Figure 8 shows the "shaft centerline" displacement as observed from the displacement sensors located on the generator side. Note the horizontal *turn* as power increases.

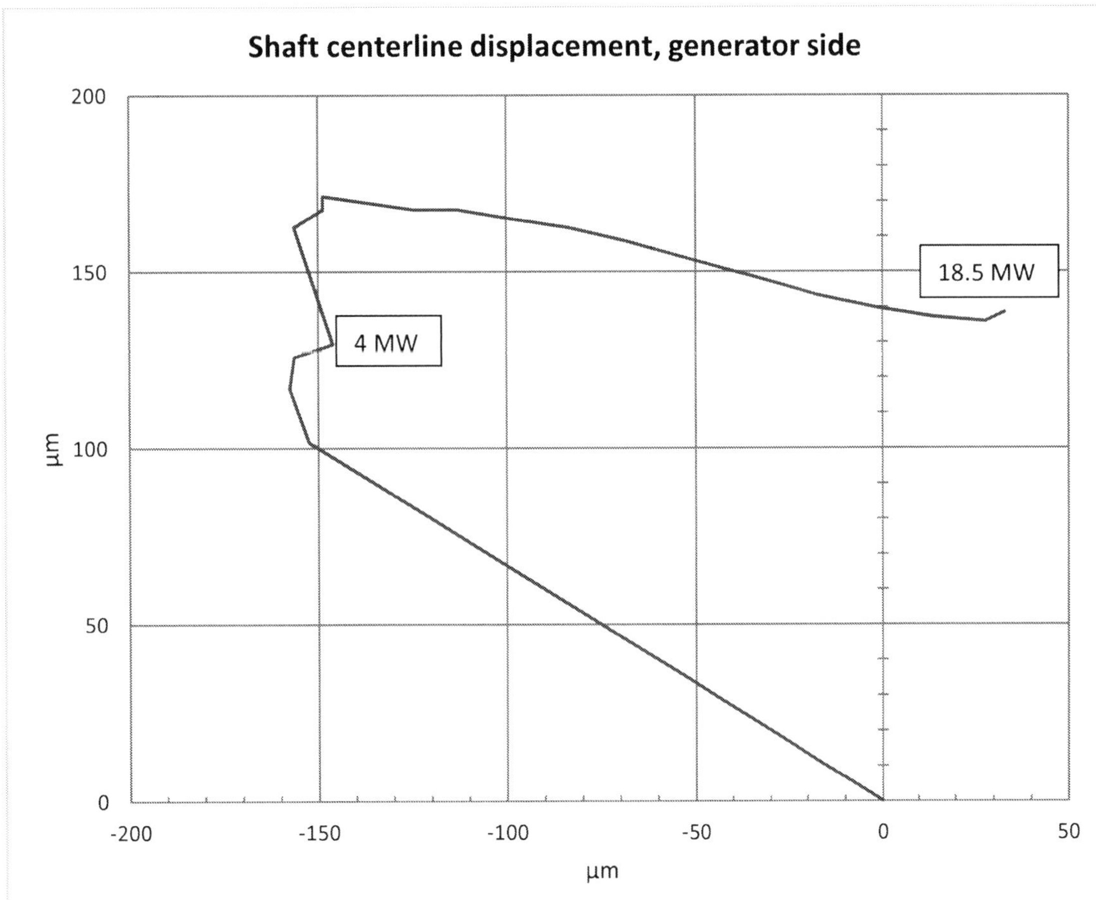

Figure 8

Figure 9 shows what the sensors are observing at the runner side.

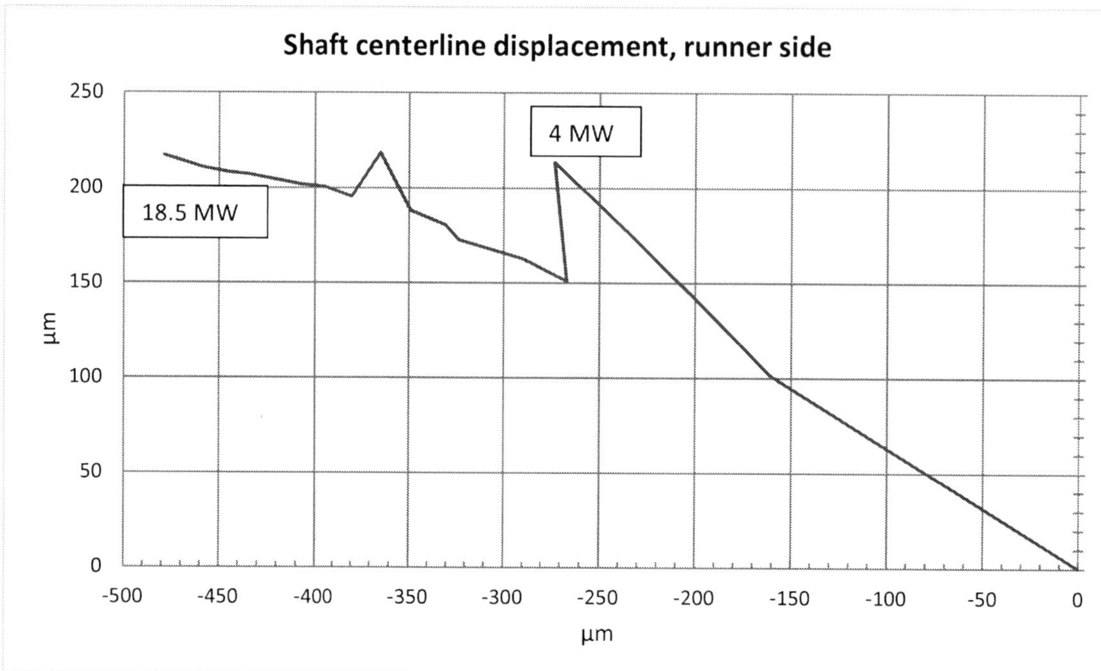

Figure 9

Diametral bearing clearance is normally calculated as 1 or 1.5 µm per mm (1 or 1.5 mils per in) In this case, 1.25 µm per mm was used, yielding 625 µm for a bearing diameter of 500 mm. A horizontal relative displacement of 480 µm can only be achieved if the bearing casing is twisting, increasing the gap at the runner side, and decreasing it at the generator side.

Combining data from figures 8 and 9, shaft centerline displacement occurring at the middle of the bearing can be calculated. Figure 10 shows the actual displacement of the shaft at the middle of the bearing. The dotted line represents half the diametral clearance. The arrow is pointing at 18.5 MW and the attitude angle (between arrow and vertical) is measured as 58.9°. The arrow itself has an angle of 211.1° Recall that the resultant force has an angle of 213°
In figures 8 and 9, a "jump" can be observed in the traces at 4 MW, this is because one nozzle is used from 0 to 4 MW, afterwards, both nozzles are used.

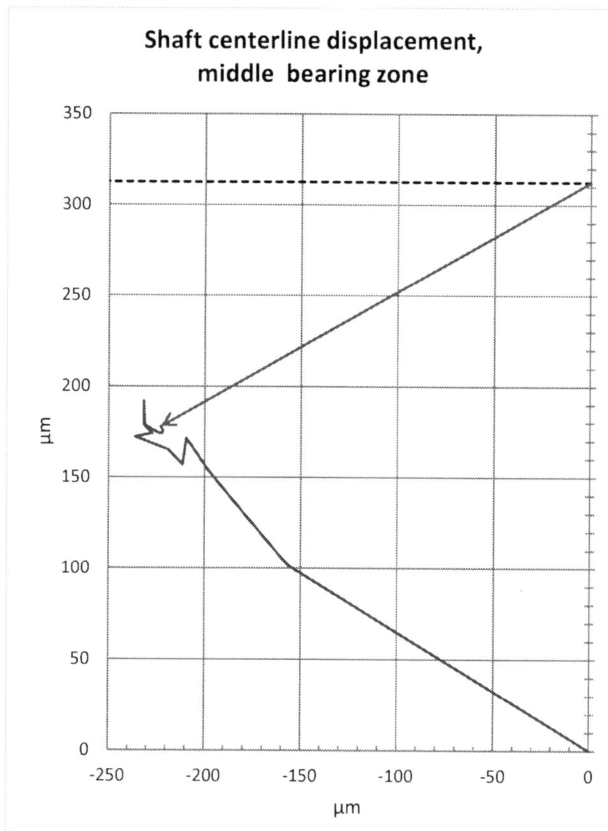

Figure 10

The specific load, W / LD (Force / Length Diameter) is calculated as 2.48 MPa (359.3 psi) According to [1], specific loads between 200 to 300 psi are considered "Heavy", above 300 psi a "Special Design" is required.

From the above analysis, it was concluded that the shaft was not operating near the horizontal and bearings failures are fatigue related. The repowering process must consider all components of the system.

HORIZONTAL AND VERTICAL DISPLACEMENT, SIMPLE MODEL:
Figure 11 shows a simple system. A motor "M" is supported inside a massless frame which can move in the vertical direction. The motor can move in the horizontal direction inside the frame. An unbalance "m" at radius "R" is attached to a disk that has a counterclockwise rotation. Two displacement sensors observe the vertical and horizontal movement.

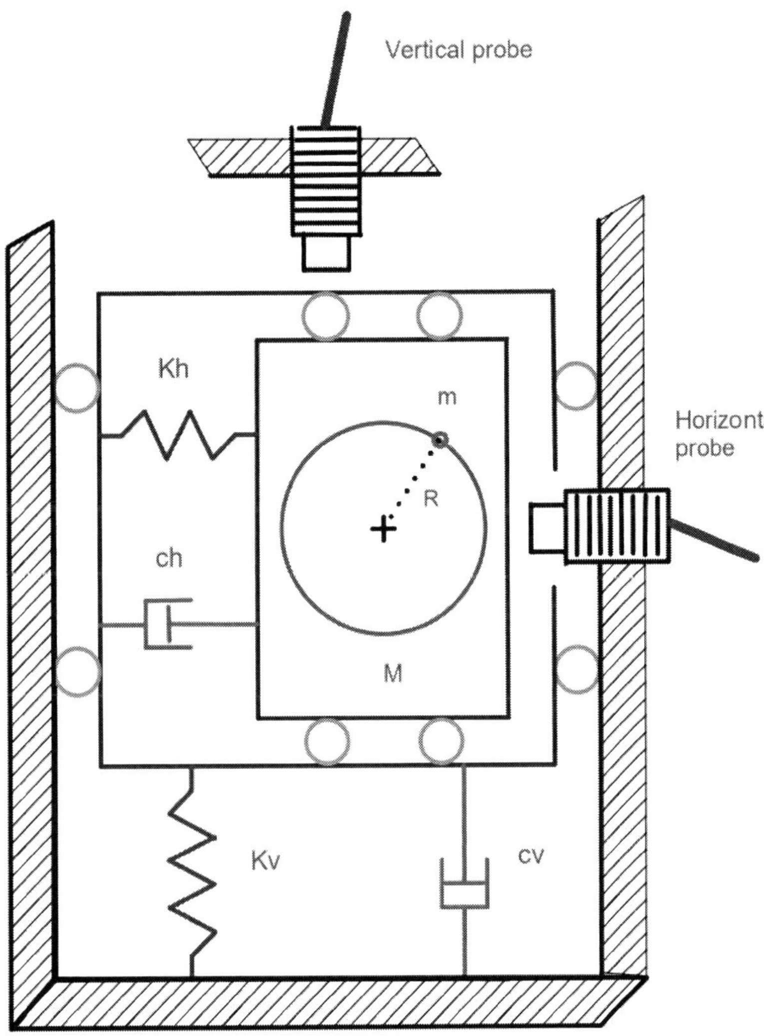

Figure 11

Figure 12 shows the Bode plots for the horizontal and vertical amplitudes vs. angular velocity (ω, rad/s) Horizontal natural frequency is 224 rad/s, vertical natural frequency occurs at 274 rad/s.

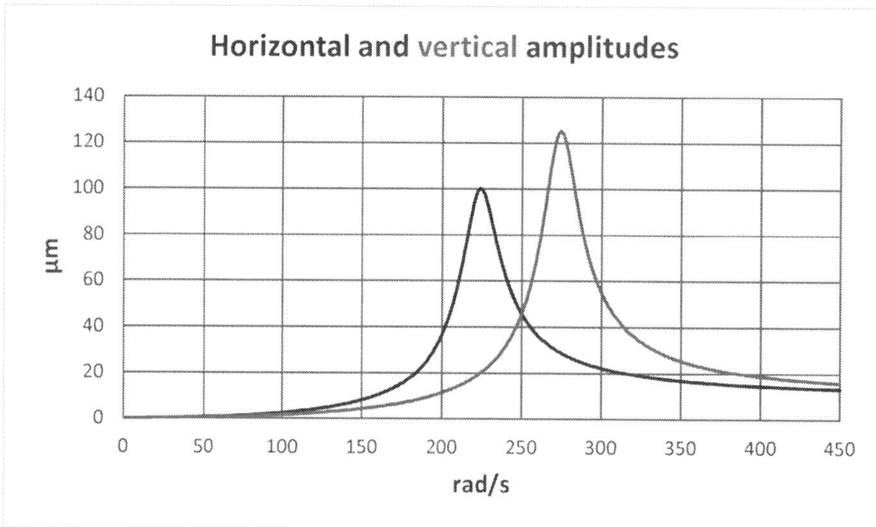

Figure 12

Figure 13 shows the complimentary Bode plot: phase vs. angular velocity. Vertical lag angle does not include an additional 90° lag due to probe mounting. Note that only the angular velocity "ω" has to be known to determine both amplitudes and both lag angles.

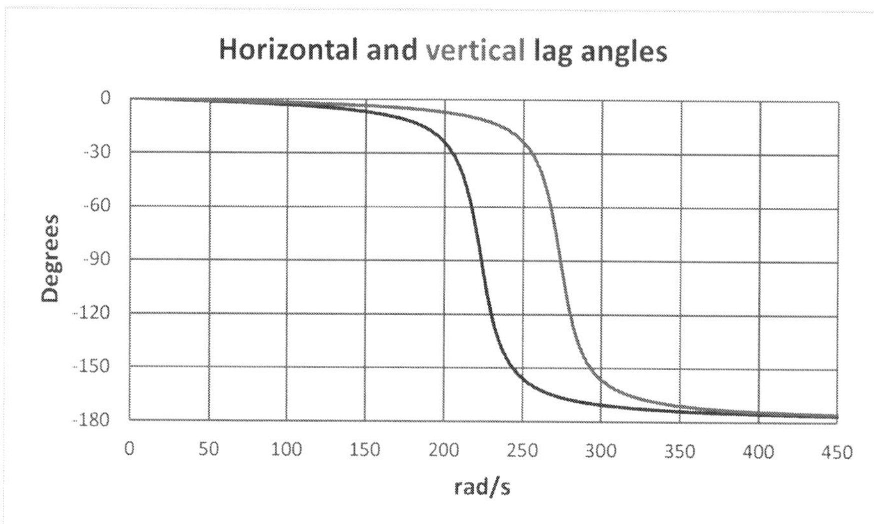

Figure 13

Damping coefficient to critical damping coefficient ratios "δ" are 0.05 and 0.04 for the horizontal and vertical directions. These low values were chosen to have a steep angular change as can be seen in figure 13.

A tachometer pulse in phase with the heavy spot (unbalance) is obtained from an optical detector placed in the horizontal positive axis. The pulse triggers data acquisition from both sensors. A dot - blank (positive pulse) convention is used to determine sense of orbit rotation. Three angular frequencies were selected to draw waveforms and orbits: 200, 250 and 300 rad/s (see figure 12)

Figure 14 shows the waveforms observed by the probes at 200 rad/s

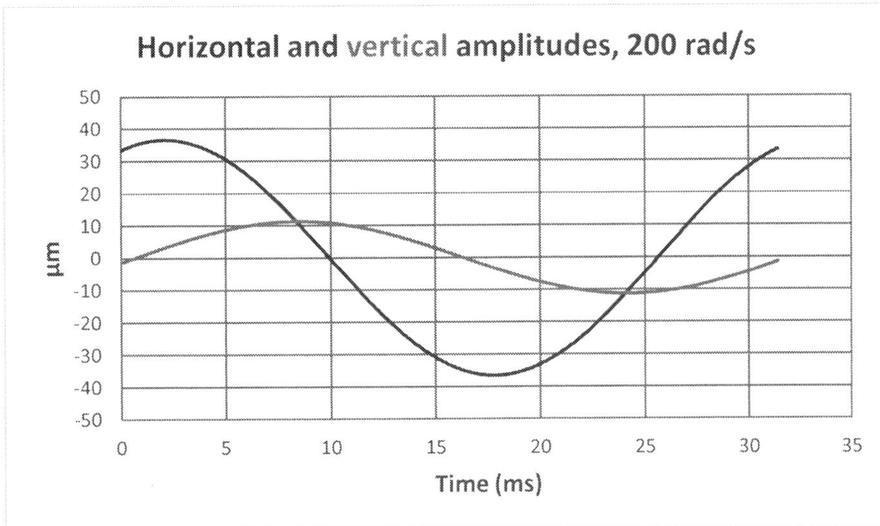

Figure 14

Horizontal high spot has a lag angle of -24°, while in the vertical direction, lag angle equals -7°. The vertical high spot will be seen by the vertical probe after -97°. Because the horizontal amplitude is larger than the vertical, orbit shape will be elliptical with main axis nearly horizontal. Figure 15 shows the corresponding orbit. Note dot – blank sequence indicating a *counterclockwise* direction.

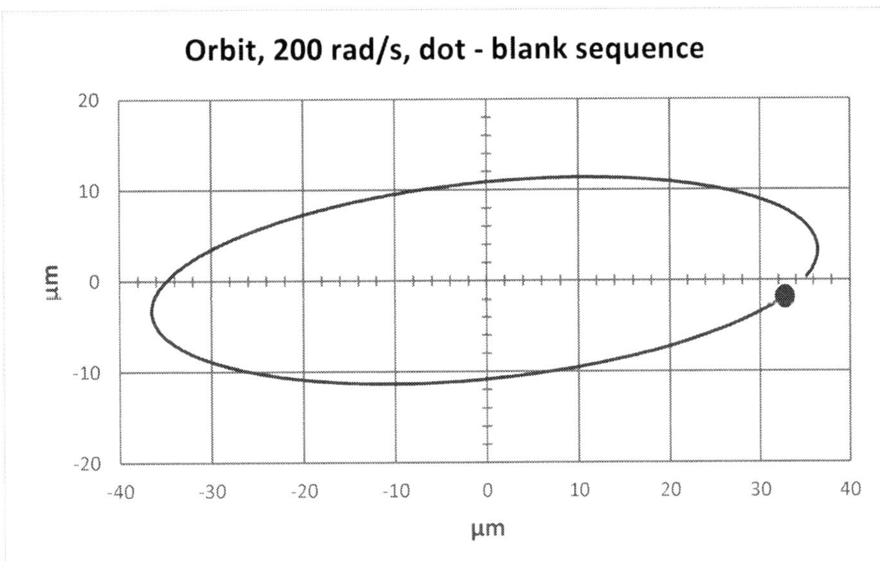

Figure 15

At a speed of 250 rad/s, the waveforms will be as shown in Figure 16. Horizontal lag angle will be -156°, vertical lag angle, including -90° is -114°. Horizontal high spot will be detected *after* the vertical high spot, so the orbit will have a clockwise rotation. The motor is still rotating counterclockwise. Although amplitudes are nearly equal, angle difference is 42° so the orbit will not have a circular pattern as can be seen in Figure 17. Dot – blank sequence indicates a *clockwise* rotation.

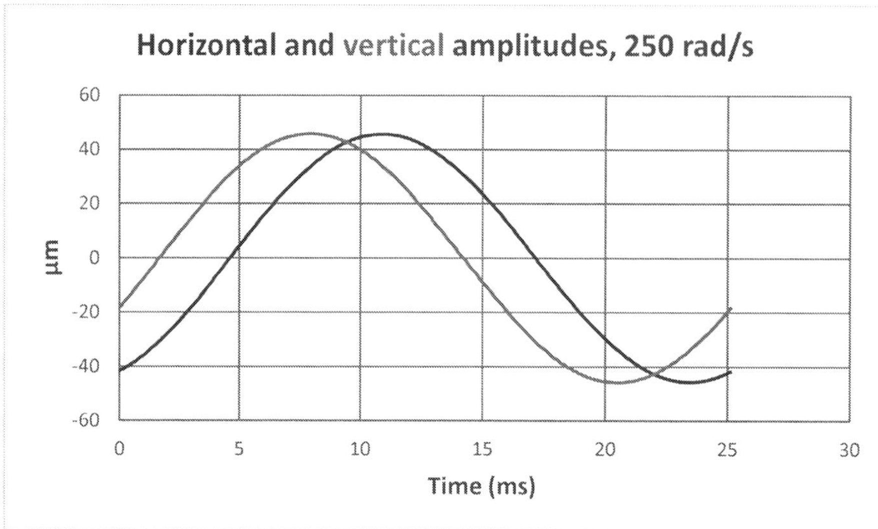

Figure 16

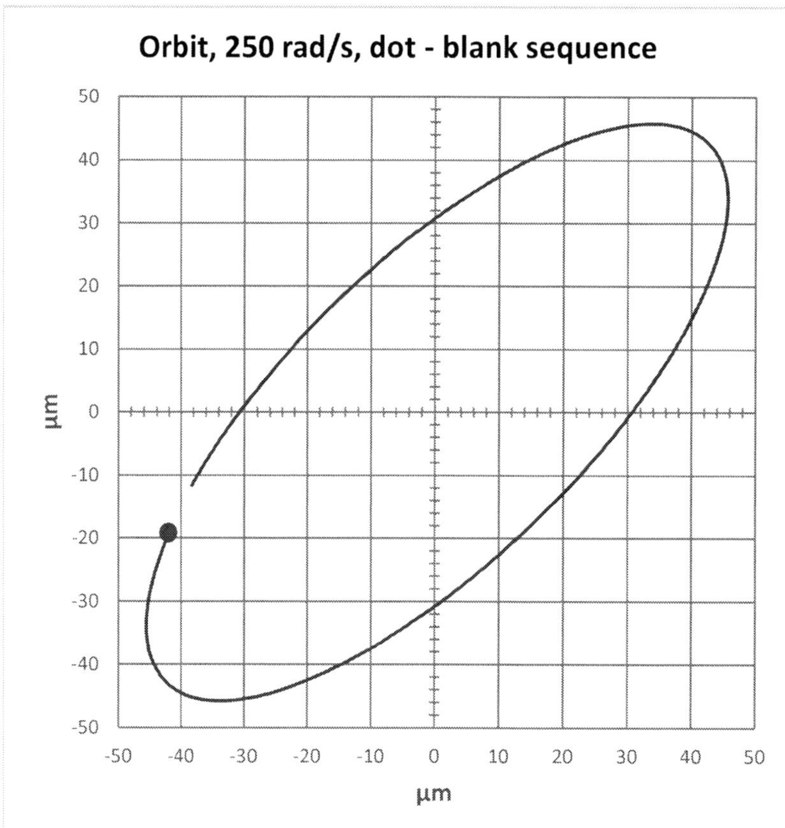

Figure 17

Increasing speed to 300 rad/s will increase the vertical lag angle to -156° plus -90° (-246°), so now, the vertical high spot will be seen *after* the horizontal high spot occurring at -170° as shown in Figure 18. Orbit rotation, Figure 19, will be *counterclockwise*. Because vertical amplitude is larger than horizontal, the orbit will be nearly vertical.

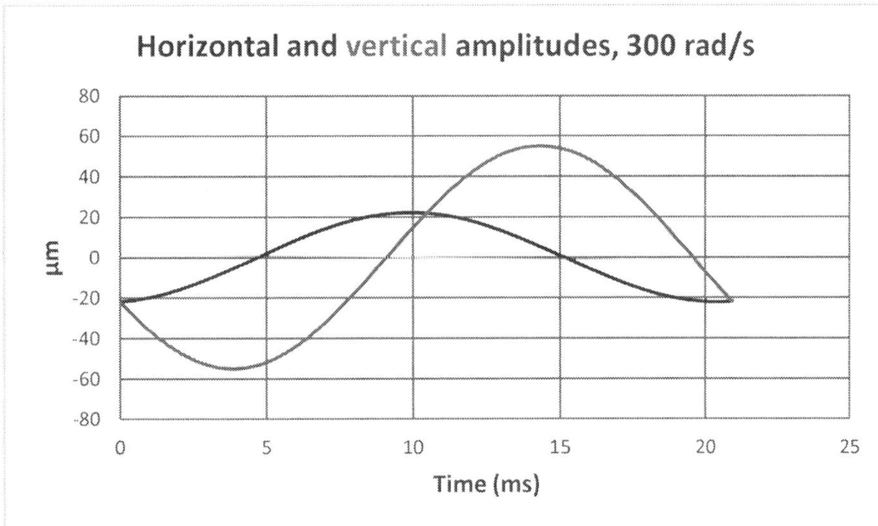

Figure 18

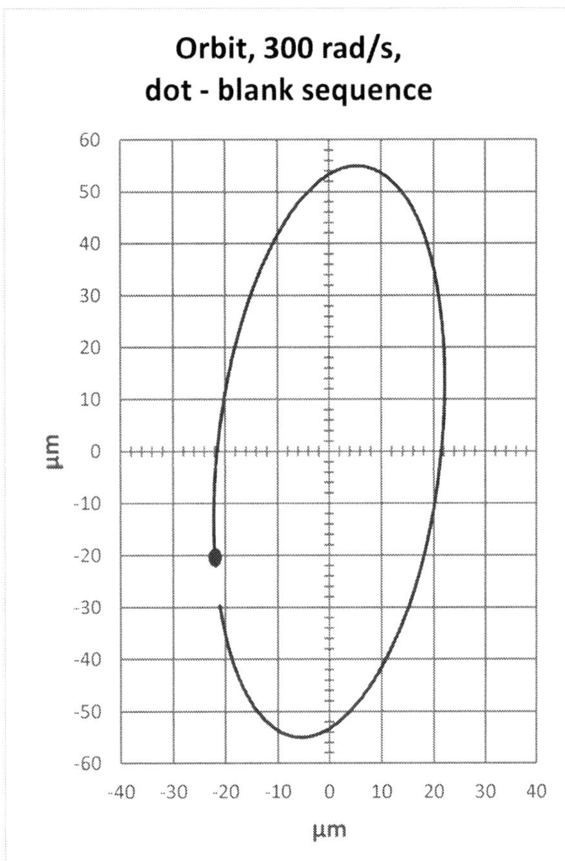

Figure 19

CONCLUSIONS

1. Relative displacement sensors are used normally to measure shaft displacement within a hydrodynamic bearing with respect to a relative frame. This frame can be the bearing itself or the bearing housing. When measuring alternating displacements, variations on gap clearance has no effect. However, if the gap clearance is to be measured, as when determining shaft centerline displacement, large errors can result if the frame of attachment is also moving as would be the bearing housing and probes mounted on the sides.

2. Correct determination of the shaft centerline requires probes mounted at the middle of the bearing or the use of four probes: two on each side of the bearing housing.

3. When repowering a system, attention to all critical parts is required as they could have been designed with no upper margin.

4. Bearing specific load calculation is useful when making a root cause failure analysis.

5. In the simple model with horizontal and vertical displacements, we saw that it is possible to change orbit direction of rotation without changing the system direction while operating above a natural frequency and below another one.

REFERENCES:

[1] Leader Malcolm, "*Understanding Journal Bearings*"

VIATC TRAINING CONFERENCE & EXPO

NIAGARA FALLS, NEW YORK JULY 26 -28, 2023

TWO CASE STUDIES USING AMPLIFIED VIDEO FOR VIBRATION TROUBLESHOOTING

Han-Chian Gee, P.E.
Chief Reliability Engineer (Rotating department)
Formosa Plastics Corporation, Point Comfort, Texas

Han-Chian Gee is a chief reliability engineer at Formosa Plastics, Point Comfort, Texas. He has over 30 years' experience on rotor dynamics and other reliability related technologies. He provided training & consultant services in Taiwan to industries, including nuclear power plant, refinery & petrochemical, etc. He is an ISO 18436-2 Category III certified analyst and a registered PE in Texas. Before his career in vibration application, he worked at AIDC (Aerospace Industrial Development Corporation) at Taichung to develop metal fuselage, composite graphite speed brake, honeycomb stabilizer and rudder design for AT-3 trainer & IDF (Indigenous Defense Fighter).

ABSTRACT:
Amplified video remotely records motion beyond equipment which makes a traditional tough job (hard, dangerous, or impossible to access) becomes easier & safer should gain more application in the future.

Normally, the image is taken on "**external**" surface, after magnification and slow down to allow the machine speak for itself and let the audience enjoy a "seeing is believing" experience which is an ideal scenario. However, the broken shaft in case #1 which was "**invisible**" since it was installed inside a stainless enclosure. Even it was "**ABSENT**" in the video, however, its failure mechanism still could be accurately forecasted & verified after tearing apart.

In case #2, motor soft foot (0.71 mils ~ 35% of average hair) was visible after been amplified. Before it was corrected, online vibration trending with associated DCS data was studied to tell **how vibration is affected by process loading variation** (viscosity, feed rate & start up, etc.) which provides valuable lessons to improve equipment reliability & operation practice for reference.

KEY WORDS: Amplified video; DCS; Feed rate; Forecast failure mechanism; ODS; Overshoot; Process loading variation; Startup; Viscosity.

PREFACE:

Vibration is a useful tool which proves its effectiveness on mechanical reliability, from detection, analysis to correction for decades. In normal practice, few spots on bearings are selected to collect data, however, when shaking is spread all over and not limited to machine itself, either hard, dangerous or impossible to reach, the job scope is beyond routine route check, hundreds or even thousands points to cover all suspected area to get an idea how each spot is moving related to each other like ODS (Operation Deflection Shape) is not un-usual which is time and labor consuming, let alone the proprietary software in need.

Vibration analysis data is valuable and meaningful for professionals, however, for managers without complicated training, technical information like spectra, waveform and phase is overwhelming and not what they are interested in. Fortunately, dynamic amplified video brings vivid & convincing field status to office environment and provide an option to let decision making become easier.

For general application, dynamic video taken on "**external surface**", after been magnified, slowed down and let the machine speaks for itself is usual, on the other hand, if the defective piece is "**behind an obstacle without transparence**", no comment can be made, however, in case #1, the failure mechanism of the "**internal**" component, which was **absent** in the video, still can be accurately forecasted before tearing apart. In case #2, soft foot less than hair line is visible after been amplified, before this deficiency was corrected, online vibration trending with associated DCS data was studied which provided valuable information to facilitate the study on how equipment vibration behavior can be affected under different process conditions.

CASE #1:

Seeing is believing, what if you cannot see it? This case was **rotating speed related** with two major challenges:

(Fig-1)

(1) The centrifuge dryer rotor was installed inside a stainless cabinet, with its majestic height which is hard to reach (Fig-1), let alone a secured door for safety concern (Fig-2). As a result, no video could be taken with a solid obstacle (the cabinet) between the rotor and the lens. Unless a sensor was pre-mounted, no vibration check spot could be reached except the motor.

(2) The motor was VFD (variable speed driver) type, its rotating speed, the two pulley diameters, the distance between the pulley centers, and belt length, all were un-known. Fortunately, the rotating speed of the centrifuge dryer rotor provided the key clue for troubleshooting was wired to DCS in control room.

(Fig-2)

In 2021/May, scary shaking from the roof, the columns to the floor was noticed, and no one could point out the source. Eventually, the dryer centrifuge shaft was broken and shut down the production.

With challenges as mentioned, **the failure mode of this invisible broken shaft** (Fig-3) **still could be forecasted and verified after tearing apart**. The three basic vibration characters, frequency, amplitude and phase are enumerated one at a time, and the inference is elaborated step by step.

(Note: Below description is based on the magnified video which will be presented at conference for better and easier understanding).

(Fig-3)

Frequency:
After requesting the unit operation, DCS indicated the dryer's rotating speed was ~ 400 cpm which matched spectra, both from magnified video and portable meter (Fig-4).

Amplitude:
The near-by structure was the most eye-catching object in this case, whereas the stainless cabinet had much smaller amplitude in different range; even though both had same finger print at dryer 1x RPM rotating speed (~ 400 cpm).

Rotating vs. bending mode:
It is well known if the clothing is not evenly distributed, the washing machine will shake. The dryer's function is to get rid of water by spinning, for same reason, un-symmetrically agglomerated puff could induce unbalance force (especially at discharge on top as the worse scenario) got the best chance, then through a cantilever beam (the shaft) which created both rotating moment & rotating shear force as counteraction at the bottom, where bearing, shaft and keyway met together with stress raiser. Other defects also could be candidates; however, no matter unbalance, miss alignment or looseness, all are **rotating in nature**. On the other hand, the near-by structure with much stronger amplitude never occurred in the past was **moving back and forth which should be a bending resonance response** to the unbalance.

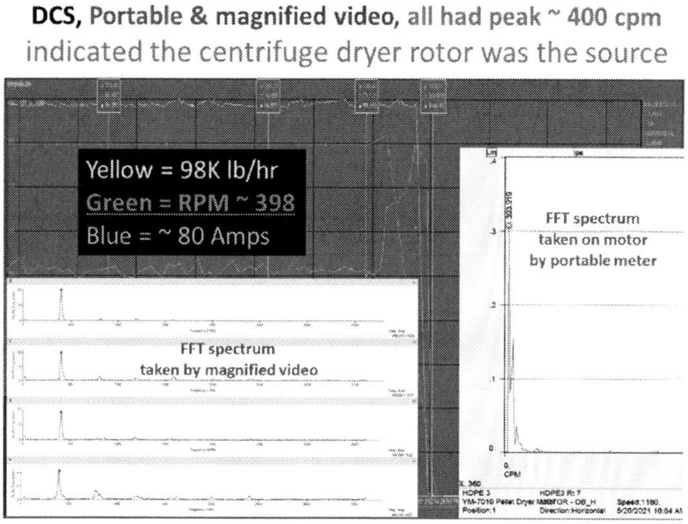

Phase: With much smaller amplitude and not synchronous with the bending structure, while the phase lags either not stable or not matched physical shift as expected, the dynamic image taken on the cabinet surface was influenced by combing both **rotating defects inside the cabinet & bending resonance of the near-by structure** was a reasonable inference.

(Fig-4)

After been dissembled, the rotor was shipped to shop for service, in the meantime, an independent lab was contracted to examine the broken shaft (Fig-3) to investigate the root cause.

For routine lab job, the "corpse body" which was the broken shaft and specification sheet on paper for each component was available, without surprising and exception; all was "**static**" based. With damaged parts and technical data on hand, the "forensic", lab analyst with years' experience could perform his job. However, such traditional lab work mode is focused on information after facts and is easy to fall into the "**See the tree but miss the forest**" trap. On the contrary, magnified video brought the "**dynamic**" image from the field and let all involved **witness the live scene at last stage before failure**, which might not be comprehensive, but enough for right direction.

The lab analyst surely can do precise diagnose, however, as a local staff on site, with better understanding equipment function and witnessed what the machine was doing before failure, homework still can be done, since it is convenient to navigate internet for answer. After diligent searching & comparison, "**bending plus rotational failure mode**" was forecasted and best matched the broken shaft section (Fig-3).

(Note: To avoid taking space in this short article, the other damages will be presented at conference, and of course, only on "**static**" photo after tearing apart).

The invisible rotating dryer behind the cabinet should be the principle dancer, although we don't know how it was moving or how much it was moving inside the cabinet. On the other hand, even the surrounding structure caught the most attention; however, actually was only a supporting dancer.

The video's hyperlink was shared with the unit manager, manufacturer and lab analyst while all still busy in communicating before the analysis report was ready, most importantly; after been put on same page, everyone expressed a "Wow!" as the common response and made agreement for prompt action.

CASE #2:

<u>To cultivate positive application in negative situation</u>. This case was **<u>process loading related</u>** and DCS data was required to verify vibration behavior affected by process.

A three shafts polymer (polypropylene) extrusion gear box had permanent online system (16 channels with one as spare) installed. After a major T/A (output speed upgraded from 226 to 265 RPM), a weird vibration pattern on motor inboard along vertical (Z) direction was developing since baseline (Fig-A), its trending stepped over wide severity range - from smooth, good, fair to rough, which brought up attention and curiosity.

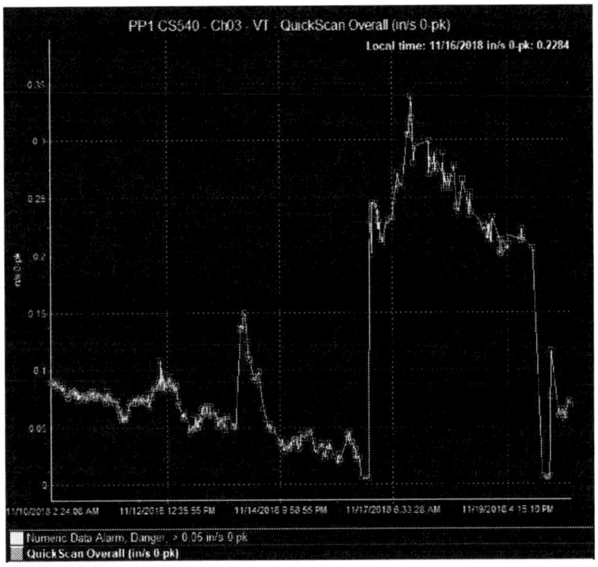

(Fig-A)

FRF (Frequency Response Function) analysis indicated minor motor 1x RPM peak (at 1200 cpm) with out of phase reading among all four corners was noticed, and soft foot was highly suspected. However, visual inspection cannot find any loosing bolts or gap under each foot.

Eventually, displacement (0.748 mils) on motor foot along vertical (Z) direction which is 37.5% of average hair thickness (2 mils) was verified after taking an amplified video. Imperceptible crevice open & close activity became visible & obvious (Fig-B).

(Fig-B)

For audience without polymer extrusion background, basic information should be illustrated first. More than rotating speed, to setup a proper alarm, the loading capacity (if variable & available), always needs to be considered. However, to meet ever-changing customized specifications, frequent viscosity (MI or Melt Index in polymer industry) transit is another factor can never be neglected which surely affects loading as well.

To squeeze polymer passing through a filter to remove impurities and thoroughly mix with chemical (like pro-oxide) to get homogenous resin ready for pelletizing is critical for quality control. The difficulty to insert a thumbtack into the ultra-fine screen (just barely held at position and not drop!) explains the hardship (Fig-C). In worst scenario, the screen might crash if it is overloaded and needs to be replaced.

(Fig-C)

Extruder gear box is surely the core of process, let alone bearing defects, or broken teeth, auxiliary issue (like lubricant pump low pressure, steam ventilation clog, etc.) might cause shut down, not only affects extruder reliability, but also motor rotor bar crack as consequence of over current in electrical armature circuit after frequent startup in long term. Before the motor soft foot was corrected, vibration was used as an indicator and related its response to DCS data; a more in-depth study was explored which provided insight, both improve operation practice and troubleshooting skills for future reference. After this hair line soft foot deficiency was fixed, there was no weird trending pattern any more. This 0.748 mils soft foot was subtle, although easy to fix, however, it was definitely hard to detect.

Understand the basics

To meet various customized specifications (especially viscosity), which impact the extrusion gear box loading and power consumption (KW), polymer industry uses another term, MI (Melt Index) and is measured differently from the other most popular kinematic unit (cSt),_ centistokes (Table-1). To understand MI (Melt Index) is necessary in advance, and five keys: MI (Melt Index or g/10 minutes), capacity (PPH or lbs. /per hour), electrical power (KW), startup & overshoot (both during startup and online) are selected.

Table-1 (the difference between viscosity and Melt Index)

Viscosity	MI (Melt Index)
"The **numbers of seconds** the liquid takes to flow through a calibrated glass tube".	"In fixed **10 minutes** duration, the **mass (grams)** can pass a capillary (specific diameter & length) by pressure."
▪ **cSt (centistokes)** is the Kinematic viscosity unit	▪ **g/10 minutes** is the MI unit
- Fewer seconds (faster) means lower viscosity.	- **Higher MI** means lower viscosity & less viscous.
- More seconds (slower) means higher viscosity.	- **Lower MI** means higher viscosity & more viscous.

It is emphasized that only 15 minutes cycle overall velocity trending is used in this case study, although twice a week spectra and waveform is available, both analysis are for reference only. The other data source is from DCS which is updated every few seconds. Even though both data are not synchronous; however, as long as we pick up an adequate time frame, our goal still can be focused: "**How does vibration respond to each DCS respective variable**?"

Fig-D1 (nine days duration in macro) including all factors is selected to relate MI, PPH & KW with associated vibration response which is pickup for its weird pattern at first glance, most importantly, **<u>every factor occurred only one at a time</u> which makes it easier to compare its individual effect on vibration.**

<u>Fig-D1 (for fundament illustration)</u>

Four vibration peaks on Fig-D1 with associated chronical DCS events is listed in Table-2:

(1) **<u>11/12/2018</u>** (green color highlighted):
MI (or viscosity) online transit from 12.1 to 2.2 (became thicker)
while PPH stayed stable (~ 28088);

(2) **<u>11/13/2018</u>** (orange color highlighted):
PPH online transit **<u>without overshoot</u>** (+4004) from 29985 to 33989
while MI stayed stable (~ 2.2);

(3) **<u>11/16/2018</u>** (red color highlighted):
PPH startup transit **<u>without overshoot</u>** (+35985) from zero to 35985
while MI stayed stable (~ 2.2);

(4) **<u>11/17/2018</u>** (red color highlighted):
PPH online transit (+831) from 35985 to 36816
while MI stayed stable (~ 2.2).

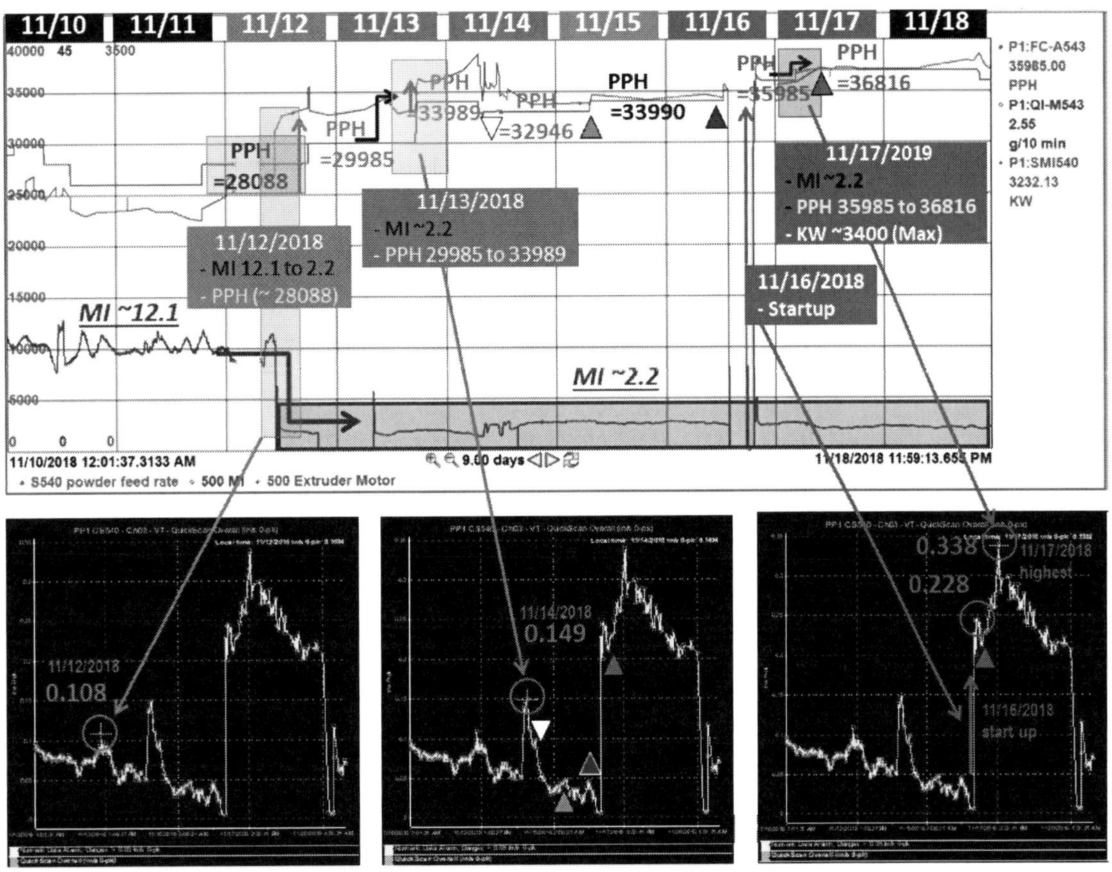

Fig-D1 (Overview macro 11/10/2018-11/18/2018)

Table-2 for Fig-D1 (11/10/2018-11/18/2018)

Date	Various DCS events & related vibration response
11/11/2018	High MI (12.1) while PPH stayed ~26000 then rose up and stayed ~28088 on 11/12/2018.
Only MI transit 11/12/2018	High MI (12.1) transit to low MI (2.2) - Dark blue hazy box (horizontal/bottom) on DCS indicated MI transit from high to low (or viscosity low to high), then stayed the same (11/12/2018 to 11/18/2018) & harder to process. - Green hazy box (horizontal & vertical) with green arrow indicated KW skyrocketed while PPH still stayed ~28088. - Green circle on trend indicated vibration climbed to 0.108.
Only PPH transit 11/13/2018 (raise)	MI 2.2 (or higher viscosity) kept the same - Orange solid box on DCS indicated PPH jumped from 29985 to 33989 and was highlighted with "re-direct arrow". - KW (green arrow) jumped (but < 11/12 during MI transit). - Orange circle on trend indicated vibration jumped to 0.149.
Only PPH transit 11/14/2018 (drop)	MI 2.2 (or higher viscosity) kept the same - White triangle on DCS indicated PPH dropped from 33989 to 32946. - White triangle on trend indicated vibration sharply dropped.
Only PPH transit 11/15/2018 (raise)	MI 2.2 (or higher viscosity) kept the same - Blue triangle on DCS indicated PPH rose from 32946 to 33990. - Blue triangle on trend indicated vibration was climbing.
Only PPH transit 11/16/2018 (raise)	MI 2.2 (or higher viscosity) kept the same - Purple triangle on DCS indicated PPH rose from 33990 to 35000. - Purple triangle on trend indicated vibration was climbing.
Only trip & re-start 11/16/2018	MI 2.2 (or higher viscosity) kept the same while tripped & re-start - Red vertical arrow on DCS indicated PPH jumped from zero to 35985. - Red circle on trend indicated vibration first jumped to 0.228.
Only PPH transit 11/17/2018 (raise)	MI 2.2 (or higher viscosity) kept the same - Solid red box on DCS indicated PPH rose from 35985 to 36816 & was highlighted with "re-direct arrow". - Red circle on trend indicated vibration first attenuated from 0.228, and then kept further climbing to 0.338.

Not like launching from idle to designated constant 1200 RPM within seconds, instead of as a response while passing critical speed which requires real time sampling and can never be captured by a 15 minutes scanning cycle overall velocity, let alone network latency for whatever reasons. **Peak at 0.228 (inch/second) on 11/16/2018 startup was definitely a feedback to extruder loading status in nature and not rotating speed relevant**.

Fig-D2 is a ten hours micro for the Fig-D1 nine day's macro to confirm both online and startup transits **without overshoot** as below:

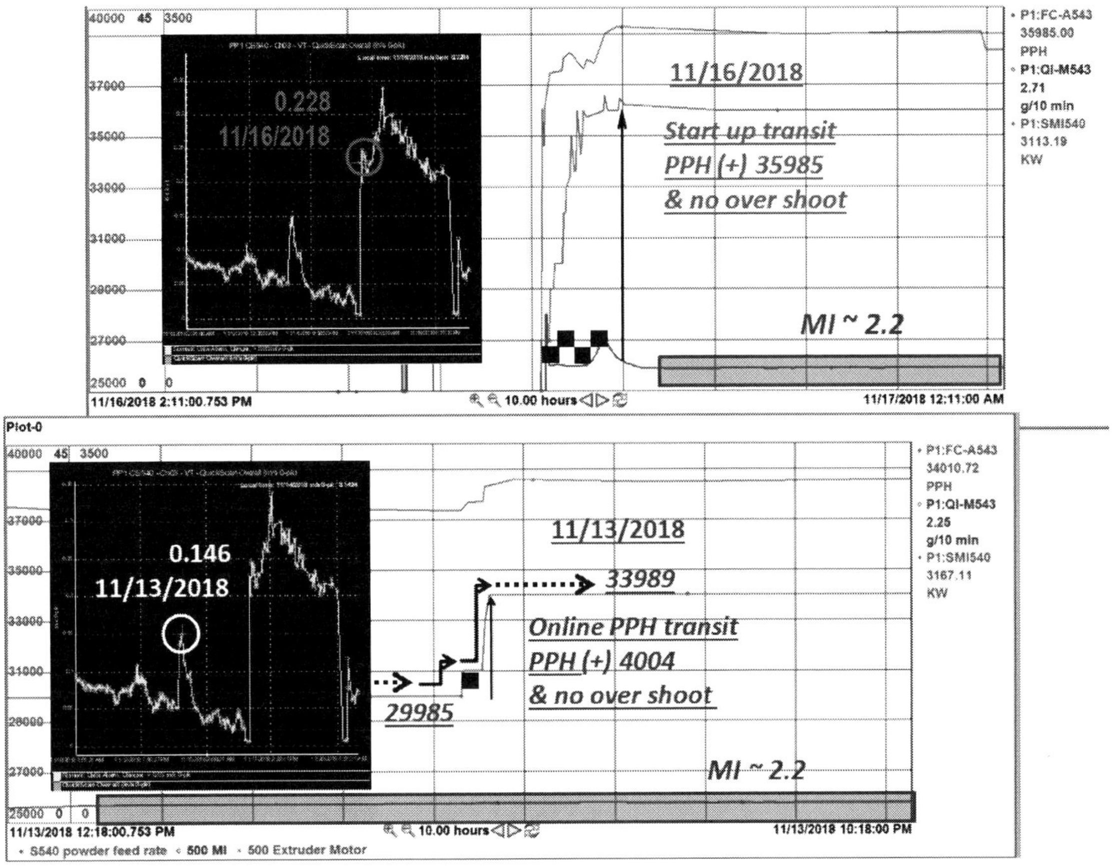

Fig-D2 (DCS after zoom in)

Those small, solid, black blocks were used as reference to get a general visual comparison to tell how long to complete the process procedure.

For example, both were **without overshoot** at the **same MI (~ 2.2)**:

(1) **11/16/2018** (upper Fig-D2)
Startup PPH transit jumped +35985 (from zero to +35985) with much slower (~ 5x blocks) duration.

(2) **11/13/2018** (lower Fig-D2)
Online PPH transit jumped +4004 (from 29985 to 33989) with much faster (~ 1.5x blocks) duration.

From Fig-D1 & Fig-D2, we can prioritize vibration sensitivity (from high to low) to DCS events at this moment:
(1) Startup transit (**without overshoot**) while MI stayed same;
(2) Online transit (**without overshoot**) while MI stayed same;
(3) MI transit while PPH stayed same.

Electrical KW sharply increased as a response to MI transit from 12.1 to 2.2 (or became thicker & more viscous) which can be clearly recognized on 11/12/2018 by the green arrow and was proportionally variable and depended upon PPH capacity on the rest days.

Fig-E1 (introduce overshoot, an new factor, both during online & startup)

Since production was scheduled in advance & operation has no active initiation to support this study. In the meantime, **single sample is not convincing unless its repeatability is confirmed**. To search one or two similar cases is not difficult; however, to include all same events is not easy. After diligent investigation, day by day, Fig-E1 (another nine days in macro) was selected for its meeting the criteria to cover all four factors, plus an extra value to introduce an new factor, **overshoot**, which not only confirm the repeatability of the study, but also extend the exploration, and its associated description is chronically listed in Table-3:

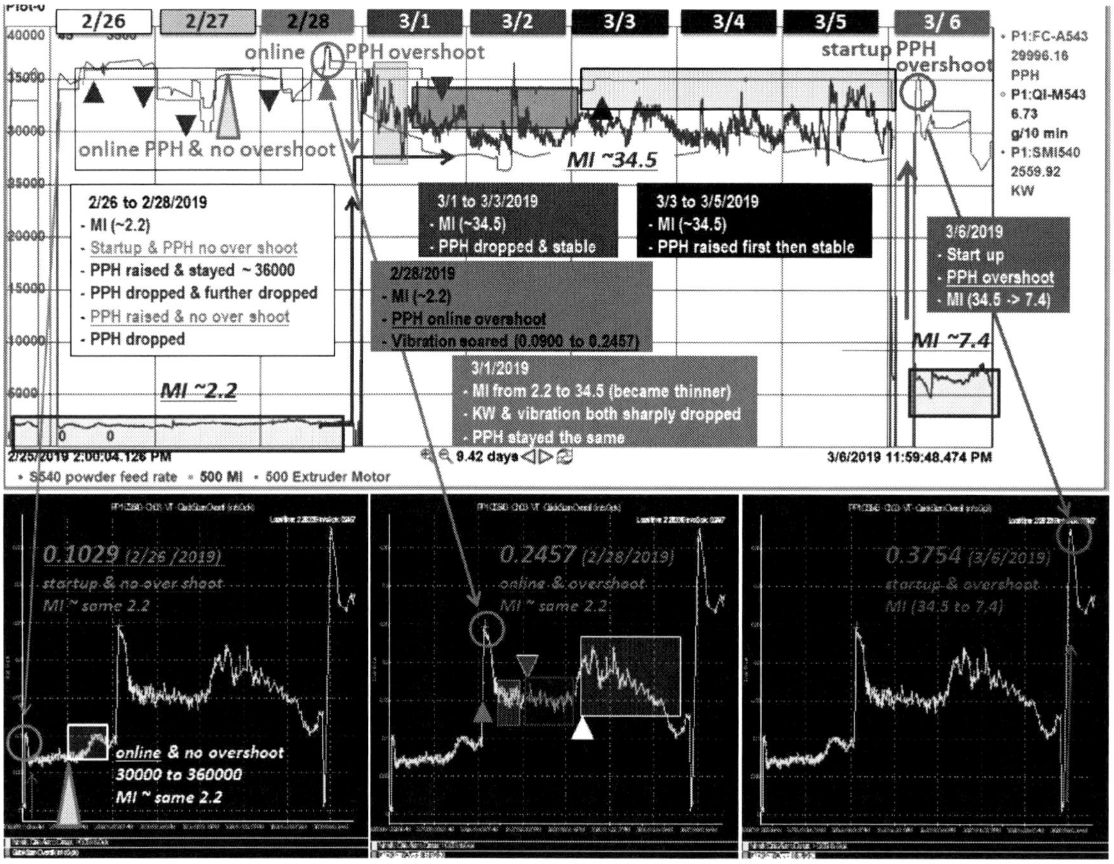

Fig-E1 (Overview macro 2/26/2019-3/6/2019)

Table-3 for Fig-E1 (2/26/2019-3/6/2019)

Date	Various DCS events & related vibration response
2/26/2019 Start up	Low MI 2.2 (or higher viscosity) startup **without overshoot**. - Blue circle & blue arrow
2/26- 2/28/2019	Same MI 2.2 (or higher viscosity) - White box on DCS indicated PPH raised & dropped, and Orange hazy long triangle (2/27/2019) indicated PPH online raised **without overshoot**. - Hazy white box on trend indicted vibration fluctuated up & down.
Only PPH transit online 2/28/2019 (**overshoot**)	Same MI 2.2 (or higher viscosity) - Orange circle on DCS indicated PPH first **online overshoot** (+3000) from 35000 to 38000, then drop (-2000) to 36000. - Orange circle on trend indicated vibration soared from 0.0900 to 0.2457.

Only MI transit 3/1//2019 (from 2.2 to 34.5)	After PPH **online overshoot** on 2/28/2019, MI transit from 2.2 to 34.5 (viscosity high to low), while PPH kept the same ~36000. - Hazy green vertical box on DCS indicated motor electrical KW immediately nosedived once viscosity became softer. - Hazy green vertical box on trend indicated vibration sharply fell first, and then followed by vibration minor fluctuation.
Only PPH transit 3/1 to 3/3/2019 (drop)	PPH dropped & MI kept the same (~34.5) - Hazy purple box & solid triangle on DCS indicated PPH dropped first, then kept stable & MI (~34.5) was fluctuating. - Hazy purple box & solid triangle on trend indicated vibration was fluctuating.
Only PPH transit 3/3 to 3/5/2019 (raise)	PPH raised & MI kept the same (~34.5) - Hazy black box & solid triangle on DCS indicated PPH jumped, then kept stable. - Hazy white box & solid triangle on trend indicated vibration climbed first then declined afterward.
Combined 3x factors - startup, over shoot & MI 3/6/2019	Extrusion processes tripped then re-start (**with overshoot**), plus a MI transit to 7.4 (which was between 2.2 and 34.5). - Red circle & vertical arrow on DCS indicated PPH jumped from zero while startup, with an **overshoot** to 35000 first, then dropped back. - Red circle on trend indicated vibration soared from 0.0061 to 0.3754.

In Fig-E1, six events are noticed and paired into three groups (startup, online PPH and MI) for two purposes & is summarized in Table-4:

- To examine if all phenomena in Fig-D1 are **repeatable (on left column)**;
- To add a **new factor, overshoot (on right column)** and confirm **all four plots (Fig-D1, D2, and E1& E2) complement each other**.

Table-4 (To confirm repeatability in Fig-D1 & introduce overshoot factor)

<table>
<tr><th>Without overshoot</th><th>With overshoot</th></tr>
<tr><td>Startup on 2/26/2019 (without overshoot) repeatedly indicated a peak at 0.1029.</td><td>Startup on 3/6/2019 (with overshoot) indicated a peak at 0.3754 (> 3.5x times higher than 2/26/2019).</td></tr>
<tr><td colspan="2">• Note#1: PPH on 2/26/2019 was 34000, whereas PPH on 3/6/2019 was overshot to 35000 first, dropped to 32000, then increased to 33000, both were within similar PPH range, except with & without overshoot.
• Note#2: MI on 3/6/2019 was 7.4, which was ~3x times softer than MI (2.2) on 2/26/2019.
• Since both PPH were around similar range, to run against softer resin with better fluidity, ideally, should be easier; however, vibration did not behave as our intuition which proved overshoot do create more violence.</td></tr>
<tr><td>Online PPH transit on 2/27/2019 (+6000 without overshoot, 30000 to 36000). With double PPH, however, did not indicate any obvious peak.</td><td>Online PPH transit on 2/28/2019 (+ 3000 with overshoot, 35000 to 38000 first, then dropped to 36000).</td></tr>
<tr><td colspan="2">• Note#3: Both PPH were targeted at 36000, however, less PPH +3000 with overshoot on 2/28/2019 indicated an outstanding peak at 0.246.
• Note#4: Both PPH events occurred while MI stayed stable (~2.2).</td></tr>
</table>

<table>
<tr><td colspan="2">• Note#5: 2/27/2019 online PPH (+6000 ~ 2x times +3000), was completed in much longer period & proved a positive influence (lower vibration) by a graceful operation manner.</td></tr>
<tr><td>Online MI transit (from 2.2 to 34.5, became thinner) on 3/1/2019 without overshoot while PPH stayed ~ 36000 & KW sharply dropped.</td><td>Startup MI transit (from 34.5 to 7.4, became thicker) on 3/6/2019 with overshoot) while PPH was 35000 & KW increased during startup.</td></tr>
</table>

Fig-E2 is a ten hours micro for the nine day's macro (Fig-E1) to confirm four events, both online & startup (**with and without overshoot**) as below:

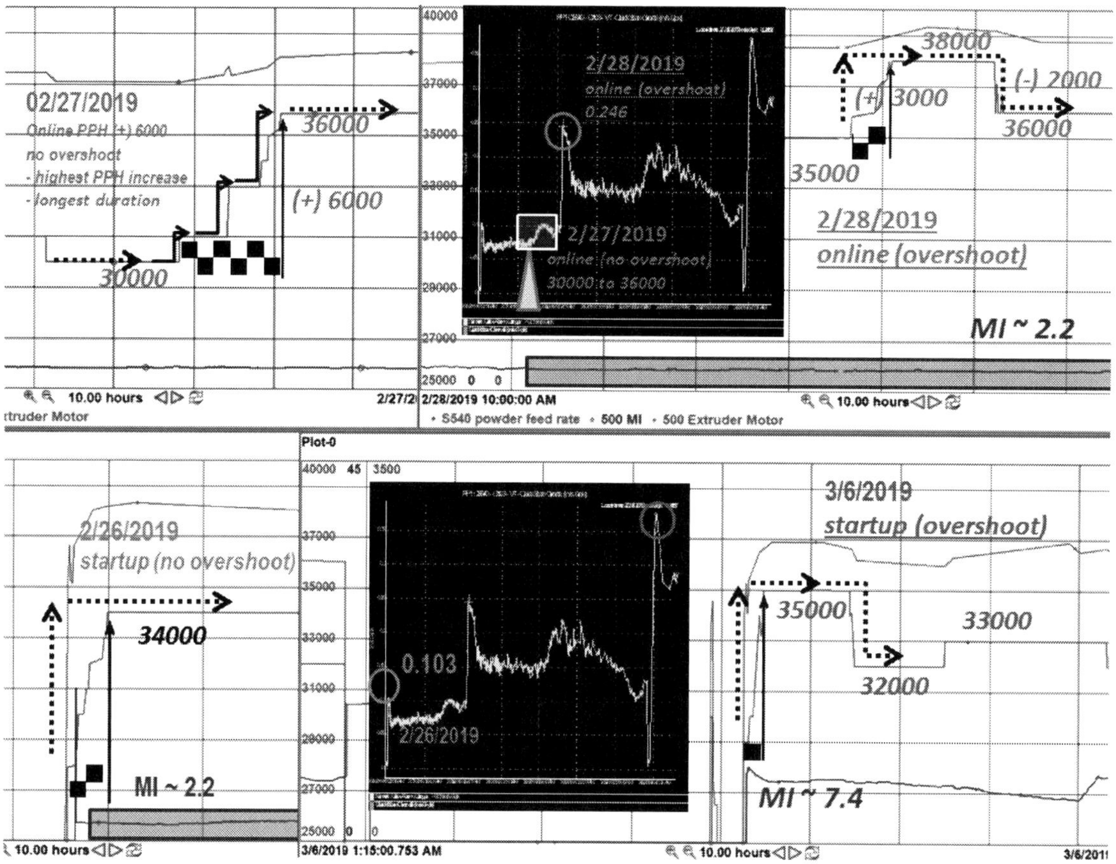

Fig-E2 (DCS for 4x events after zoom in)

(1) **2/27/2019 (left/upper corner)**
Online PPH transit (+6000) **without overshoot**, from 30000 to 36000 in **much longer duration (~6x blocks)** & proved a positive influence (lower vibration response).

(2) **2/28/2019 (right/upper corner)**
Online PPH transit (+ 3000) **with overshoot**, from 35000 to 38000 first, then dropped to 36000.

(3) **2/26/2019 (left/lower corner)**
Startup **without overshoot** repeatedly indicated a minor peak at 0.1029.

(4) **3/6/2019 (right/lower corner)**
Startup **with overshoot** with **the shortest duration** (~ 1x block) indicated a peak at 0.3754 (> 3.5x times higher than 2/26/2019).

CONCLUSION:

(1) Magnified video definitely saves time; lower the cost for labor, scaffolds and proprietary software, especially when the job scope is bigger and beyond normal route check. Most importantly, the dynamic image can be shared without explanation; it is surely a great tool for reliability application in the future.

(2) Magnified video can help looking for defects beyond human vision limitation, and its value is proved by numerous examples posted on internet. When the photographers proudly present those amazing dynamic images, most stories stop after all the "witness" give off their surprising admiration "Wow!", whereas it is just the beginning. There still can be more topics worthy of exploration and this article intends to start further discussion to find its new application and case#1 is an example for failure mechanism while the major actor been damaged was absent in the video.

(3) With good sense of reliability, like a machine will surely be affected by the manner how it is operated by different practice, or how the vibration is affect by different process factors, however, our intuition is proved by real data in case #2.

AKNOWLEDGEMENTS:

(1) For case#1, the author would like to appreciate the great job done by Seth Rozner and Greg Adams who are my colleagues at Formosa Predicative Maintenance Rotating department.

(2) For case#2, the author would like to appreciate the effort by Nick Ma who is the operation engineer in PP-1 unit to extract all the DCS plots, without his contribution, the study would not be informative. In the meantime, I want to thanks Andy Lerche and Chad Pasho from MSI (Mechanical Solutions, Inc.) who took video to discover the hair line soft foot and solved my curiosity - why the FRF (Frequency Response Function) analysis did not match my searching around the motor feet? Without their help, I might have no confidence to relate the DCS data for further exploration so far.

TRAINING CONFERENCE & EXPO

NIAGARA FALLS, NEW YORK JULY 26 -28, 2023

VALUE FROM YOUR REMOTE CONDITION MONITORING PROGRAM ON ROTATING MACHINES: PARAMETERS TO MEASURE

Akshat Sharma *is the Global Monitoring & Controls Manager at ITT Corporation and is based out of Houston, TX. He holds a B.S. in Electrical Engineering from the Georgia Institute of Technology. He has held positions in operations, engineering, R&D, field service and sales support throughout his career. Throughout his professional life he has spent time working on instrumentation, controls and wireless communications products with over a decade of experience in the pump & seal industry.*

Jeremy Sheldon *is Director Asset Management and Analytics at Poseidon Systems, LLC (based in Victor NY). He has over twenty years of experience in the Condition Based Maintenance and Prognostic Health Management domain with a strong background in signal processing, data analysis, health monitoring system implementation, testing, vibration diagnostics. In his current role, Jeremy manages Poseidon's condition monitoring team while simultaneously contributing to technical and programmatic aspects of several programs.*

ABSTRACT:
Remote monitoring enables users not collocated with the machinery to detect or infer the health of machinery using various sensors and analysis methods. Temperature, current monitoring, vibration, oil analysis and sensing, loads monitoring, and many other sensing methods can be used to measure parameters correlated to machinery health and machinery failure modes. The authors cannot possibly cover every sensor technology available, rather will focus on the most critical or the sensors that provided the most benefit.

KEY WORDS: condition monitoring, diagnostics, machine health, oil analysis, oil condition, pumps, reliability, remote monitoring, sensors, vibration,

INTRODUCTION:
In recent years, remote monitoring and analysis have become increasingly important for ensuring the reliability and diagnostics of rotating equipment. Rotating equipment such as turbines, pumps, compressors, and generators are critical components in many industrial processes, and it is crucial to ensure their proper operation and avoid unexpected failures. This paper aims to provide a comprehensive overview of best practices for remote condition monitoring and analysis for rotating equipment reliability and diagnostics. The paper will cover topics such as sensor design features, measurement configurations, signal processing, data analysis, as well as benefits and limitations of remote monitoring.

Remote monitoring enables users not collocated with the machinery to detect or infer the health of machinery using various sensors and analysis methods. Temperature, current monitoring, vibration, oil analysis and sensing, loads monitoring, and many other sensing methods can be used to measure parameters correlated to machinery health and machinery failure modes. The authors cannot possibly cover every sensor technology available, rather will focus on the most critical or the sensors that provide the most benefit.

One of, if not, the main goals of condition monitoring systems is to detect faults as early as possible in their failure timeline to enable the lowest cost repair possible. Each sensing technology and its interpretation offer varying fault detection and isolation performance. As shown notionally in Figure 1, various sensing technologies offer various detection horizons. Those implementing a condition monitoring system (CMS) must evaluate the sensors, fault detection performance, cost, and other aspects before selecting a CMS. A short discussion of many of the key sensing methods relevant to rotating machinery is presented below.

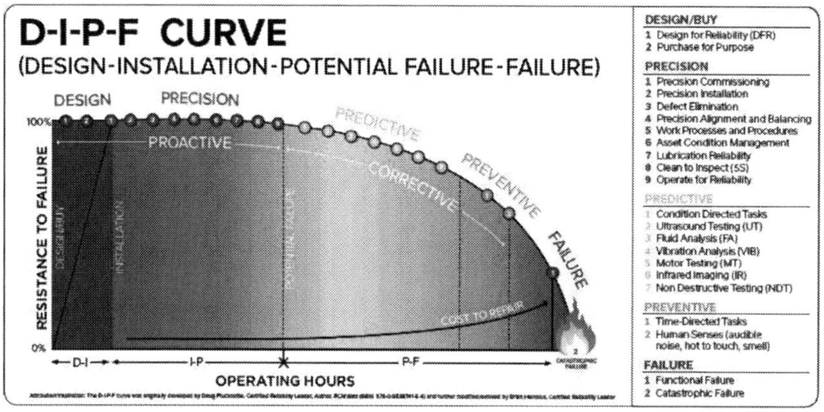

Figure 1 Progression of Fault and Detection Methodologies [0]

VIBRATION MEASUREMENT:

One of the most common parameters to measure on rotating equipment is vibration. Machines produce vibration that can be indicative of running condition, upcoming faults and component failure. Vibration is also a parameter that is well suited to measuring and displaying remotely as one of the first indicators of machine health.

Rotating equipment vibration is typically measured by the oscillatory motion on a stationary part, such as the bearing housing. The closer to the source of vibration, the more accurate the measurement will be.

Vibration can be measured in displacement, velocity or acceleration. Using the frequency of the vibration energy, one domain can be converted to another utilizing integration or differentiation. The type of vibration and sensor used often depends on the source frequencies of interest. Acceleration is more sensitive to higher frequencies while displacement is more sensitive to lower frequencies. Velocity, the most common unit used in industrial applications, provides good sensitivity in both high and low frequencies.

Figure 2 Wireless vibration sensor mounted on a pump bearing housing

VIBRATION SENSOR DESIGN FEATURES:

A critical aspect of remote vibration monitoring is the design of the sensors used to measure vibration. There are several factors to consider when selecting a sensor, including sensitivity, frequency range, environmental conditions, and cost. The sensitivity of the sensor is crucial because it affects the accuracy of the measurement, and the frequency range determines the range of vibrations that can be detected. Environmental conditions such as temperature and humidity can also impact sensor performance, and the cost of the sensor will influence the number of sensors that can be deployed.

One of the most commonly used sensors for remote vibration monitoring is the accelerometer. Accelerometers are relatively inexpensive and can measure a wide range of frequencies, with the acceleration signal typically used to convert into velocity as well. Accelerometers are also relatively easy to install, and can be used in harsh environments. Another option is the proximity probe for displacement, which measures the distance between the probe and the surface of the rotating equipment. Proximity probes are more expensive than accelerometers, but they offer higher accuracy and a wider frequency range, which is especially beneficial for lower speed machines and lower frequency measurements.

In addition to the basic type of sensor, there are some additional features and characteristics that are important when selecting a remote monitoring system:

- Accuracy: One of the most important, and often overlooked, factors in choosing a sensor is the accuracy of the data it can measure. Accuracy is often a combination of the sensing element as well as the packing of the whole sensor, combining designs in the electronic and mechanical fields. In order for a remote monitoring system to deliver value, its first job is to deliver accurate data that can be trusted for monitoring the true condition of rotating machinery.
- On-board Processing Capabilities: Some remote monitoring sensors have on-board processing capabilities that can be used to pre-process the data generated by the sensors before it is transmitted to the central monitoring system. On-board processing can include data reduction, data compression, data filtering and data analysis, among other things. On-board processing can reduce the data transmission requirements and minimize the risk of data loss, as well as provide a more detailed representation of the data generated by the sensors. Greater processing power on-board the sensor allows it to perform more analysis locally and provide more advanced evaluations as well as better adjust sensor behavior based on machine condition (eg. Take data more frequently or send out alarm notifications based on true machine condition rather than simple thresholds). This is important when scaling a remote monitoring system to hundreds or thousands of machines to avoid data overload or congestion with less valuable measurements.
- Data Storage Capabilities: having the ability to store data on-board sensors is important so that the data can be transmitted to the central monitoring system at a later time, when a wireless connection is available. Data storage capabilities are also useful for redundancy where the wireless connection may be intermittent or unreliable (as is often the case in industrial facilities). Sensors with larger memory on-board can handle larger amounts of data.
- Power Requirements: The power requirements and onboard power of remote monitoring sensors are important, as they determine the power consumption of the sensors and the length of time that the sensors can operate on a single battery charge. Selecting sensors that are power-efficient with long battery lives, maintain battery power over a wide range of environmental conditions and allow for easy battery changes or re-charging will minimize the risk of data loss and maintenance required for the monitoring system when the battery is depleted.

By considering the above factors, it is possible to choose sensors that are best suited for the intended application, in terms of vibration measurement as well as performance, reliability and utility of the remote monitoring system.

VIBRATION MEASUREMENT CONFIGURATIONS:

Once the appropriate sensors have been selected, the next step is to determine the measurement configuration. The measurement configuration determines the placement of the sensors and the type of data that will be collected. The most common measurement configuration is the triaxial configuration, in which three sensors are placed at 90-degree angles to each other to capture the full range of vibrations.

Another common measurement configuration is the single-axis configuration, in which a single sensor is used to measure vibrations in a specific direction. This configuration is less expensive and requires fewer sensors, but it provides less information about the vibration patterns.

Sensors with advanced capabilities will allow the capture of triaxial vibration accurately from a single unit, simplifying installation and lowering cost.

The measurement configuration will also depend on the type of rotating equipment being monitored. To better determine the condition of the full machine train, sensor placements may be required on multiple machines (eg. Pump and motor) as well as multiple locations (e.g., Inboard and outboard bearings) to better capture machine conditions and understand issues related to mating of the machines (e.g. Misalignment).

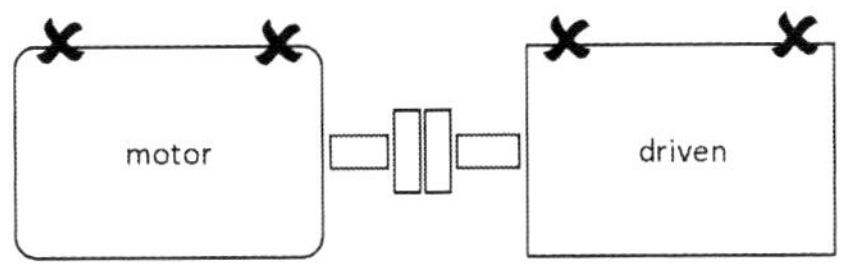

Figure 3 Sensor installation locations on a motor close coupled to a rotating machine

VIBRATION SIGNAL PROCESSING & DATA ANALYSIS:

After determining the measurement configuration, the next step is to process the raw data collected by the sensors. Signal processing is a crucial step in remote vibration monitoring and analysis because it helps to extract meaningful information from the raw data. The signal processing techniques used will depend on the type of vibration data being collected and the goals of the analysis.

One common signal processing technique is the Fourier transform, which converts the raw vibration data into its frequency components. The Fourier transform is useful for identifying the dominant frequencies present in the vibration data and for determining the root cause of a vibration. Another signal processing technique is the wavelet transform, which can be used to detect and analyze transient events in the vibration data.

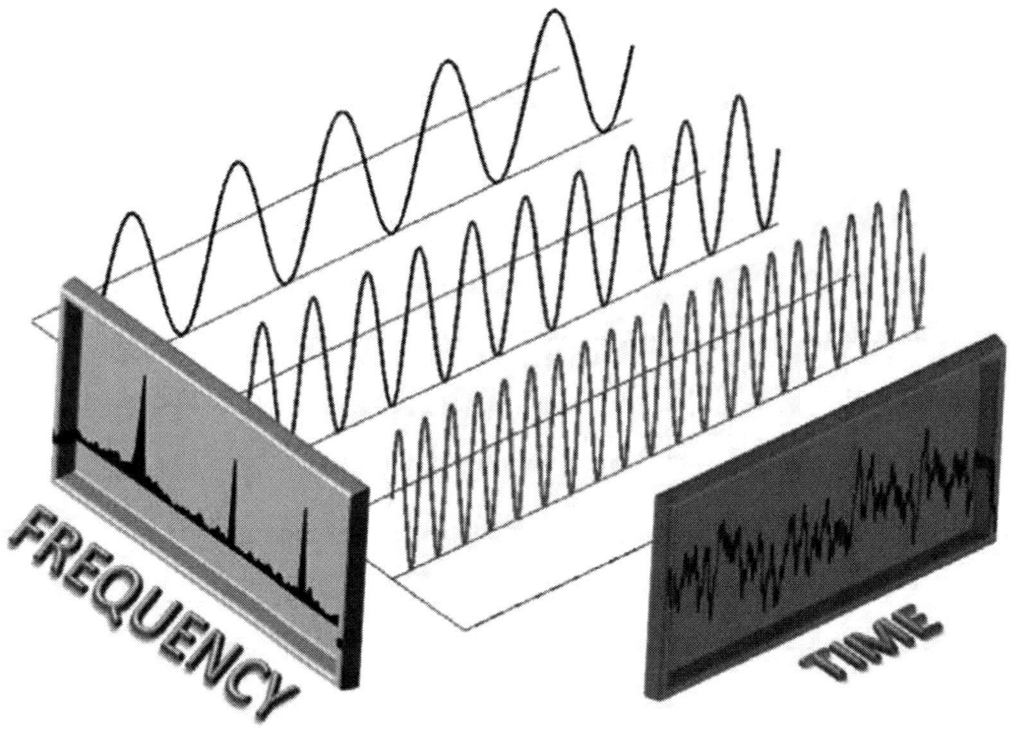

Figure 4 Example vibration signal with 3 distinct frequency components

Increasing the amount of signal processing done at the sensor, or at the edge, will limit the amount of information needing to be sent to the cloud or other central processing system. This is very beneficial as scaling remote monitoring systems plant-wide can quickly overwhelm communication systems, especially if they are wireless. Additionally, the more advanced signal processing and data analysis that is done at the sensor level, the more actions the sensor can take to respond to machine conditions (e.g. By taking additional data to be used by personnel for analysis or by alerting someone to a true machine condition based on data analysis).

After processing the vibration data, the next step is to analyze it to determine the health of the rotating equipment. There are several approaches to data analysis, including time-domain analysis, frequency-domain analysis, and modal analysis.

Time-domain analysis involves analyzing the raw vibration data over time to identify trends and changes in the vibration patterns. Frequency-domain analysis involves analyzing the frequency components of the vibration data to determine the dominant frequencies present and to identify the root cause of a vibration. Modal analysis involves analyzing the vibration patterns to identify the natural frequencies and modes of vibration of the rotating equipment.

All of these analyses make it easier for vibration analysts to identify what is happening with a machine and any fault conditions that may exist. Some automated data analysis can be beneficial as vibration analysts' time is usually in short supply. Care should be exercised when selecting and implementing automated data analysis to ensure it is reliable and fast. A system that generates large amounts of fault analysis can quickly result in time spent chasing non-existing issues and cause personnel to lose trust in the remote monitoring system overall. Additionally, if the data analysis is not done accurately in a short period, it can slow down the process and may not provide much benefit over existing manual analysis. Automated analysis that requires a learning period on the machines may result in long time periods before it is useful and have less than desired accuracy.

The best use of automated data analysis is in conjunction with accurate sensor data to augment an existing vibration analyst workforce by making them more efficient and allowing focus on solving more advanced issues.

MECHANICAL WEAR DEBRIS:

Online wear debris sensors provide the unique capability of constantly monitoring the levels of ferrous and nonferrous wear debris particles within a mechanical system. Accelerating wear generation rates in particle counts can be easily recognized and acted upon before a critical gearbox failure occurs. The utility of online wear debris monitoring as a means of detecting mechanical component failures and continually assessing their severity makes them an attractive monitoring option in many applications, especially oil-lubricated gearboxes.

Various online debris monitoring technologies are available. Inductively Coupled Plasma Atomic Emission Spectroscopy (ICP-AES) or analytical ferrography methods are commonly used in laboratory settings [0]. Optical based particle counters are common in labs but can also be found in certain field applications. Magnetic-based chip collectors and detectors are typically used in gearbox monitoring applications [0]. Another technology used for debris sensing is based on an inductive sensor created by one or more coil windings around a non-conductive fluid passage (sensing bore). The coil is excited with a high-frequency oscillating voltage and the resulting magnetic field couples with any metallic particle present within the field. The particle's passage results in a measurable impedance change in the sensing coils providing information about the particle's size, type (ferrous or nonferrous), and speed. Unlike many other sensing technologies, the output of wear debris is a direct measurement of mechanical damage. In addition, the results are easy to interpret with little expertise needed to analyze and understand.

Mechanical faults can propagate quickly, developing from a state of low risk of major component failure to high risk in a matter of days, if not hours. Online metallic debris monitoring enables these changes in severity to be observed and tracked in real-time, allowing adjustments in operation to prevent catastrophic failure and extend operating life until a repair or replacement can occur. Furthermore, wear debris data can be correlated with machine usage, temperature, vibration, and other sensor data for further insight into the health of the machine. For example, a large wear debris increase could correspond to a severe load event. It is often useful to summarize oil debris sensor outputs in two measurements: rate of material loss (mass per unit time/flow, sometimes called concentration) and fault severity (cumulative mass). An example from a failing gearbox is shown Figure 5. The rate of material loss is sporadic as wear material generation rates change with gearbox loads and speeds, while the cumulative mass steadily increases over time. Larger peaks in the concentration level are indicative of more severe wear events and increasing slope of the cumulative mass also indicates increase fault severity.

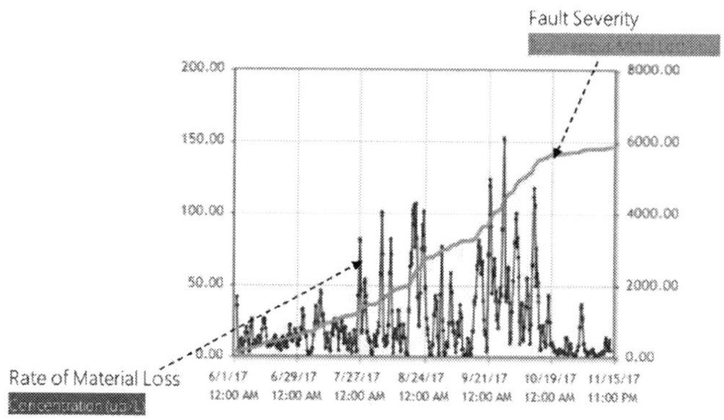

Figure 5 Example Oil Debris Trend

In another example, an inductive-based oil debris monitor was installed on several large industrial gearboxes at a customer site. Although several gearboxes were monitored, only one is shown herein for brevity. The ferrous wear debris concentration for a 24-hour interval data from several months is shown in Figure 6. The concentration levels steadily increase from February through June at which point they reach warning levels. The alarm and warning levels were statistically derived from a large data set that included healthy and known faulted gearboxes and is recommended best practice when sufficient machine populations are available.

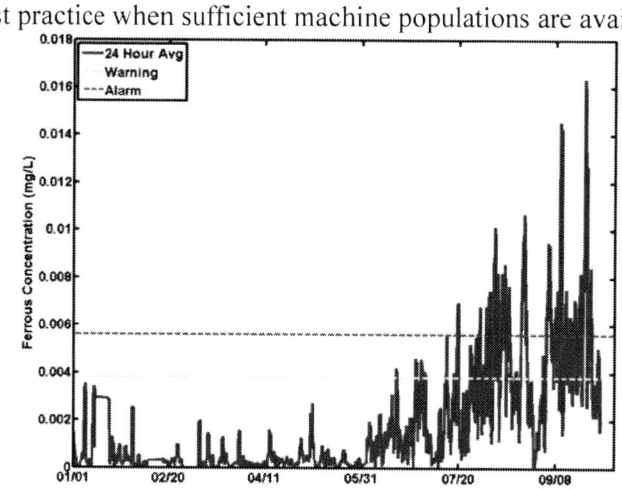

Figure 6 Industrial Gearbox Wear Concentration Trend

When selecting a debris sensor, the ability to count and size the sensed debris in a sequence of size ranges is very beneficial. Understanding the sizes of the particles, both in counts and size distribution, provides valuable information that can be used to characterize fault type, fault severity, and even fault location in many instances. Figure 7 shows the number of particles observed in each month in four key size ranges. During fault initiation (February through May), the wear generation is almost exclusively sub-200μm particles. As fault propagation increases, (June through September), the distribution shifts to include larger particles, with an increase in the smaller size debris as well. These distributions and their evolution over time are important fault signatures that can be used to diagnose fault severity and location, for example, a gear fault signature looks different than a bearing fault signature. Based on the trend and the distribution the operator was notified of a progressing fault.

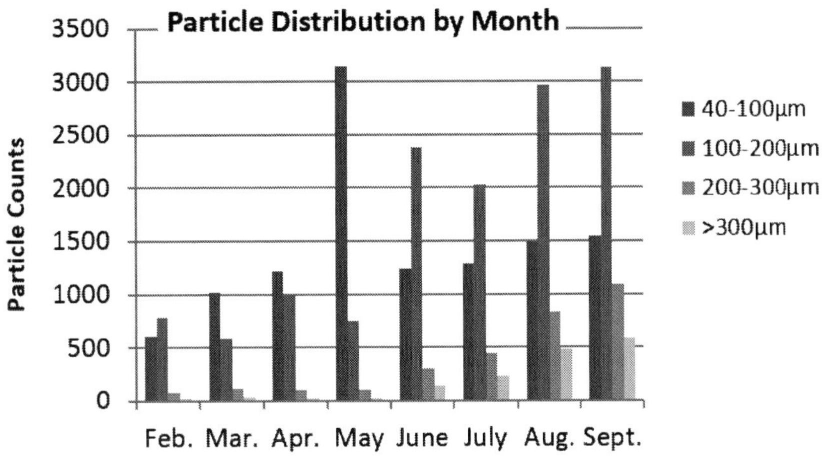

Figure 7 Industrial Gearbox Wear Size Distribution

The resulting borescope inspection revealed a fractured tooth on a helical pinion, as shown in Figure 8. The maintenance team was able to perform an in-situ repair to prolong the life of the gear. Fault identification occurred early enough to prevent secondary damage to the mating bull gear, significantly reducing the overall cost of repair. Post repair debris data was used to confirm the repair and monitor the gearbox for the remainder of its life ensuring safe operation.

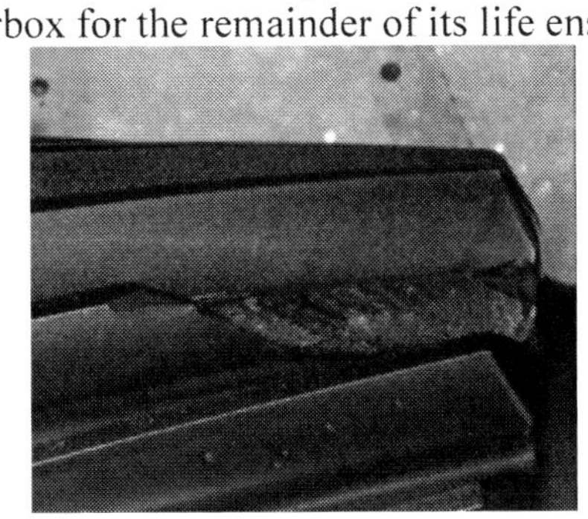

Figure 8 Industrial Gearbox Damage

OIL CONDITION:

Oil sampling analysis is the backbone of most oil reliability programs. From early analysis technology used by oil labs to the more modern miniaturized benchtop equipment, manual oil sampling is critical to most rotating machine preventative maintenance. The latest advancements are new in-line, real-time oil quality monitoring sensors that deliver many of the benefits of manual oil sampling in an online monitoring capability.

The value from manual oil samples is centered around oil health itself. Properties such as oxidation, TBN, TAN, additive packages, viscosity, water, fuel, soot, etc. are all used to assess the remaining-useful-life (RUL) and identify preventative, as well as corrective actions. A few notable challenges with oil sampling are the time delay between sample and results, access to remote and mobile assets, infrequency of sampling compared to reliability events, and high risk of human error. Because of these challenges, many operators are moving towards online fluid monitoring. Utilizing these advanced sensing technologies is leading to higher uptime through lubrication health tracking and event detection, enabling informed, timely preventative maintenance. In turn, online oil condition monitoring is lowering maintenance costs as operators can reduce failures and take corrective action earlier.

Typically, online sensors are electrochemical impedance, dielectric, conductivity, or permittivity measurement devices detecting degradation of the oil. Some methodologies can detect degradation of not only overall quality, but estimate percent soot, varnish, total base number, relative humidity, additive depletion, etc. Regardless of sensing method, effective oil condition sensors must detect most, if not all, key oil events and have an output that trends with the life of the oil while the asset is in operation.

While these sensors cannot duplicate lab analysis results, they can provide the necessary insight to make preventative maintenance decisions before damage occurs. Online oil quality sensors are key enablers to shift oil sampling from periodic to condition-based sampling and remote monitoring. To completely monitor the condition of an oil, multiple parameters are required. At a minimum a combination of sensors to monitor oil level, track overall oil health, and monitor water activity are recommended. With these types of sensors operators are now able to know the current health of their oil, trend degradation, project fluid RUL, and significantly reduce overall oil consumption as part of a truly condition-based fluid change program.

Laboratory oil analysis consists of many tests specifically selected to provide adequate insight across the range of lubricant performance requirements so that a lubricant analyst can determine whether the lubricant will be able to continue performing as intended. Exactly reproducing traditional oil analysis parameters (TAN, TBN, viscosity, etc.) is not as important for an online device as is the ability to provide the same lubricant condition assessment that a lubricant expert, using those laboratory results, would make. An oil condition sensor should be able to determine the difference between a healthy oil and a degraded oil. Additionally, the real-time analysis methods must be sensitive to a sufficient range of degradation modes for any lubricant. Most online oil condition sensors are not designed to measure, infer, or output any laboratory oil analysis parameter directly, rather are designed to provide users with real-time insight into lubricant health and to provide alerts when abnormal conditions are detected.

Most oil sensors are best used to determine anomalous conditions and when a sample of the lubricant should be taken and sent to a traditional oil-lab for a more traditional oil analysis before any further action is taken or the output can be used to directly determine what action should be taken when significant changes are detected. To illustrate, a series of laboratory sub-scale testing was performed to determine the performance of a commercially available sensor at varying oil contamination levels. Two separate testing phases were performed to assess the sensor response to coolant contamination and fuel contamination. sensors at varying oil contamination levels. Coolant and fuel were used as they are common sources of contamination, however the sensor performance is expected to be similar for other fluid contaminations, for instance with incorrect fluid addition. In the first test, coolant was added incrementally. One of the measure parameters trended very well with both coolant and fuel were added, as shown in Figure 9. In addition, the parameter also responded proportionally when increasing amounts of contaminates were supplied.

Consistent and proportional response to oil condition of change of interest is a critical aspect to assess when selecting an oil condition sensor. The sensed parameters must behave in a proven way to allow confidence in the resulting decisions. In addition, one must ensure the sensor is sensitive to the failure modes of the target machine.

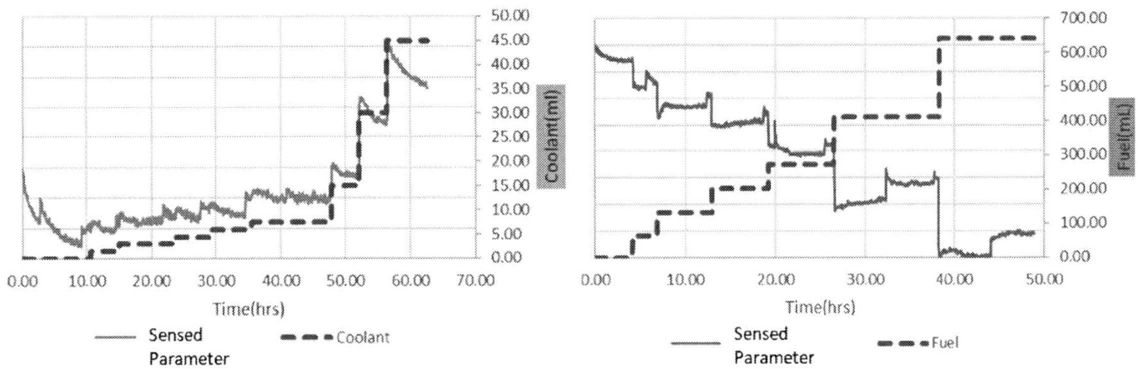

Figure 9 Example Oil Condition Sensor Data

RELATIVE HUMIDITY OR WATER CONTENT:

Like water is detected in the atmosphere by monitoring relative humidity, relative humidity sensors in oil provide a measure of the dissolved water present. While in the dissolved state water likely poses relatively low risk to the system; however, as the level approaches the oil's saturation point the risk of free water and emulsion formation greatly increases. When interpreting the data and setting limits one must consider the risk the free water poses to the specific system.

In most laboratory reports water content (both free and dissolved) is expressed in parts per million (PPM). The acceptable PPM levels vary greatly by oil type because the saturation points of oils are highly variable. Oil saturation points vary due to the oil chemistry, additives, age, depletion, and other influences. Saturation points also vary with oil temperatures, meaning that for fixed PPM limit the actual saturation level of the oil will change with temperature. A more relevant to machine health way to express water content in oil is as a percentage of saturation, or relative humidity. Expressing water content in relative humidity accounts for the changing saturation point and at any given time directly indicates the risk of free water without needing to know the often difficult to find oil specific saturation curve.

Oils can typically absorb more water with increasing temperature. So, for a given fixed PPM limit at lower temperatures the oil could be close to or above saturation limit while at higher temperatures the oil could be relatively dry. A notional example is shown in Figure 10. In this example an assumed 1000 PPM of water, as shown as grey line, is present in a hypothetical oil. For that fixed amount of water the lubricated system is exposed to a variable risk of free water as the oil temperature varies (x-axis). At high oil temperatures the system is relatively dry with low RH (orange line) resulting in no free water to cause damage. However, as temperatures decrease, the system is exposed to free water and increased risk of damage, highlighted by the red box. When implementing a remote monitoring solution specific to water content in oil, one must consider the entire operating range of the system to ensure the measured parameters provides accurate insight to the damaging conditions such as free water.

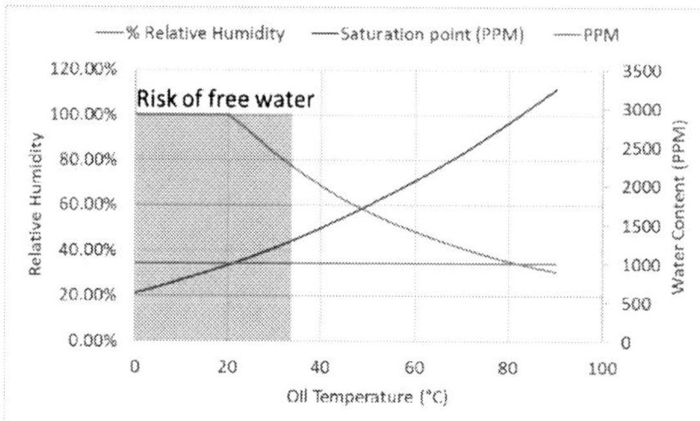

Figure 10 Notional RH vs PPM Comparison

FLUID LEVEL:

When implementing a condition monitoring solution, fluid level is often overlooked. Insufficient lubrication is obviously detrimental to the machine, but it can also more rapidly deplete whatever amount of oil is present. Various studies report inadequate, or no lubrication is responsible for greater than 30% of bearing failures [0]. Examples of low oil or oil out events causing massive gearbox damage are all too common. Many of these would have been avoidable had the actual oil level been actively monitored.

Therefore, monitoring the level of the fluid is a critical aspect to overall machine health. An integrated level sensing mechanism is built into the developed sensor. Online real time monitoring of oil level offers many benefits over traditional manual/visual readings. Monitoring continuously allows trending of oil level with machine operation to confirm adequate lubrication at all operating conditions. Alarms and notifications can be raised at the instant oil levels drop avoiding the time delay associated with manual readings that may occur days apart. In addition, online monitoring of level removes the need for manual checks that can be expensive and dangerous depending on the location of machinery.

TEMPERATURE:

Although seemingly trivial, machine temperature (bearing, housing, surface, or oil temperature) is a critical part of condition and machine monitoring. Measurement of machine internal or surface temperatures can provide an overall indicator of machine health, even if the indication may be late in the fault progression. Additionally, temperature measurements can often provide insights into secondary system performance, such as cooling systems. Prolonged periods of high oil temperatures can indicate poor lubrication condition, moderate to severe mechanical damage, or other abnormal operating conditions that require remediation. Oil condition can more rapidly degrade under abnormal temperatures. In addition, some of the parameters measured by sensors are temperature dependent and a real time oil temperature reading is necessary to ensure analysis conclusions are accurate. Therefore, temperature measurements are a valuable addition to any remote monitoring system.

SENSOR COMMUNICATIONS:

Wireless sensing is an attractive option for many sensors and remote monitoring systems. Therefore, the wireless part of remote monitoring sensors plays a critical role in the overall performance and success of a remote monitoring system. When selecting wireless sensors, it is important to consider the following features:

- Wireless Range: The wireless range of the sensors determines how far the sensors can transmit data from the equipment being monitored to the central monitoring system. It is important to choose sensors with a wireless range that is sufficient for the intended application, with consideration of the size of the facility, the type of equipment being monitored, and any potential obstacles in the signal path.

- Wireless Frequency: It is important to choose a wireless frequency that is appropriate for the intended application, taking into account any potential interference from other wireless systems in the facility. Depending on any existing wireless infrastructure in the plant and geographical location, this may need to be adjusted to account for interference and government radio licensing requirements.
- Compatible Communications Protocols: It is important to choose sensors that support compatible communications protocols with the central monitoring system. This could mean utilizing a wireless communication protocol already in the plant or using a remote monitoring system that will output data over a protocol that can be easily incorporated into the data storage and viewing system. Keeping this in mind will allow for easier integration of the sensors with existing systems.
- Wireless Data Transfer Rate: The wireless data transfer rate determines the speed at which data is transmitted from the sensors to the central monitoring system. Sensors with a fast data transfer rate can help minimize the risk of data loss and ensure that the monitoring system can keep up with the data being generated by the sensors. As a remote monitoring system scales, this becomes a crucial consideration when the number of sensors in the facility is high and data collection intervals are short. Even small or medium amounts of data collected often and by thousands of sensors can quickly add up to congest a wireless communication system, leading to data losses, repeat transmissions and reduced battery life.
- Battery Life: For wireless sensors, battery life is an important consideration, as it determines how long the sensors can operate without needing to be recharged or replaced. It is important to choose sensors with a long battery life to minimize the risk of data loss and to minimize the maintenance required for the monitoring system.
- Wireless Security: The wireless security of the sensors is critical to ensure that sensitive data is protected and that unauthorized access to the monitoring system is prevented. It is important to choose sensors with strong wireless security, such as encryption and secure authentication, to protect the data being transmitted. Although monitoring systems typically do not perform machine control functions, depending on the facility, they may have enough information on proprietary processes and plant capacities to warrant data protection.
- Robustness and Reliability: Wireless sensors need to be robust and reliable to ensure that they can withstand the harsh industrial operating environments in which they are used. If sensors are not designed for the desired operating environment, this will lead to high failure rates and increase time and cost for replacements.
- Data Integrity: The data generated by remote monitoring sensors must be transmitted accurately and without corruption to be useful for diagnostic and predictive maintenance purposes. Sensors built with data integrity in mind, and with features such as error correction and data redundancy to automatically account for inevitable data transmission challenges will allow for easier communications and reduce time spent designing and maintaining the wireless infrastructure.
- Network Scalability: The remote monitoring system may need to be scaled up or down over time, as the number of sensors and the volume of data generated by the sensors changes and plant capacity changes. Planning ahead by choosing sensors and a

monitoring system that are scalable makes this process easier so that the system can be easily adapted as required.

- Real-time Data Availability: Wireless remote monitoring systems always have a trade-off between data availability and battery life. Remote monitoring systems must provide near real-time data to support predictive maintenance and diagnosis within a reasonably short timeframe.

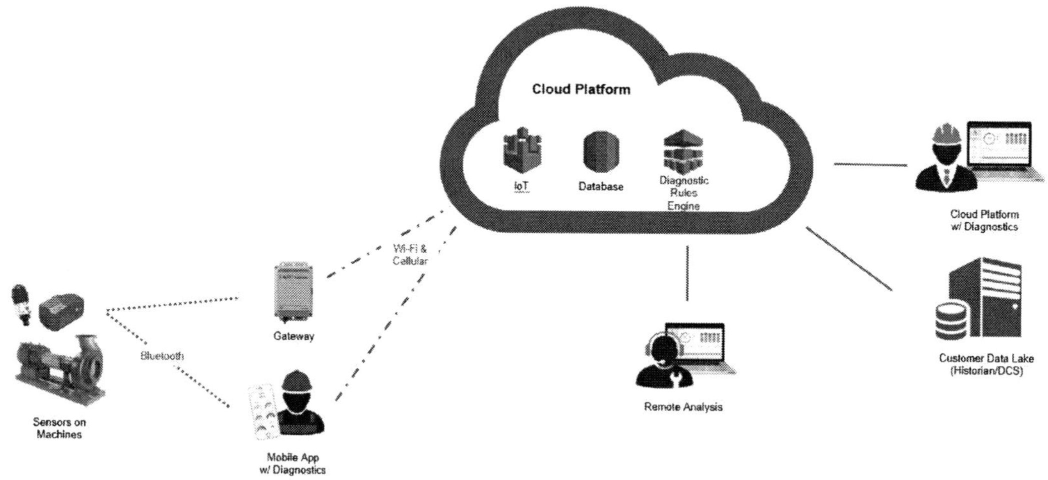

Figure 11 Architecture of a wireless remote monitoring system incorporating multiple communication features

By considering the above sensor communication features, it is possible to choose sensors that are best suited for the intended application, and that will provide the highest level of performance and reliability for the remote monitoring system.

REMOTE MONITORING:

Remote condition monitoring is an important tool for ensuring the reliability and diagnostics of rotating equipment. Monitoring equipment remotely makes it possible to alert personnel of potential issues and detect and diagnose problems before they lead to costly breakdowns or unplanned downtime. This helps minimize the risk of equipment failure, extend the lifespan of machines, and reduce the cost of required repair or maintenance work.

Reducing equipment failure and downtime will result in increased machinery uptime. This leads to improved productivity and reduced operating costs for facilities.

One of the main benefits of remote monitoring is the ability to continuously monitor the health of rotating equipment. This allows for early detection of potential problems when compared to historical performance, letting them be addressed before they escalate into more significant issues. For example, if a bearing is beginning to show signs of wear, remote vibration monitoring can detect the increase in vibration levels and alert maintenance personnel. This allows for prompt maintenance and repairs, which can prevent a catastrophic failure with unplanned downtime, emergency repairs and potential damage to other parts of the machine.

Remote monitoring also provides valuable data for predictive maintenance. Predictive maintenance is a maintenance strategy that uses data and analytics to predict when equipment is likely to fail. By monitoring data remotely, it is possible to collect data on the performance of rotating equipment over time, which can be used to predict when maintenance is required based on past performance as well as analytical capabilities of the monitoring system. This helps reduce maintenance costs and increase equipment uptime by scheduling maintenance only when needed.

Another benefit of remote monitoring that is often overlooked is safety. By being able to collect a majority of data remotely, personnel are put in hazardous areas less often just to collect data. Personnel may now only be required to visit machinery in areas of heightened risk for further data collection or to resolve issues. Additionally, by reducing the risk of unplanned machine failure, this also reduces the risk of injury to personnel onsite and the potential for unplanned environmental releases of hazardous fluids. These considerations are particularly important in industries where equipment failure can have serious consequences, such as in the chemical, oil and gas, and nuclear industries.

In addition, remote vibration monitoring is a non-intrusive method of monitoring rotating equipment. This means that it does not require the equipment to be taken offline or disassembled, which minimizes downtime and reduces the risk of damage to the equipment. The sensors used in remote vibration monitoring can be selected for taking data often while being easily installed with minimal required maintenance, which helps to reduce maintenance costs and improve equipment reliability.

Although remote condition monitoring can be a powerful tool for improving equipment reliability and diagnostics, it is not without limitations.

One of the first limitations of implementing a remote condition monitoring system can be the financial cost. This includes not only the sensors' cost, but also the cost of supporting data infrastructure (wired or wireless) and software licenses required to utilize those sensors. Additionally, the cost of implementing and maintaining a remote condition monitoring system can be a significant barrier for some organizations. It is essential to understand and plan for the true cost of remote condition monitoring versus the benefits it can bring. While a remote condition monitoring system can often easily pay for itself within several months, tracking the financial benefits and value brought by the system must be done to ensure the return on investment is healthy.

Technical limitations such as equipment interference and data quality must be considered. Interference from other equipment can affect the accuracy of the vibration measurements. This can include mechanical or electrical interference from other equipment, as well as environmental factors such as temperature and humidity.
The quality of the data collected through remote condition monitoring depends on the sensors' accuracy and reliability. If the sensors are not calibrated properly, are damaged or have poor accuracy in general, the data collected may be inaccurate or unreliable. This can very quickly lead to an inefficient monitoring system with little tangible benefits but large amounts of required effort, thereby eroding trust in the monitoring program as a whole.

The accuracy and reliability of remote condition monitoring depends on the proper installation, maintenance, and interpretation of the data collected. If a proper plan is not made and implemented for all system elements, including accounting for personnel training and required expertise, it can result in inaccurate data, incorrect diagnoses, abnormally high maintenance of the system or premature failure of equipment.

Another challenge can be presenting data from remote monitoring in the right place, at the right time and in an easy-to-use fashion. If the remote condition monitoring data is not properly integrated into existing systems, especially if the systems are outdated or not compatible with the remote condition monitoring software, this can lead to data loss or knowledge silos, preventing the value of the program from being fully utilized. Examples include integrations into plant data historians, data dashboards or computerized maintenance management systems.

With the increased amount of data collected through remote condition monitoring, it can be challenging to manage and analyze the data in a meaningful way, especially when compared to much smaller amounts of data that were previously collected manually. This can lead to data overload, making it difficult to identify meaningful trends and patterns in the data as well as actions to take from it. It is important to select a system with robust data analysis features built-in to enable spending time on higher-level analysis and actions. Additionally, planning for and designing internal processes to account for the increased amount and varied types of data available through remote condition monitoring will help ease the cultural transition within the organization.

CONCLUSION:

In conclusion, remote monitoring is a valuable tool for ensuring the reliability and diagnostics of rotating equipment. By detecting and diagnosing problems early, it is possible to minimize the risk of machine failure, extend the lifespan of the equipment, improve machine reliability and uptime, reduce maintenance costs and reduce plant safety risks. While considering remote condition monitoring, developing a plan for implementation is important. To maximize the benefits, it is important to carefully select the right sensors and software, properly install and maintain the system, and clearly understand the data being collected and analyzed. By investing in well-suited remote monitoring paired with a strong implementation and usage plan, plants and factories can benefit from a more efficient and effective maintenance program, resulting in improved productivity and reduced operating costs.

REFERENCES:

[1] Reliabilityweb.com. Copyright 2016-2018. NetexpressUSA Inc. d/b/a Reliabilityweb.com, accessed January 9, 2020.
[2] Booser, E. Richard, "Tribology Data Handbook: An Excellent Friction, Lubrication, and Wear Resource," CRC Press, Boca Raton, FL 1997, pg 900-907.
[3] "Engineered Sensors Tedeco Products Lubrication System Diagnostics Sensors Magnetic Chip Collectors," http://www.tedecoindustrial.com/mag.htm.
[4] SKF, "Bearing damage and failure analysis," SKF Group, June 2017.

VIATC TRAINING CONFERENCE & EXPO
NIAGARA FALLS, NEW YORK JULY 26 -28, 2023

INVESTIGATION OF A PREMATURE FAILURE OF A LARGE VERTICAL PUMP

James Chin Yut
Senior Technical Specialist - Bruce Power
Tiverton, Ontario, Canada

James Chin Yut, P Eng, is employed at Bruce Power as a Senior Technical Specialist. Current duties include the life cycle management of rotating equipment- large pumps and motors. His career experience spans various roles in maintenance, inspection and engineering in aviation, petrochemical and nuclear power generation. Mr. Chin Yut is a graduate of the University of the West Indies with a B Sc. – Mechanical Engineering (Hons) and holds a Cat III certificate with the Vibration Institute (VI) and Canadian Machinery Vibration Association (CMVA).

ABSTRACT:

This case study will focus on a failure of a Bruce Power Bruce B low pressure service water pump (LPSW), the investigation, findings and rebuild. The pump was overhauled in the summer of 2021 using Green Tweed ARHT (Abrasion Resistant High Temperature) bearing material. It was removed from service in Feb 2022 after less than a year in service. The Green Tweed ARHT bearing replacement was intended to be more durable and resistant to fouling, thereby improving the mean time between pump overhauls. This Case Study will highlight the findings after the pump was disassembled, inspected and steps were taken to prevent repeat occurrence.
.

KEY WORDS: Centrifugal pump, composite bearings, fouling, large vertical pump, low bearing flow, misalignment

SECTION 1: Pump data

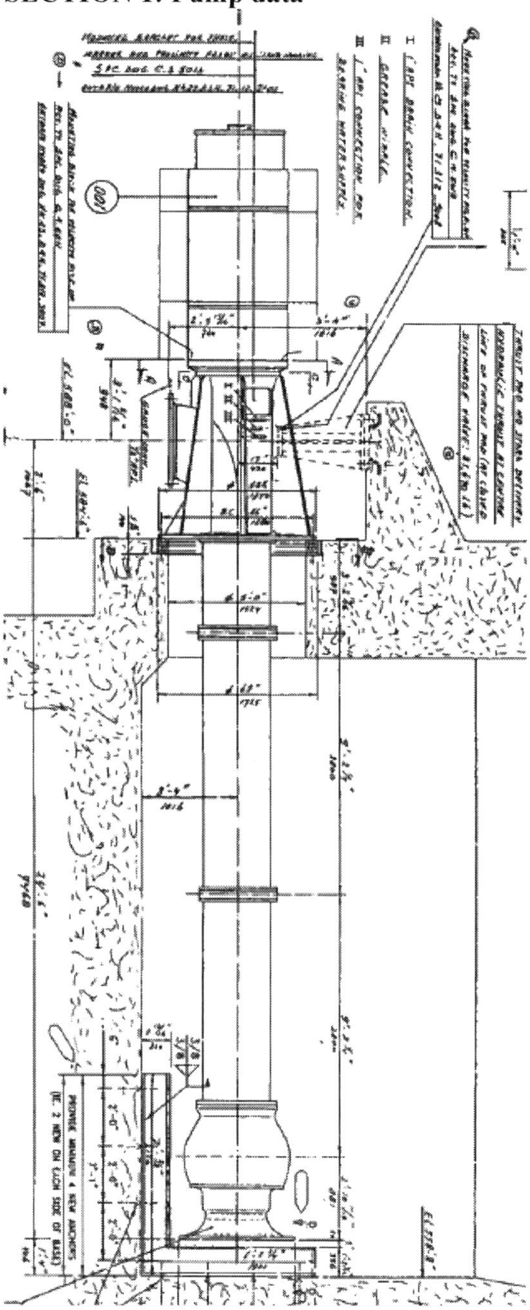

Figure 1: Pump and Motor Arrangement

The unit Low Pressure Service Water (LPSW) system provides strained lake water for cooling and other services in each of the four Bruce B units. The water in each unit is pumped through four vertical turbine type pumps (71310-P1 through P4) in the unit pump house (elevation 591 ft-0 in). A maximum of three pumps operate in parallel (3 X 33.33%) with the fourth pump on standby. Two pumps have their power supplied from a Class III bus and the other two from a Class IV bus. Class III pumps are provide water to safety related loads in the event of a loss of grid.

8-71310-P1- Low Pressure Service Water (LPSW) pump was removed from service due to low bearing injection flow, this resulted in reduction in redundancy of the number of U8 available LPSW pumps with the potential impact to production if other LPSW pumps /motors became unavailable in the summer months.

Pumps are Stork Pump Type VOA 85-75 – single stage, vertical pump, original head, 174 ft and flow 30,000 USGPM driven by Reliance Type P, 1500 HP motor, 900 rpm, 4160V. Original impeller – Aluminum Bronze (ASTM B 148 Alloy 995), Pump shaft- stainless steel (ASTM A 276 Type 420) and shaft coupling- stainless steel (ASTM 276 Type 420).
These pumps were originally commissioned in the 1980's and had a history of low bearing flows, the major reason for overhauls of the pump.

SECTION 2: Failure history

8-71310-P1 was overhauled in summer of 2021 using Green Tweed (GT) Abrasion Resistant High Temperature (ARHT) bearing material. This new configuration was recommended in consultation with the OEM after repeat and premature fouling events were observed using the previous bearing material (Tenmat Ferroform T-12). The GT ARHT bearing is a modern composite type bearing and was expected to be more durable and resistant to fouling, thereby improving the mean time between pump overhauls.

Stork Type VOA 85-75 is a large vertical pump designed with a protective pipe, enclosing the shaft and directs centrifuged process water to the bearings from the discharge piping through a cyclone separator to lubricate the bearings. The normal flow of water injected to the pump is

0.5litres/sec, some small percentage leaks upwards through the packing to keep it cool and lubricated; this leakage is piped back to the sump. The larger percentage of the water flows down firstly through the throttle bushing located at the stuffing box then through two intermediate sleeve bearings and finally through the sleeve bearing located at the impeller. The water rejoins the process water though holes in the impeller. The upper, intermediate and lower (pump) individual shafts are coupled together to form the pump shaft. The two couplings perform double duty as a shaft coupling and as journals for the two intermediate sleeve bearing.

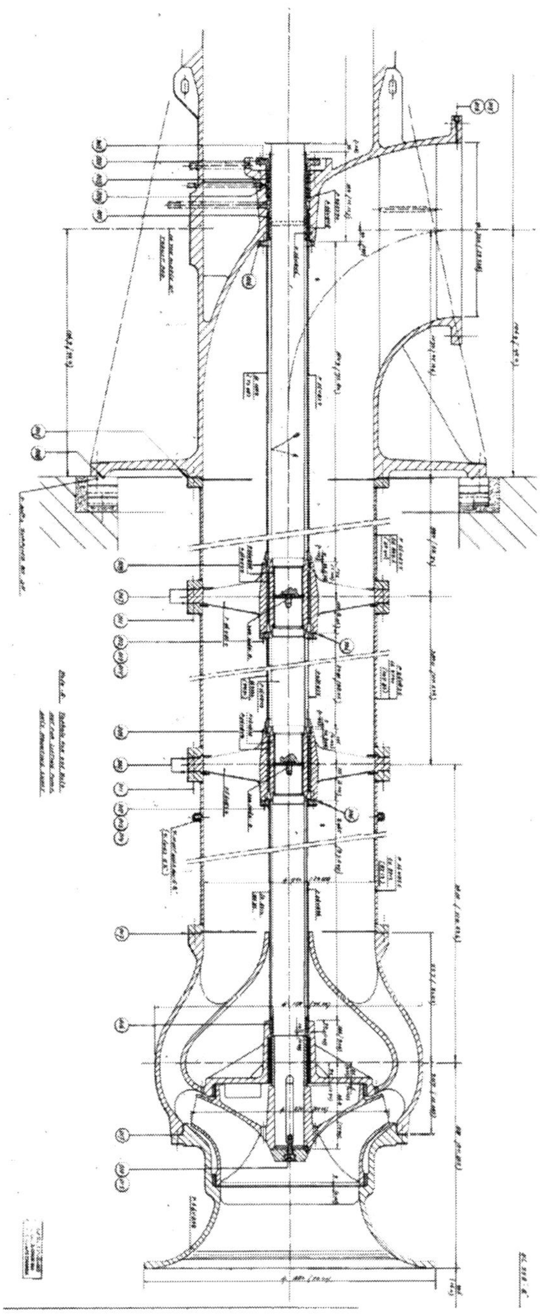

The injection flow is monitored when the pump is in operation and the alarm is set at 0.25litres/sec. During the failure sequence operations reported that the injection flow fell from 0.5litres/sec to 0.0litres/sec. Overhauling a LPSW pump requires the removal of the roof of the pump house and due to the length, the use of two mobile cranes to tail the vertical pump in a horizontal position and avoid the overhead high voltage power lines that exit the power station, so all attempts were made to return the pump to service. After troubleshooting and attempts to reestablish injection flow was exhausted, a decision was made to go intrusive and remove pump for disassembly and overhaul.

There was a concern that the premature failure of the pump was due to GT ARHT installation and the extent of condition implications of other pumps where the bearings were installed. A station condition report (CR) was created to document the premature failure of 8-71310-P1 and the recommended action to perform an equipment failure checklist (EFC) assigned **Figure 2: Pump cross section** to component engineering. The EFC is an internal investigation tool that identifies gaps in process, procedures, design, supply chain and practice under subsections- prevention, detection and correction. Forensic analysis and 5 Whys method was used to complete the EFC.

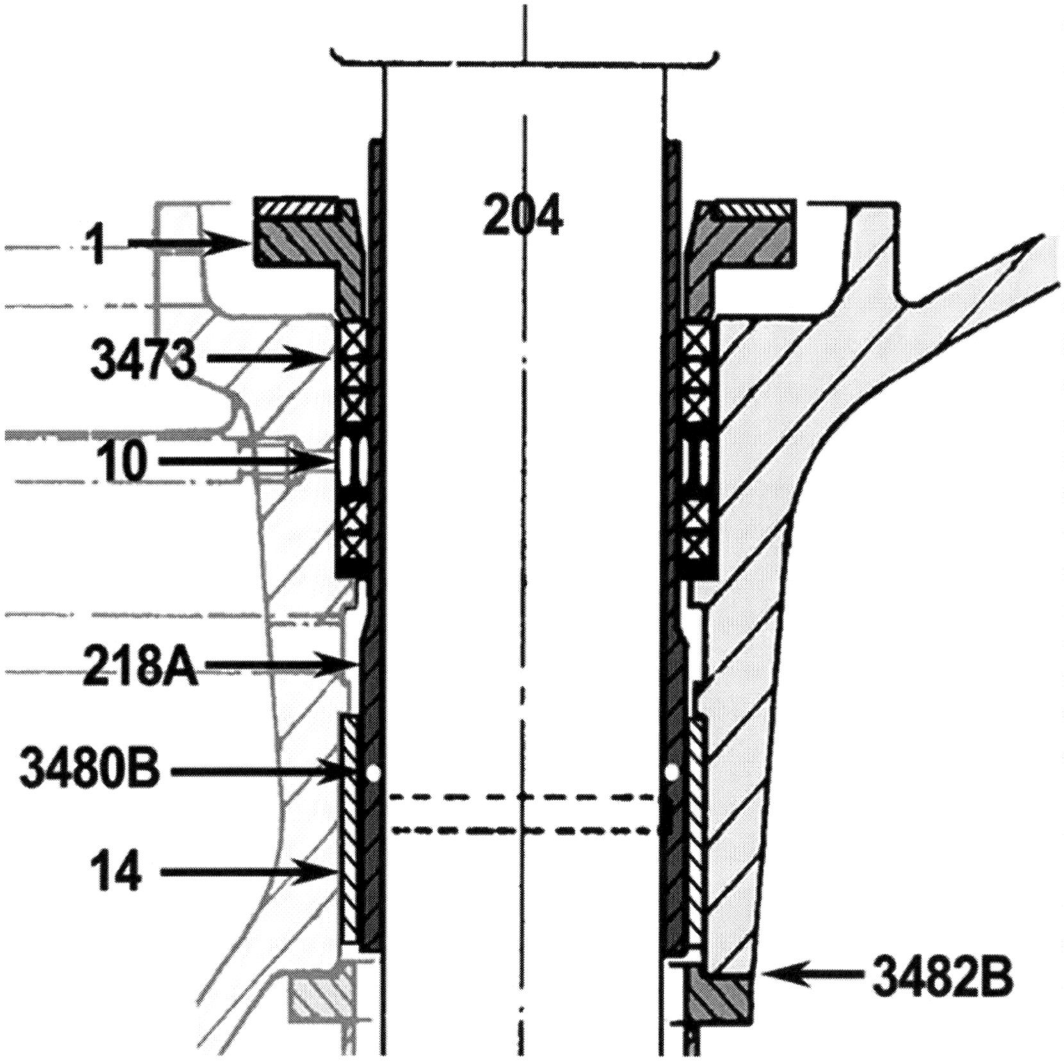

Figure 3: Stuffing box detail

SECTION 3: Disassembly – initial findings

8-71310-P1 was disassembled and fully inspected. The as-found alignment to the stuffing box was 0.005" TIR, the specification is 0.002" TIR. Although out of specification the condition of the stuffing box was pitted so may have affected the accuracy of the measurement. The throttle bushing Figure 3- Item 14 located in the stuffing box was found to be the cause of the pump failure. It was a relief to the organization to discover that the failed throttle bushing was manufactured from Tenmat Feroform T-12 material and not replaced with the GT ARHT during the last overhaul. The other 3 pump bearings manufactured from GT ARHT were found to be in acceptable condition. ARHT bearings is retained with an interference fit of 0.003-0.010" and recommended to be used with a sleeve of minimum hardness of 40RC.

Running clearances	Original bearings	Green Tweed ARHT

Stuffing Box bushing	0.012-0.015"	0.013-0.017"-ID 6.707/6.712"
Intermediate bearings	0.012-0.015"	0.013-0.020"-ID 6.707/6.712
Lower bearing	0.020-0.025"	0.013-0.019"-ID 6.109/6.114"

Table 1: Bearing running clearances

Figure 4: Throttle bushing lubrication groove plugged

The lubrication grooves in the throttle bushing was blocked with debris that prevented the flow of the injection water to the lower bearings. The debris was smeared and caked on the internal diameter (ID) of the throttle bushing and had a mud like appearance. Although it was obvious what caused the pump failure there was no clear answer at that point on why it became plugged.

Historically, low injection flow failures were attributed to plugging of the lubrication grooves by external organic material and silt which bypassed by the cyclone separator before injection into the pump. After review and discussion, it was decided to engage Canadian Nuclear Laboratories (CNL), an external vendor that shared a business relationship with Bruce Power to perform a forensic analysis on the throttle bushing. The intent of the forensic analysis was to determine the cause of the throttle bushing failure and incorporate findings in the pump reassembly to prevent a repeat failure.

Internal misalignment created during assembly of the three piece pump shaft was investigated as a possible cause of the throttle bushing failure. The shaft coupling is constructed with fine internal threads and the shafts are screwed into coupling to allow assembly. While the forensic analysis was being conducted on the throttle bushing, in parallel, maintenance conducted detailed inspection and as-found measurements to support or refute the hypothesis of internal misalignment.

The pump upper, intermediate and lower shafts were inspected and run out measured. Obtaining accurate run outs of the long shafts can be challenging as run out can be introduced by the inappropriate location of the supports when taking the measurements. Initially, run out criteria of 0.002" TIR was used to determine the suitability of the shafts and only the top shaft met the criteria. A decision was made to fabricate new intermediate and lower shafts. After the new intermediate and lower shafts were fabricated and measured, it was found not meeting the 0.002" TIR criteria. Discussions with third party pump vendor suggested the criteria of 0.0005" per foot of shaft be applied. This criterial allowed a run out of 0.005" TIR for a 10ft shaft; both new shafts met this criterion. The pump shaft was reassembled using the original upper and lower shaft with a new intermediate. Internal misalignment was ruled out as the being the primary cause of the throttle bushing failure.

SECTION 4 Forensic findings:

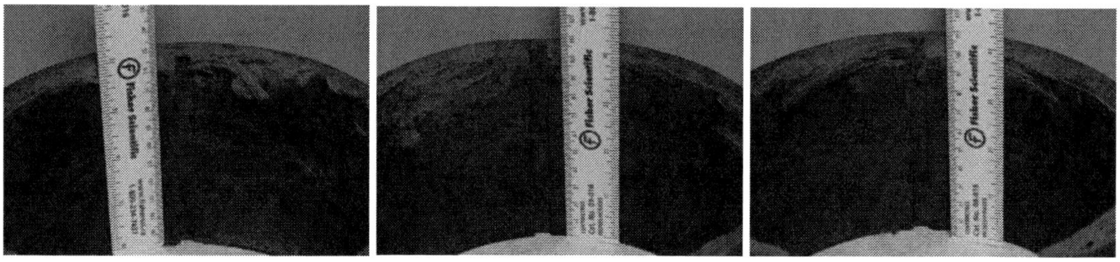

Figure 5: Initial visual inspection

At Canadian Nuclear Labs, the throttle bushing was inspected using visual, optical and scanning electron microscopes (SEM). Visually it was observed that the deposits were heavier on the upper throttle bushing closer to the bearing injection and less to the bottom and smeared around the circumference.

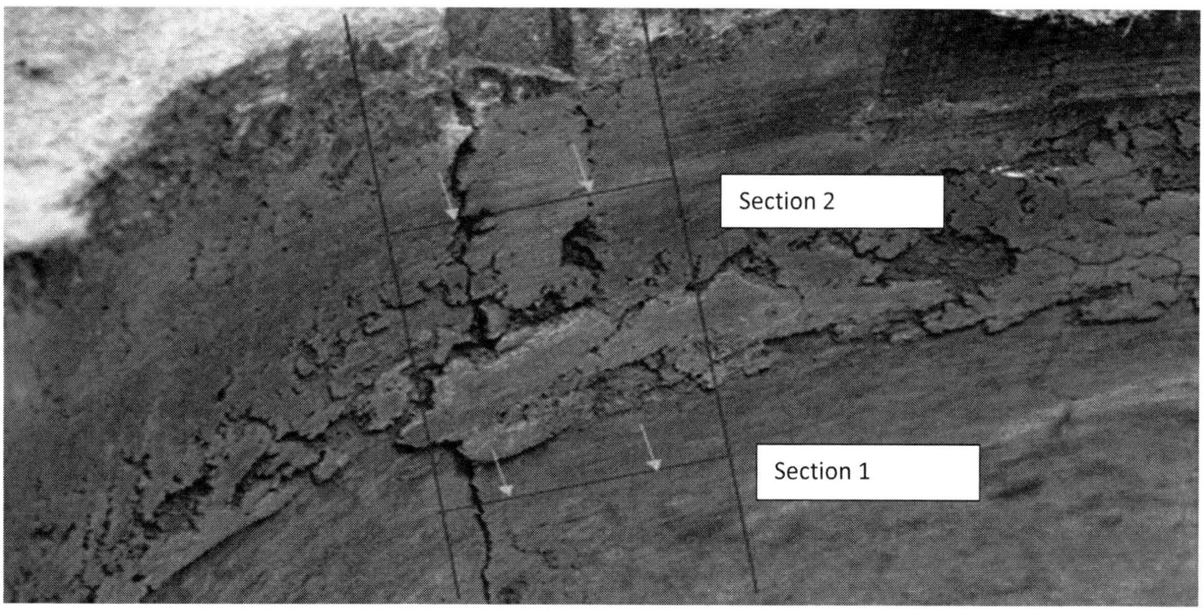

Figure 6: Throttle bushing sectioned in two areas

Figure 7: Section 1- Optical Microscope

Figure 8: Section 2- Optical microscope

Under the optical microscope examination of the sectioned throttle bushing showed that the deposit had the same microstructure of the bushing. SEM inspection indicated that the deposit was composed of the two main phases constitutive of the bushing matrix without presence of other elements. This indicated that the deposits plugging the lubrication grooves was bearing material and not silt or external materials.

The report from Canadian Nuclear Labs concluded:

- 8-71310-P1 throttle bushing degradation was induced by mechanical wear
- The deposits are bushing wear products and not a foreign material
- Potentially misalignment could be responsible for the wear

The CNL failure analysis findings indicated that the pump lost the 0.012"-0.015 "diametrical clearance between the shaft sleeve and throttle bushing. This prompted a review of the mechanical maintenance procedures (MMP) used in the overhaul and installation of the LPSW pump and motor and further discussions with maintenance. EPRI TR-112449 –Shaft Alignment Guide- 1999, suggest in the absence of other documentation maximum allowable shaft alignment tolerances for a 900 RPM machine of 6 mils –offset and 1.0 mils per inch – angularity. The Stork Instruction and Operation Manual (IOM) was available and required aligning the motor shaft to stuffing box to a maximum 0.002 "TIR. The MMP included this alignment instruction.

Courtesy of Ludeca, Inc.

Tolerances for Shaft Alignment									
	Short Couplings							Spacer Shafts	
	Excellent			Acceptable				Exc.	Accpt.
	Offset	Angularity		Offset	Angularity				
RPM	mils	mils per inch	mils per 10"	mils	mils per inch	mils per 10"		mils per inch	mils per inch
600	5.0	1.0	10.0	9.0	1.5	15.0		1.8	3.0
900	3.0	0.7	7.0	6.0	1.0	10.0		1.2	2.0
1200	2.5	0.5	5.0	4.0	0.8	8.0		0.9	1.5
1800	2.0	0.3	3.0	3.0	0.5	5.0		0.6	1.0
3600	1.0	0.2	2.0	1.5	0.3	3.0		0.3	0.5
7200	0.5	0.1	1.0	1.0	0.2	2.0		0.15	0.25

Table 2: Alignment tolerances- EPRI TR-112449

Maintenance reported that the corroded state of the stuffing box and limited access to the measuring point created challenges to executing the MMP instructions. An alternative method by centralizing the pump shaft in the pump stuffing box and indicating off of the shaft or pump coupling was proposed and agreed between engineering and maintenance.

The EFC was completed and two actions created to close the identified gaps:

- Revise the MMP to include explicit instructions of the alternative method to allow more accurate alignment between the motor and pump
- Resume the planned refurbishment program on the BB LPSW pumps to reestablish registered fits and renew major subcomponents to improve pump reliability

SECTION: Installation and post maintenance test

Maintenance reassembled and installed pump and motor and aligned motor to pump by first centralizing the pump shaft in stuffing box then setting dial on motor to the pump shaft to obtain 0.002" TIR on the shaft run out. The pump was released for post maintenance test. The gland flow was 0.375litres/s and the vibration recorded AN-S- 2.2mm/s and AE-W: 1.40mm/s overall measured at the non drive end of motor. Attached is the vibration spectrum at the top of motor in north- south and east west direction. The pump was returned to service with unrestricted operation.

Figure 9: AN-S- Motor non-drive end (mm/s)

Figure 10: AE-W- Motor non-drive end (mm/s)

AKNOWLEDGEMENTS:

Nicolas Huin-CNL, Mike Husk, Mike Sanders, Chris King, David Durance, Zbigniew Komosa-Bruce Power

REFERENCES:

(1) 153-32710-REPT-002- Bruce Unit 8-71310-P1 Throttle Bushing examination- Huin, Nicolas- Jul 2022.
(2) NK29-MMP-71310-00021- Stork VOA 85-75 Pump Overhaul
(3) NK29-MMP-71310-00022- Reliance Electric Pump Motor Overhaul
(4) EPRI- TR-112449- Shaft Alignment Guide

BIBLIOGRAPHY:

NK29- D0H-71312-9003- Assembly Drawing Machinefabriek Stork- P568570
NK29-D1H-71312-9002- Dimension Print Drawing Machinefabriek Stork- P567983
Green Tweed ARHT- product brochure

VIATC | TRAINING CONFERENCE & EXPO

NIAGARA FALLS, NEW YORK JULY 26 -28, 2023

MOTION MAGNIFIED VIDEO APPLIED TO VIBRATION MODAL TESTING

Bill Marscher, CEO
and
Andy Lerche, Senior Staff Engineer
Mechanical Solutions, Inc.
11 Apollo Drive
Whippany, NJ 07981
(001) 973-326-9920
AHL@mechsol.com

Bill Marscher, P.E. founded MSI in 1996, which has grown to a consulting firm of 30. Bill has been an MFPT and VI participant since the 1980's, and is a long-standing member of the MFPT Board, and past chair. He has BS and MS degrees in Mechanical Engineering from Cornell University, and an MS in Mechanics from RPI. Bill previously worked at Dresser Pump and Pratt & Whitney. He is past president of the Society of STLE, and is on the Advisory Committee of the Texas A&M Pump Symposium. He is a member of the ISO TC108 Machinery Standards Committee, and is past Vice Chair of the Hydraulic Institute Vibration Standards Committee.

Mr. Lerche has more than 20 years of professional experience in the areas of design, testing, analysis (fluid and structural), and troubleshooting of mechanical equipment and machinery such as pumps and turbomachinery. He is experienced in the areas of turbomachinery, high pressure fluid control equipment, and shipboard equipment. He has extensive experience with flow-induced vibration of turbomachinery blades, as well as shock and vibration testing and analysis of fixed mechanical and electrical equipment. Mr. Lerche has authored and presented several technical papers in the area of turbomachinery and holds three patents related to industrial compressors and valves. He holds a B.S. and M.S. in Mechanical Engineering from the University of Texas at Austin and the University of Texas at San Antonio, respectively.

ABSTRACT:
Vibration detection and display using magnification of high-speed video (Motion Magnified Video, MMV) has been demonstrated to be a very useful tool. The authors have developed an approach that applied MMV in the determination of natural frequencies and mode shapes. Experimental Modal Analysis (EMA) vibration testing detection is typically performed by a dual (or more) channel spectrum analyzer, an instrumented impact hammer, and an accelerometer. Some form of modal testing has been able to be performed since engineers and technicians gained access to the Fast Fourier Transform (FFT) starting in the 1960's, since the FFT breaks vibration down into its frequency components. Multiple channel FFT analyzers have facilitated obtaining data in the manner from many locations simultaneously. This had permitted the visualizing of the vibration of the machine and its attached system, including the foundation and any piping networks. When plotting the mode shape of a given natural frequency, hundreds of phase-linked impact response data sets may be superimposed onto a CAD model, and then scaled-up vibrations can be animated at frequencies of interest. This visualization process provides valuable insights, but is time-consuming and therefore expensive each time it is applied by experts, and it is prone to user error. An alternative method has been developed that is based on evaluation of MMV high resolution/ high speed videos. The method provides information equivalent to a high-sensor-count EMA accelerometer-based test, by treating each video scene pixel as an accelerometer. The pixel's light intensity modulation is used to translate information statistically available from the video frames into vibration motion able to be observed and interpreted by human investigators. This video scene amplification method was named by researchers at MIT as Motion-Magnified Video (MMV).

KEY WORDS: Motion Magnification; Experimental Modal Analysis; Rotating Machinery; Machinery Diagnostics; Video Amplification

1. INTRODUCTION:

Turbomachinery, pump, and motor vibration is often useful in determining whether or not a machine is operating properly, and for diagnosis of problems if the operation appears improper, or if reliability issues have been experienced (e.g. fatigue cracking, or premature failure of bearings and seals). An important part of such an investigation is the determination of natural frequencies and their mode shapes, in a process known as Experimental Modal Analysis (EMA). For several decades, visual methods have animated a CAD model to exhibit mode shapes in an understandable manner. This has been an important tool in getting a complete and simultaneous view of the vibration of the machine and its system (e.g. piping, foundation, and driver or driven machine), so that the potential importance of any given mode shape is given proper context. The resulting animations represent, mode by mode, the motion implied by phase-linked signal spectra from the group of accelerometers. Such accelerometers typically have been located over hundreds of test points while (at least in single input techniques) the impact hammer acts at a consistent location and direction. All impact response data is superimposed onto a simplified CAD model of the machine and system, and then the exaggerated (but to-scale) vibrations are animated as "cartoons" at frequencies where the response is sufficient to be of interest to the troubleshooter. This process may take one or more days for complex machinery and systems.

An innovative vibration detection and display method (MMV) has been developed that is based on evaluation of high resolution/ high speed video taken of the operating machinery. A modal test version of the new technique begins by being excited by an instrumented impact hammer, as in conventional EMA tests. Following this, however, the MMV method provides information equivalent to a high-sensor-count EMA test, by treating each pixel in a video scene as an accelerometer, using the pixel's light intensity modulation statistics to determine local vibration displacement frequency spectrum. From this information, realistic magnification of slow-motion video footage permits microscopic vibration to be magnified and thereby observed and interpreted by the human troubleshooter. This video technique that is the core of the method was originally called Motion-Magnified Video (MMV) by an MIT research team, as provided in the references. The MMV-based modal method is much faster than classical EMA, yet provides similar information at many more locations (at least in directions in the 2-D field of view), with less opportunity for error, and in a manner that is more credible to both technical as well as non-technical decision makers.

2. BACKGROUND:

Classical EMA Methods:

A cartoon depiction of the modal test concept and process is shown in Fig. 1. Around 1980, some researchers developed a methodology for "seeing" EMA vibration, including the complete mode shape response (as discussed in Ref. [1]. Their approach was to acquire data at many locations and directions on a vibrating structure, and then allocating those motions to a stick-figure CAD model. This model was then animated on a computer video screen, or the extremes of the vibration in the model were plotted by a computer printer or plotter.

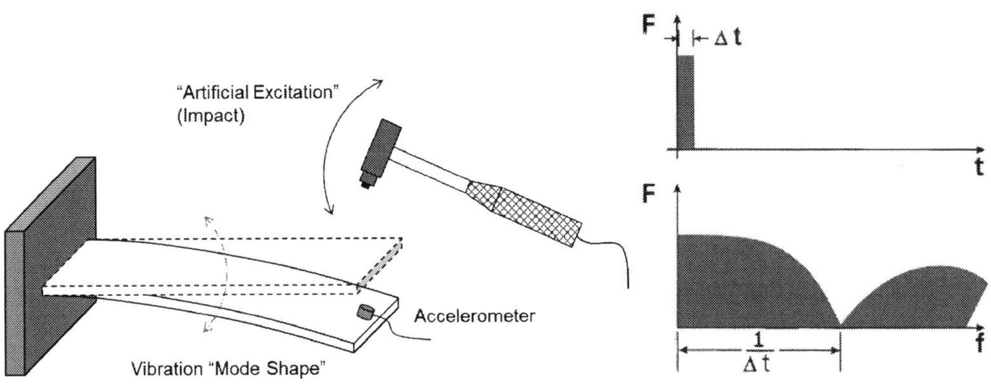

Figure 1: Simple example of a classical EMA test for natural frequency. A very brief impact force is applied over a time period Dt, to excite a broad range of frequencies f.

The following outline summarizes the EMA technique:

- It is based on the "impact excitation" frequency spectra of the rotating equipment (variations can also be based on excitation by a shaker)
- The User acquires vibration data from various locations and directions on the machine and system (eventually, the method evolved to permit/encourage hundreds of vibration measurements)
- Typically, the impact is always kept at the same location and direction, to provide a consistent reference signal as other probes are "roved" (multiple input methods are also available, but not typical)
- A large database is built up of amplitude vs. frequency and phase angle
- Stick figure representations have evolved to realistic 3-D CAD model constructions, which are assigned motion from each individual vibration data point
- Software is used to amplify the impact test results filtered at specific frequencies of interest, creating mode shape animations of the CAD model of the equipment and related system, including piping and foundation.

EMA has been a powerful method for understanding the nature of natural frequencies in machines and their foundations and systems. Used properly, it also often provides visual "hints" concerning options to avoid resonant amplification, and thereby reduce excessive vibration. However, classical EMA has its drawbacks as well. Some of the pros and cons of classical EMA are summarized in the list below:

- ✓ Powerful and intuitive diagnostic tool
  - Can clearly demonstrate vibration natural frequencies & mode shapes
- ✓ Proven over decades of application
- ✗ Time consuming
  - Data acquisition – hundreds of data points
  - Post processing – careful compiling of database to match model points
- ✗ Potential for bookkeeping error
  - Match all gathered data points to appropriate place and direction on the model
- ✗ Response sensors typically require attachment – not appropriate for all cases
  - High temperature, radiation, shock, accessibility/scaffolding requirements

Motion-Magnified Video (MMV) Based Method:
Many of the drawbacks can be avoided if an actual visual scene is able to be evaluated concerning the vibration of components within the scene. Academic researchers, some of whom are listed in the References [2], [3], [4], had this thought as much as 30 years ago (albeit not for the purpose of modal testing at the time), and have been gradually perfecting techniques. Basically, these techniques fall into two categories: 1) tracking of specific points, edges, or (as machine-vision scientists call them) "blobs", and 2) performing statistics, including signal vs. time as well as signal FFT frequency spectra, on the individual independent pixels. The former are called "Lagrangian Methods" by many

researchers, and the latter are called "Eulerian Methods". A rich literature basis exists for these methods. This paper's references emphasize the Eulerian technique.

The Motion-Magnified Video (MMV) method may be summarized as follows:

- Uses high-speed, high-resolution video
  - Eulerian method is equivalent of millions of accelerometers, 1 per pixel
- Analyzes/quantifies displacement motion, in 2-D
  - FFT is used to break motion into individual frequencies
- Algorithms amplify motion to human visual threshold
  - Filterable by desired frequencies, high resolution (up to 4+ kHz in the system as developed by the authors).
  - Amplifies up to 1000+ times the actual motion, using a statistical algorithm that can perceive and magnify motions down to less than 0.1% of pixel. At a 10 foot field of view, motions as small as 20 millionths of an inch (half a micron), peak-to-peak, can be reliably detected, quantified, and magnified, based on both controlled laboratory as well as practical field tests.

History:
This has led to various researchers, in the US as well as Europe, to develop such video acquisition and evaluation processes to study other dynamic phenomena. In fluid dynamics, laser-doppler velocimetry (LDV) has been able to track streamlines, and similar processes have been applied to mechanical vibration. In a lower-tech method, strobe lights have been used on machinery for many decades to "freeze" motion. These methods have been useful for interpreting dynamic behavior if relatively large motions are involved. However, displacement magnification was not achieved.

In the 1920's, advanced research organizations used relatively high-speed cinematography to guide development and to evaluate the validity of new vibration analysis procedures. A good example is the work of Wilfred Campbell at GE. At the time, steam turbines driving generators were growing larger, and as they did, they encountered unexplained fatigue failures. Campbell's work, as exemplified by the photo in Figure 2 (a single frame of a "movie" or video taken by his cameras), determined that bladed disks had much more complicated natural frequencies than had been predicted up to that time, including zero-vibration nodal lines in their mode shapes, now known as nodal diameters and nodal circles. When the associated natural frequencies matched a strong excitation frequency (such as the number of stator nozzles times running speed), and when simultaneously the mode shape of the natural frequency matches the lobes of a circumferential static pressure pattern (such as associated with the number of blades versus number of stator nozzles), a strong resonant response was seen to occur.

Based on this insight, GE and eventually others were able to avoid damaging resonances by either adjusting natural frequencies, or by adjusting problematic lobe patterns of the nozzle pass frequency. The Campbell Diagram, well known to turbomachinery engineers, became a graphical method of codifying this procedure.

Figure 2: Campbell's bladed disk video from 1924

This range of efforts, and others, led scientists and engineers to an appreciation for the power of video to uncover useful information in dynamic scenes, assuming that appropriate algorithms, implemented by software, were able to be used to evaluate the video footage.

In the US, several universities, including MIT [2], [3], have been actively pursuing MMV. At MIT, the particular technology which became the core of their research was an Eulerian approach [4], which essentially tracks the variation of individual pixels over time, and then exaggerates those differences. To the human eye, no matter how long and hard a person stares at a machine with vibration at, say, several times the ISO limits, you would struggle to detect any motion. For a computer, however, the tiniest per-pixel fluctuations (between white and slightly-off-white, say) are easy to detect and quantify. MIT originally developed their software to measure the vital signs of neonatal babies without physical contact, but they realized that there were other, far-ranging applications. For biological applications, Eulerian MMV detects changes in skin color, as well as exaggerates movements. More lately, MIT researches have applied their Eulerian approach to modal test as well [5]. The steps and hardware associated with Eulerian MMV are illustrated in Figure 3.

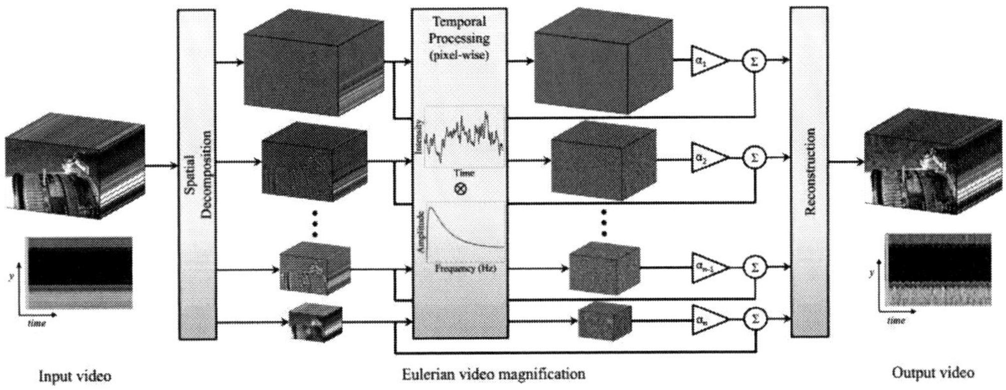

Figure 3: Illustration of the MMV process, as implemented by MIT [3]

Research by the Authors:
The authors' organization initiated its detailed research into engineering uses for high speed video in 2003. The US Department of Defense contracted the authors' group to study how to make non-lethal projectiles operate reliably, ensuring that they functioned without harming humans or animals. Applications included various forms of "soft bullets". High speed video, up to 20,000 frames per second, documented their transient behavior during flight and impact.

A later application (2005) involved an inexpensive video sensor to use pattern recognition software to rapidly identify rocket-propelled grenades (RPGs) fired by a belligerent, so that defensive steps could be taken. Following this was use (2011 to present) of high-speed video to determine the moment-to-moment effectiveness of passive counter-measures to fool MANPADS missiles fired by adversaries to bring down (for example) an airliner, providing feedback facilitating improved countermeasure design. More recently (2015 to present), the authors' group was contracted to use high-speed video and pattern recognition to rapidly characterize lethal ranges of various munition types, so friendly military forces are better able to ensure sufficient buffer zones for civilians, schools, and hospitals.

During this process, the authors began applying Eulerian algorithms to measure and amplify motions of individual pixels [1]. Different procedures were tried during both government and IRAD-funded testing, and after several years a successful method was achieved. Transient motions and vibration levels as low as tenths of a mil (i.e. several microns) of machinery surfaces were demonstrated first in the laboratory, and soon after in on-site machinery troubleshooting.

The authors' methods are based on the Eulerian approach, which observes the intensity variation of individual pixels, combining their effects statistically into a full scene. Alternative methods typically keep track of motion pixel-to-pixel, the so-called Lagrangian techniques. Variations include feature tracking, and an approach called "optical flow". These Lagrangian methods are good for amplifying motions that occur over multiple pixels, e.g. motions of more than 2mm (0.08 inches) for a 3 m (10 foot) field-of-view.

However, these approaches are limited in how small a displacement they can detect for a given field of view and a given camera resolution, in that interesting vibration levels can be well below such limits, particularly at higher frequencies of vibration. If higher camera resolutions are used to overcome this limitation, the computational overhead becomes excessive. One application mediates this by selecting only a greatly reduced number of pixels in a scene.

Eulerian methods have been shown to work best for detection of small displacements, since motion detection as low as 1/1000 of a pixel have been demonstrated. Typically, as shown in Fig. 3, Eulerian methods build a time history for each pixel, and then apply signal processing techniques common to more conventional vibration detection methods. Such conventional methods include digital filters, time and frequency domain digitization (Laplacian and Fourier), and amplification of the subsequent amplitudes by simple multiplication. In implementations such as video-based EMA modal testing, an FFT is useful to separate motion from individual frequencies and/ or vibration modes, to make vibration mode shapes more evident. Multiple frequency animations can be made from the same ten second video strip.

The benefits of MMV performed for modal testing are that it becomes a powerful and intuitive procedure for displaying complex mode shapes during EMA testing able to animate detailed mode shapes in a relatively short time-frame. It realistically animates modes of vibration in a manner that is easy for non-experts to believe and understand. Importantly for certain applications, the technique does not require contact, so that surfaces above 1000 F, or that are radioactive, or that are (for example) 50 feet vertically above in a ceiling, can be observed without difficulty. Furthermore, items such as instrumentation wires or lubrication lines, which would change their behavior if mass-loaded with even a small accelerometer, can be evaluated in a manner that does not alter their behavior. The results are obtained quickly, and at modest cost and effort.

However, there are also several potential problems in applying MMV to EMA modal testing. A summary of these is:

- If ring-down vibration is too low, getting "clean" response FFT is challenging. Data may stay buried below "noise floor".
- So: *Need to detect displacements as small as possible!*
- "Time-Averaged Pulse" using triggered averaging is very beneficial in this regard. The conceptual approach of Time-Averaged Pulse is illustrated by Figure 4.

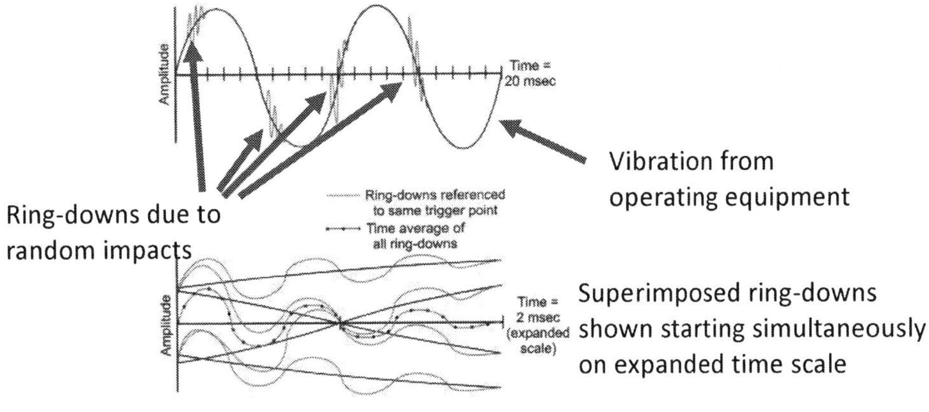

Figure 4: The use of time-averaging of multiple impact ring-downs, triggered consistently with the hammer impact

The concept of triggered averaging is common in EMA testing, in that a threshold value of the signal from an instrumented hammer is used to trigger response acquisition by an FFT spectrum analyzer. Typically, this process is performed many times, with the output of the response sensors (usually accelerometers) averaged together, typically as separate real and imaginary vectors, therefore statistically enhancing the ratio of ring-down response versus response from sources (e.g. imbalance) that are not correlated with the hammer impact. The authors often use this technique, but have found that cumulatively averaging the response data in the time domain prior to ultimately performing an FFT provides superior frequency response functions with fewer impacts. Therefore, the "Time-Averaged Pulse" technique illustrated in Fig. 3 has been implemented by the authors in the described MMV EMA method, as outlined below:

- Modal hammer and triggering enhancement kit provide reference signal
- Multiple video clips are "averaged" together
  - Reduces level of vibration not coherent with the "ring-down"
  - Reduces camera sensor white noise effects
  - Lowers vibration detection threshold for same duration video
  - Results in a novel technique (Patent Pending)
- Provides modal frequencies AND comprehensive mode shapes, quickly

Applications and Case Histories:
Some examples will be presented as videos during VIATC2023.

High noise in an industrial fan:
An industrial fan was producing extremely high noise (120 dB) due to blade passing frequency (8 blades, with a frequency of about 480 Hz). Everyone including the OEM was baffled because accelerometer measurements on the fan casing, motor, and piping indicated surface velocities were too low to result in such high acoustic response. The MMV EMA

video, taken at a frame rate to permit full-screen video up to 640 Hz, tuning in the natural frequency peak as excited by white noise turbulence near blade passing frequency, clearly showed that the problem was a natural frequency at close to 480 Hz. The mode shape is easily visible and "intuitive" once observed, as exhibited by strong "pulsing" in-and-out of the upper sidewall in Fig. 5. As can be seen in the animated video, the pulse direction was opposite on the left side versus the right side, with no movement in the vertical centerline (a "node line"). This was the source of the manufacturer's confusion and why testing with probes on the centerline did not give useful results. A frame from the video is shown in Figure 5, which animates a 1/4 mil (7 micron) motion, such that it appears to be ¼ inch (7 mm) of concave/convex motion of the sidewall.

Figure 5: The mode shape obtained by the MMV process applied to an industrial fan

Example of Large Vertical Wastewater Pump and Motor:
A vertical pump and piggy-back motor was tested using the author's MMV EMA approach, with excitation and associated triggering provided by an instrumented 12 lbm (5.4 kg) plastic-headed hammer. The pump flow rate was 45,000 GPM, at a speed of 594 RPM (9.9 Hz) and power of 1000 HP. Some tests involved exciting pumps in the direction parallel to the horizontal discharge line, and some perpendicular to the discharge direction, since separate and orthogonal natural frequencies in such pumps tend to occur in these directions. Fig. 6 shows a frame from a the MMV video, exhibiting the perpendicular-to-discharge mode at 7.5 Hz. The results for the parallel to discharge direction was found to be 7.8 Hz, with a similar "reed" frequency mode shape.

Figure 6: The EMA MMV process applied to a vertical pump and motor, Determining "reed" natural frequency and mode shape at 7.5 Hz

Compressor Impeller Modal Test, Detecting High Frequency Modes:
Figure 7 shows a high-speed centrifugal compressor impeller being tested for its natural frequencies and mode shapes. The MMV EMA procedure was applied, using an 0.75 lbm head hammer s the excitation and triggering source, and resulted in detection and display of its modes up to 4000+ Hz. Fig. 7 shows a frame of one of the animations exhibiting a "bladed disk" natural frequency mode shape at 2180 Hz. This type of testing on a small impeller is particularly difficult because there can be hundreds of modes, many closely spaced. Furthermore, detecting the mode shape with accelerometers is impractical because the accelerometers "mass load" the blades, significantly affecting the results. Proximity probes are also a problem, because of the curved nature of the blade surfaces, and lasers struggle because of most areas not having good line-of-sight. Line-of-sight is also an issue for video, but much less so, with motion at oblique angles typically being easily detected and displayed, in the experience of the authors. However, the vibration displacements that can be excited, without damaging the impeller, are very small. This issue was overcome in the author's MMV technique by averaging a large number of impacts, using the time-averaged pulse averaging method.

Figure 7: The EMA MMV process applied to a compressor 6.5 in. diameter impeller, evaluating mode at 2180 Hz, the shape of which is shown in the lower picture.

3. Future Research:

The following summarizes the research currently underway at the authors' organization:

1. Improving MMV capability using low-cost COTS cameras
2. Continuous improvement of displacement sensing accuracy
3. Acquiring mode shapes of rotating components
4. Processing and displaying 3-D mode shapes

4. Conclusions:

As presented, the new video-based motion magnification procedures and equipment now on the market can be very useful for determining natural frequencies and mode shapes of rotating machinery and associated piping and systems. The new method has advantages over the conventional EMA method in many instances. MMV EMA also typically takes much less time and "logistics" than conventional EMA to implement.

MMV benefits applied to EMA are:

- ✓ It is a powerful and intuitive diagnostic tool
  - Realistically demonstrates modes and vibration natural frequencies
  - Easy for non-experts to understand (management, etc.)
  - Able to observe, and display, the entire system of interest
- ✓ It does not require "contact" – perfect for restricted areas
  - Avoids heat, radiation, accessibility/scaffolding issues
- ✓ It is fast and cost-effective
  - Millions of data points ready to evaluate in minutes, not days

Experimental Modal Analysis using Motion-Magnified Video represents an important addition to a vibration diagnostic tool kit.

REFERENCES:

[1] Marscher, W.D., and Onari, M.M. (2018), Video Motion Magnification vs. Operating Deflection Shapes for Machinery Diagnosis, MFPT2018, Virginia Beach, May

[2] Timoner, SJ, and Freeman, DM (2001), "Multi-image gradient-based algorithms for motion estimation", Optical Engineering 40.9, pp. 2003-2016

[3] Liu et al (2005), "Motion magnification", ACM Transactions of Graphics (TOG) – Proc. of ACM SIGGRAPH 2005, TOG Homepage, v. 24, Issue 3, July

[4] Wu, H-Y et al (2012), "Eulerian video magnification for revealing subtle changes in the world", ACM Trans. of Graphics (Proc. SIGGRAPH 2012), v. 31, August

[5] Chen, J.G., et al, "Modal Identification of Simple Structures with High-Speed Video Using Motion Magnification", J. of Sounds & Vibration, vol. 345, 9 June 2015

VIATC TRAINING CONFERENCE & EXPO

NIAGARA FALLS, NEW YORK JULY 26–28, 2023

VIBRATION SPECIFICATIONS AND STANDARDS AS PART OF EPC CONTRACTS

Kaitlin Spak, Ph.D., P.E.
Principal Engineer, Exponent
kspak@exponent.com

Daniel J. Mennitt, Ph.D., P.E.
Managing Engineer, Exponent
dmennitt@exponent.com

Dr. Spak is a Principal Engineer at Exponent in the mechanical engineering practice. She specializes in the analysis and testing of systems and structures, with specialized knowledge of vibration characterization and broad experience in analysis of machines and dynamic systems. As the highest level of certified vibration analyst, she has extensive experience in instrumenting and conducting dynamic tests and analyzing vibration and operational data, and has testified domestically and internationally for vibration-related issues.

Dr. Spak has applied her engineering skills to projects involving pipeline joint failures; drilling and mining operations; industrial machinery; heavy equipment; engine failures; commercial and passenger vehicle design, analysis, accidents, and failures; bearing analysis; operational data analysis; and consumer products and appliances.

Dr. Mennitt is a Managing Engineer at Exponent in the mechanical engineering practice. He specializes in acoustics and is skilled in the characterization, design, and modeling of acoustical environments and devices. His primary areas of expertise include noise control, environmental sound, and signal processing. Dr. Mennitt has applied his expertise to noise and vibration control of machines, medical devices, consumer products, and other equipment. He has also provided engineering support for litigation involving acoustic and vibration matters.

ABSTRACT:
Large-scale engineering efforts such as power plants and processing plants require huge investments of time and resources, and the development, construction, commissioning, and operation of these facilities almost always include considerations for pumps, compressors, or other equipment that is a source of vibration. The requirements for vibration associated with these machines are generally set out in engineering, procurement, and construction (EPC) contracts, but disputes can arise between the facility owner and engineering contractor if the contract is not adequately clear, specific, or thorough.

One area in which the authors have experienced vague or ambiguous specifications is the topic of vibration criteria. This paper illustrates the value of standards and vibration limits for acceptance testing specified in the original EPC contract, some of the commonly used standards and guidelines for vibration of pumps and other processing equipment, considerations for the applications of those standards, and approaches to evaluating vibration when no standard or other guidance is specified.

KEY WORDS: EPC contracts, acceptable vibration, standards, pumps, machinery vibration, acceptance criteria

INTRODUCTION:

A contract is intended to outline what each party will contribute to an agreement; typically, one party provides goods or services and the other party provides payment for those goods or services. The most effective contracts provide specific details regarding the nature and quality of the goods or services in terms of dimensions, throughput, efficiency, or other relevant characteristics. Such contracts may reference standards, include guidelines to be followed, or provide numerical limits that must be met.

Contracts for everything from buildings to processing plants to consumer products may have requirements for vibration. Too often, those requirements are vague, using only phrases such as "minimal vibration" or "vibration not to exceed industry standards," instead of specifying an acceptable vibration limit and measurement procedure or identifying a specific standard. Particularly in the case of new, large-scale construction projects, nonconformances (i.e., deviations from the contract or contract requirements that are not met) can arise that must be resolved between the parties to the contract [1]. When process machinery shakes apart or causes excessive vibration to surrounding structures, opposing parties may select different references, criteria, or methods that they believe prove their case, especially if sufficient guidance was not identified in the contract. This paper provides an overview of common vibration standards in industry along with caveats for their use, particularly as they relate to engineering, procurement, and construction contracts, or "EPC contracts."

EPC CONTRACTS AND VIBRATION:

EPC contracts are used for large-scale projects such as power generation facilities and manufacturing plants. EPC contracts typically indicate that one party will be responsible for the design of the facility, acquiring all materials to build the facility, and constructing the facility to completion. Subcontractors are often used because these efforts can be massive, and often require many different types of specialized expertise.

Most power generation facilities or manufacturing plants have some type of rotating or reciprocating machinery, and these are the machines that often are prone to vibration problems. Such problems can affect the machine itself as well as surrounding equipment and processes. Requirements for these machines are typically provided in the EPC contract or in sub-contracts between the EPC contractor and its subcontractors.

Many EPC contracts require payments in stages as different milestones are met. Milestones may include completion of the facility design, acquisition of equipment, construction of the facility to some percentage, commissioning and startup, and operation of the plant to meet expected performance levels successfully. There may be a warranty period as well, during which the EPC contractor is responsible for correcting any defects or contract nonconformances that are found. Standards and other specific guidelines are vital for demonstrating the completion of milestones and correction of defects, which often requires testing and evaluating results according to criteria.

Vibration issues can be particularly problematic for EPC contracts as there are multiple ways in which vibration issues can manifest, with not all issues immediately apparent:

- Initially high vibration from a piece of equipment. These issues are the easiest to diagnose and consider as a contract nonconformance because the problem is immediately apparent when the machine starts up. Often, the equipment manufacturer is likely still involved with the construction and can (and should!) assist in diagnosing the issue, particularly if the manufacturer has internal standards for acceptable vibration. This is a good reason to have factory acceptance testing, in which the manufacturer must ensure that the equipment it offers meets provided specifications before the equipment is delivered to the site for construction. These types of vibration issues are typically found in the commissioning process as the facility is brought online (especially if factory acceptance testing was insufficient) and can usually be corrected at the machine level (for example, balancing a motor or correcting a soft foot condition of a piece of equipment that is wobbling on its base). Note that if process requirements have changed, such as the flow rate through a pump, equipment may need to be reconsidered or resized to ensure that negative repercussions do not occur from the design change.

- Initially high vibration that affects surrounding equipment or components. These issues can be simple to identify and diagnose but may be difficult to resolve in a dispute. Although vibration criteria are often determined and provided for vibration generating equipment, criteria may not be well defined for equipment or infrastructure that is located nearby and prone to damage or other adverse effects from excessive vibration. Thus, a machine that meets its specified vibration limit, but causes proximate components to resonate or otherwise vibrate excessively, may be difficult to define as a defect, according to the contract. However, if vibration standards are included in the contract, those standards can be applied regardless of the source of the vibration. If the expected vibration response, particularly for long flexible materials such as piping, is calculated and evaluated at the design phase, many types of transmitted vibration issues can be avoided in the construction and commissioning phase. There are useful guidelines to assist in this type of analysis, as presented in the next section.

- A step change or progressive increase in vibration during the warranty period. These issues present a particular challenge for diagnosis because the problem would not have been present during the startup or commissioning period. A root cause analysis must consider what has changed and whether the change is due to a material selection or equipment issue (which would likely be the responsibility of the EPC contractor), or to a process or operation issue (which would possibly be the responsibility of the facility owner, particularly if the equipment were operated incorrectly or outside of provided specifications).

- Fatigue-inducing vibration that may not cause early operational life failures but make the equipment or surrounding materials prone to failure before the contract life is achieved. These types of vibration issues are generally the most difficult to diagnose and the most difficult to resolve in disputes. Relatively low levels of vibration can contribute to the accumulation of fatigue damage over time (similar to bending a paper clip back and forth until it breaks), and a fatigue life assessment must be conducted to determine whether the equipment or infrastructure is at risk. Fatigue life assessments are inexact (i.e., engineers cannot predict the exact number of cycles after which failure can occur but can generally determine an order-of-magnitude likely number of cycles to failure) and depend heavily on the specific material properties, so additional testing of the material used may be necessary for accurate results.

When evaluating vibration issues related to EPC contracts, it is important to consider the history of construction and operation; equipment that was replaced multiple times during commissioning, complaints from operators about vibration or noise (particularly if only during certain operations), and cracking of material or loosening of bolts can all be signs of a vibration issue that may cause the equipment to deteriorate over time. These warning signs should be investigated to determine if a root cause can be mitigated. Sometimes a relatively simple strategy, such as adding a support to stiffen a structure or adding material to damp a structure, can significantly reduce the vibration. However, a word of caution is in order, because adding supports without adequate consideration can transmit the vibration to another sensitive location. It is recommended that a vibration analysis of any corrective actions be conducted before the corrective action is implemented.

Another consideration for EPC contracts is alleged failure to achieve contractual performance guarantees [1]. While there may not be performance guarantees directly related to vibrating equipment, excessively vibrating equipment can affect plant processes and lead to shutdowns or inefficiencies that reduce throughput or uptime and thus affect contract requirements.

RELEVANT ORGANIZATIONS:

EPC contracts often mention “industry standards” or similarly ambiguous language when referencing specific equipment requirements. Thus, it is recommended to understand the range of organizations that address vibration criteria for machinery. Many vibration standards relevant to the equipment procured and constructed in EPC contracts are developed and maintained by the organizations described here.

The standards that the International Organization for Standardization (ISO) develops are widely used for many different types of equipment; ISO standards are often incorporated (or adopted in their entirety) into national or regional standards. Particularly notable ISO standards of interest used in the authors’ experience include the ISO 20816 series (Mechanical vibration – Measurement and evaluation of machine vibration), which combines standards from the ISO 10816 series (covering bearing vibration) and the ISO 7919 series (covering shaft vibration), which in turn evolved from other prior standards. Some portions of the ISO 10816 series are still used since the ISO 20816 series does not include all the same sections as ISO 10816. ISO also recognizes that selecting the appropriate standard can be a challenge. ISO TR 19201 is a technical report that provides an overview of relevant standards on machinery vibration [2]. The scope and applicability of many standards are described, which provides a good starting point for an investigation.

Other organizations may rely on ISO or other primary standards as a starting point for their own standards with specific applications. For example, the Norwegian petroleum industry uses its own standard, NORSOK S-002 (requirements for safe working environments) [3]. NORSOK S-002 includes vibration criteria that have the same shape as the curves in ISO 10137 [4], but NORSOK S-002 specifies alternate categories and limits for specific applications (work area, exposure time, etc.).

Although a national organization, the American Petroleum Institute (API) provides comprehensive standards that are used globally for petroleum processing facilities. Many of the categories of equipment used in the types of facilities that EPC contracts cover may also be used in petroleum processing and are thus covered by an API standard. Many API standards specify vibration limits for specific types of equipment. For instance, API 610 requires a vibration velocity of less than 3 mm/s root mean square (r.m.s.) for horizontal pumps and less than 5 mm/s r.m.s. for vertical pumps when operating in the preferred operating region [5]. Identifying standards that apply to specific pieces of equipment is one approach to identify acceptable vibration levels. The API 6XX series covers a wide variety of mechanical equipment standards for refinery service [6]. API 670 covers vibration monitoring systems, which can be an important consideration in EPC contracts for evaluating machinery vibration trends to prevent catastrophic failures [7]. API also offers recommended practice (RP) documents that may define vibration requirements. For example, RP 688 provides vibration requirements specifically for reciprocating compressors [8].

Another national organization, the American National Standards Institute (ANSI), often works in partnership with other organizations such as ISO, the Acoustical Society of America (ASA), the American Society of Mechanical Engineers (ASME), and the American Gear Manufacturers Association (AGMA). For example, the ANSI/ASA S2 series on mechanical vibration and shock covers measurement and evaluation of machine vibration, viscoelastic materials, and human exposure, among other topics. ANSI standards are primarily used for US projects rather than international projects, since international projects will generally revert to the underlying ISO standard when possible. However, ANSI standards may be internationally referenced or incorporated into other national standards.

As process systems have become more complex, the failure of associated piping systems due to vibration has become more common. Although not typically included in EPC specifications, the Energy Institute's "Guidelines for the avoidance of vibration induced fatigue failure in process pipework" provide useful limits to determine whether piping is likely to be suitable for long-term operation and should be a consideration if piping vibration is suspected to be damaging [9]. These guidelines were published in January 2008 to provide guidance, assessment methods, and advice on control and mitigation measures for situations including when a vibration issue is identified on an existing plant. The Energy Institute guidelines provide frequency-dependent vibration values above which "concerns" or "problems" are identified. These guidelines can be useful for plants or facilities with piping, since they provide the limits for an initial assessment (noting that these guidelines recommend a fatigue life analysis for those measurements indicating a concern or problem).

Another specification that is not commonly called out in EPC specifications but may be useful when no other guidance is available, is military specification standard MIL-STD-810 from the US Department of Defense Military Standards [10]. This comprehensive testing document can provide useful testing approaches for components that may not be covered in other standards. ASTM (formerly, American Society for Testing and Materials) also provides guidance for vibration testing, but ASTM standards are more often used for individual component or product testing rather than process plants or heavy equipment.

Regional or national organizations can also provide standards for special application, often either providing input for cross-listing or incorporating international standards. The German Institute for Standardisation (Deutsches Institut für Normung, DIN) sets unified standards for products and processes, and has a few vibration standards that typically align or are cross-referenced with

ISO standards. Verein Deutscher Ingenieure (VDI) is another example of a localized group with potentially appropriate standards; VDI is an association of engineers in Western Europe that provides some additions to ISO standards such as the ISO 10816 series and its own standards for specialized equipment such as wind turbines [11] and screw compressors [12].

ISO, API, and ANSI are internationally recognized and cover many of the situations that arise from EPC contract disputes. However, there may be other local, regional, or equipment-specific organizations that provide vibration guidance. Generally, such guidance is aligned with existing standards but may have additional information that the organization developed for challenges specific to its region or equipment type. As an example, the Hydraulic Institute (HI) focuses on pumps and pumping systems and provides vibration standards for 11 different pump configurations. It is also common for organizations to adopt or incorporate existing standards, so some standards are cross-listed with multiple organizations. (For example, API 610 is equivalent to ISO 13709, and ANSI 9.6.4 is also used by HI as the standard for rotodynamic pumps, HI ANSI 9.6.4.)

RELEVANT VIBRATION STANDARDS:

The range of facilities covered by EPC contracts is broad, so there is a correspondingly broad range of vibration-related standards that may be applicable. This section provides a review of some standards and other guidance that may be considered both for contract requirements and resolution of EPC contract disputes. Note that this list is not exhaustive, and other references may apply to specialized equipment or may specifically be called out as a contract requirement. In particular, some of the more specialized organizations than those listed above provide guidelines for allowable vibration of specific pieces of equipment, such as cooling towers [13]. Many additional standards not listed here are applicable to very specific situations, such as vibration and shock in buildings with sensitive equipment, especially optical equipment such as scanning electron microscopes [14].

Generalized vibration severity charts can be found that may be based on existing standards but are often presented without a reference or sufficient context. For example, the precise vibration measurement value (r.m.s., peak, peak-peak, etc.) and associated characteristics such as the frequency range of interest must be specified and must match the corresponding criteria. It is easy to see why there may be disagreement about whether a certain amount of vibration should be considered acceptable: there are numerous standards (including some that overlap with other standards or supersede prior standards), and their specific applications vary by machine type, characteristics, and the effects considered.

Generally speaking, there are two distinct areas of interest regarding effects of vibration: the effect of vibration on materials, machines, or structures, and the effect of vibration on human comfort and health. EPC contracts are typically concerned with vibration that can damage equipment or components, especially in the early stages of construction. However, excessive vibration or noise that is detrimental to employees during long-term operation can also be a concern. In the case of EPC contracts, problems are generally related to the operation, reliability, and durability of equipment, so this paper focuses on vibration impacts to equipment rather than effects on humans. However, vibration can affect humans as well as structures and equipment, so, with the consideration of worker safety and the rise of acoustic regulations around the world, inclusion of vibration criteria for multiple distinct purposes is advisable. Whereas humans can perceive vibration directly and may subsequently describe vibration as "excessive," using measurements and standardized criteria to assess vibration in a repeatable way provides a more accurate and objective understanding of the risk to humans, structures, and equipment.

Table 1 provides a non-exhaustive list of standards and guidelines that may be applied to an EPC contract. Note that some standards cover a broad range of equipment generally, while others are focused on very specific types of equipment of specific sizes. Selection of the standard to be followed must be based on a combination of the contract requirements and the applicability of the standard to the specific equipment at issue. In some cases, EPC contracts will provide a list of precedents to indicate the order in which requirements must be met; for instance, many EPC contracts will list their own data sheets or appendices as the primary requirements for specific equipment, with local or international standards as secondary or tertiary considerations. Typically, international standards are highly regarded for analysis, unless a localized or specialized organization has specific requirements due to unique features of the region or industry (for instance, heightened corrosion considerations in areas of high salt content).

Table 1. Standards and guidelines from various organizations for a subset of common equipment found in EPC contract builds.

Organization	Number	Title	Equipment Covered	Use or Effects Considered
ISO	20816	Mechanical vibration — Measurement and evaluation of machine vibration	Various equipment including gas and steam turbines and generators, industrial machinery, pumps, compressors, and gear sets	General evaluation criteria for operational monitoring and acceptance testing designed for reliable, safe, long-term operation of the machine while minimizing adverse effects on associated equipment
ISO	10816	Mechanical vibration — Evaluation of machine vibration by measurements on non-rotating parts	Most equipment similar to and superseded by ISO 20816, but some parts still active, including 10816-21 which covers horizontal axis wind turbines	Evaluation of mechanical vibration from internal and external forces (in the case of wind turbines) to address operational reliability and service life
ISO	19201	Mechanical vibration — Methodology for selecting appropriate machinery vibration standards	Various machinery; provides an overview of relevant international standards to assist in selecting appropriate standards to conduct measurements and evaluate machinery	Provides guidance for selecting appropriate vibration standards for specific machine types, and thus selecting the appropriate vibration measurement and evaluation method
ISO	13373	Condition monitoring and diagnostics of machines - Vibration condition monitoring	Various equipment including gas and steam turbines, fans and blowers, electric motors, electric generators	Specifies procedures for vibration diagnostics for machines
ISO	10137	Bases for design of structures — Serviceability of buildings and walkways against vibrations	Wide range of equipment including pumps and compressors	Considers human response, effect on sensitive equipment and building structures
ISO	2631-1	Evaluation of human exposure to whole-body vibration. Part 1: General requirements	Vehicles, machinery, and industrial activities that expose humans to whole-body vibration	Considers comfort, activities, and health of humans with respect to vibration
VDI	3836	Measurement and evaluation of mechanical vibration of screw-type compressors and root blowers - addition to DIN ISO 10816-3	Screw-type compressors and root blowers	Provides criteria for evaluation of vibration and acceptance criteria for specific equipment
ANSI/ASME	B73.1	Specification for horizontal end suction centrifugal pumps for chemical process	Centrifugal pumps, typically for chemical processes	Provides maximum allowable vibration limits
ANSI/ASA	S2.27	Guidelines for the measurement and evaluation of vibration of ship propulsion machinery	Propulsion machinery for ships (rotating machinery on ships covered in S2.28 and S2.29)	Evaluation of mechanical vibration from the use of marine propulsion machinery
ANSI/ASA	S2.70	American national standard guide for the measurement and evaluation of human exposure to vibration transmitted to the hand	Vehicles, machinery, and industrial activities that expose humans to vibration, including localized vibration such as to the hand (S2.70) or to humans in buildings (S2.71)	Provides methods to characterize and evaluate vibration effects on humans

Table 1. Standards and guidelines from various organizations for a subset of common equipment found in EPC contract builds.

Organization	Number	Title	Equipment Covered	Use or Effects Considered
ANSI/ASA/ISO	S2.80/20816	Mechanical vibration — Measurement and evaluation of machine vibration (jointly adopted)	Various equipment including gas and steam turbines and generators, industrial machinery, pumps, compressors, and gear sets	General evaluation criteria for operational monitoring and acceptance testing designed for reliable, safe, long-term operation of the machine while minimizing adverse effects on associated equipment
ANSI/HI	9.6.4	Rotodynamic pumps for vibration measurement and allowable values	Rotodynamic pumps with measurements taken from the bearings	Specific maximum allowable vibration values, acceptance criteria for field and factory tests
ANSI/AGMA	6000-C20	Specification for measurement of linear vibration on gear units	Reducing or increasing gear drive machinery	Vibration limits at specific frequencies are recommended for acceptance testing.
BS	6841	Measurement and evaluation of human exposure to whole-body mechanical vibration and repeated shock	Vehicles, machinery, and industrial activities that expose humans to whole-body vibration or repeated shock	Considers human health and comfort with respect to vibration
API/ISO	610 / 13709	Centrifugal pumps for petroleum, petrochemical, and natural gas industries	Centrifugal pumps, typically for petrochemical and natural gas processes	Correct selection and reliability of pumps chosen
API	617	Axial and centrifugal compressors and expander-compressors	Axial and centrifugal compressors	Provides maximum allowable vibration limits
API/ISO	619 / 10440	Rotary-type positive displacement compressors for petroleum, petrochemical and natural gas industries	Dry and oil-flooded helical-lobe rotary compressors in special-purpose applications	Allowable shaft and bearing vibration limits provided for achievement of design life
API	541	General purpose form-wound squirrel cage induction motors—185 kW (250 hp) through 2240 kW (3000 hp)	Squirrel cage induction motors	Allowable shaft and bearing vibration limits provided for reliability and durability
API	547	Form-wound squirrel cage induction motors—375 kW (500 Horsepower) and larger	Large squirrel cage induction motors	Allowable shaft and bearing vibration limits provided for reliability and durability for larger motors
EI	-	Guidelines for the avoidance of vibration induced fatigue failure in process pipework	Machinery including pumps and compressors, fluid flow and interaction of flow with pipework components (e.g., valves, dead legs), with a focus on the vibration of pipelines themselves	Effect of vibration on pipelines and fatigue failure of small-bore connections

EXAMPLE—PUMP VIBRATION:
As a hypothetical example, consider a pump installed in 2023 as part of an EPC contract build for a diesel fuel production facility in the United States. We will assume that the facility owner would like to know whether the pump is running appropriately with respect to "industry standards" for vibration, as several operators complained during commissioning that the pump was "shaking a lot." Standards specific to pumps as of the preparation of this paper include ANSI/HI 9.6.4-2022 [15], API 610 (12th edition, 2021) [5], ASME B73.1- 2020 [16], and ISO 10816-7:2009, last reviewed and confirmed in 2020 [17].

At first glance, all four standards might seem applicable. Moreover, there is some agreement among the standards (indeed, the ASME B73.1 standard was revised to better align with the HI and API pump standard in 2012). ASME B73.1-2020 states that performance tests must include vibration measurements, and "*The vibration level measured on the pump bearing frame, when specified, at the supplier's test facility at rated condition point (speed +/- 5%, flow +/-5%) shall not exceed the allowable "factory" pump bearing housing vibration limits shown in ANSI/HI 9.6.4 for type OH1 pump (B73.1 pumps)*" [16]. ANSI/HI 9.6.4 is primarily focused on vibration analysis and includes extensive detail regarding bearing housing vibration measurements, whereas ASME B73.1 addresses pump design in general. In many cases, it would be appropriate to use ANSI/HI 9.6.4 criteria to address the question of whether a pump is running appropriately. However, first we need to consider the type of pump and additional information. ANSI pumps (built in compliance with ASME B73.1) have distinctly different construction and characteristics from API pumps (built in compliance with API 610/ISO 13709) [18]. Accordingly, API pumps and ANSI pumps also have different vibration criteria for determining acceptable performance, although there are similarities in the test methods. Therefore, the evaluator needs to determine if the pump in question was manufactured to be compliant with API 610/ISO 13709. We can assume that the pump is an API pump for our example scenario, given that it is being used in a diesel fuel production facility to pump a flammable liquid.

API 610 is a comprehensive pump standard that includes design, testing requirements, and criteria. Detailed guidance regarding how and where vibration measurements should be taken are provided for a pump mounted on a test stand (usually conducted by the manufacturer) and mounted on a foundation (typically after installation, such as in our example scenario). The API 610 standard provides vibration criteria for any flow rate within the pump's preferred operating region. An evaluator could follow this standard to collect the data needed to answer the question of whether the pump vibration is acceptable during operation. If the test results do not indicate excessive vibration, further investigation is warranted e.g., to determine if the reports of the pump "shaking a lot" occurred during operation outside of the preferred region.

The last standard in the example list, ISO 10816-7, is broadly applicable to a wide range of pumps in industrial applications. Although ISO 10816-7 could also be consulted in our example, API 610 appears to be more closely aligned (given the petroleum production facility and intended use), and it contains sufficient guidance to determine if the pump vibration is acceptable. For a complete evaluation, the evaluator should obtain additional information such as the running speed and pump bearing type. As discussed above, the EPC contract will ideally designate standards to avoid ambiguity, but this example demonstrated some of the information that can be used to identify a reasonable standard.

CONCLUSION:

Vibration analysts can provide valuable insight into the nature and extent of a vibration problem, but if clearly applicable guidance does not exist and vibration criteria have not been agreed to by both parties, disputes can be difficult to resolve. If the EPC contract does not provide a means to identify when vibration is acceptable, the risk of damage to the equipment, surrounding components, or plant operators should be investigated. Equipment type, application, location, operating conditions, and other circumstances influence which standards are appropriate to use. Even with this information available, ambiguity can still exist, especially for the unique engineering and construction typical of one-of-a-kind plants. Appropriate standards and specific vibration criteria should be included in an EPC contract or otherwise discussed by parties before the design phase so that desired conditions can be met and the facility can operate smoothly.

REFERENCES:

[1] Garner, S., Kennett, S.C., Fecke, M., Clear Documentation Helps Avoid Power Project Disputes, Law360, August 20, 2021.

[2] International Organization for Standardization (ISO). Mechanical vibration—Methodology for selecting appropriate machinery vibration standards. ISO/TR 19201:2013.

[3] Standards Norway. Working environment. NORSOK S-002:1997.

[4] International Organization for Standardization (ISO). Bases for design of structures—Serviceability of buildings and walkways against vibrations. ISO 10137:2007.

[5] American Petroleum Institute (API). Centrifugal Pumps for Petroleum, Petrochemical, and Natural Gas Industries. API STD 610, 12th Edition, January 2021.

[6] American Petroleum Institute. API Catalog, https://www.api.org/-/media/files/publications/2023_catalog/2023_api_publications_catalog.pdf. Accessed April 6, 2023.

[7] American Petroleum Institute (API). Machinery Protection Systems. API STD 670, 5th Edition, November 2014.

[8] American Petroleum Institute (API). Pulsation and Vibration Control in Positive Displacement Machinery Systems for Petroleum, Petrochemical, and Natural Gas Industry Services. API RP 688 1st Edition, April 2012.

[9] Energy Institute. Guidelines for the avoidance of vibration induced fatigue failure in process pipework. January 2008.

[10] Army Test and Evaluation Command (ATEC). Environmental Engineering Considerations and Laboratory Tests. MIL-STD-810. 2022.

[11] Verein Deutscher Ingenieure (VDI). Measurement and evaluation of the mechanical vibration of wind turbines and their components—Wind turbines without gearbox. VDI 3834 Blatt 2. 2022.

[12] Verein Deutshcer Ingenieure (VDI). Measurement and evaluation of mechanical vibration of screw-type compressors and root blowers—Addition to DIN ISO 10816-3. VDI 3836. 2012.

[13] Cooling Technology Institute. Standard 163: Vibration Limits in Water Cooling Towers. https://www.cti.org/blogs/posts/standard-163-vibration-limits-in-water-cooling-towers-copy. Accessed April 6, 2023.

[14] International Organization for Standardization (ISO). Mechanical vibration and shock—Vibration and shock in buildings with sensitive equipment—Part 1: Measurement and evaluation. ISO/TS 10811-1:2000.

[15] American National Standards Institute and Hydraulic Institute (ANSI/HI). Rotodynamic Pumps For Vibration Measurement And Allowable Values. ANSI/HI 9.6.4-2022.

[16] American Society of Mechanical Engineers (ASME). Specification For Horizontal End Suction Centrifugal Pumps For Chemical Process. B73.1-2020.

[17] International Organization for Standardization (ISO). Mechanical Vibration - Evaluation Of Machine Vibration By Measurements On Non-Rotating Parts - Part 7: Rotodynamic Pumps For Industrial Applications, Including Measurements On Rotating Shafts. ISO 10816-7:2009.

[18] Bloch, H.P. and Budris, A.R. Pump user's handbook: life extension. CRC Press. 2021.

VIATC TRAINING CONFERENCE & EXPO
NIAGARA FALLS, NEW YORK JULY 26 -28, 2023

VALIDATION OF STALL / SURGE AND MITIGATION OF ISSUES FOR CENTRIFUGAL COMPRESSOR AT TEST BED – A PROPOSAL TO OEM & END USER

Mantosh Bhattacharya
Petrofac International UAE
Mantosh.b@petrofac.com

Mantosh Bhattacharya is designated Subject Matter Expert on Turbomachinery in Petrofac Intl. located in UAE. He is a chartered engineer CEng (MIMechE-UK), Intl. PE (UK), a member of ASME, Vibration Institute and ISCM – (BINDT UK). He is a member of the board of directors of MFPT (Society for Machinery Failure Prevention Technology) USA. He is member of advisory committee of ATPS [TEES – Texas A&M] He is the member of HI 9.6.4 , 9.6.5 and 14.5.1 standards .He holds 3 U.S. patents related to structural dynamics. He has published and presented several technical papers in journals and conferences.

ABSTRACT:
During centrifugal compressor performance test as per ASME PTC 10 test code , importance is given only for validation of thermodynamic performance of compressor. The tests for surge are conducted in a bit implicit manner as data taken as fluctuations in various process parameters only . In recent years, some more works have been done to co- relate the compressor stall / surge with machinery vibrations and validated at internal tests carried out by some of the OEM (Original Equipment Manufacturer) of centrifugal compressors.

This paper provides a road map to identify the source and mitigate such aerodynamic induced vibration. For mission critical high-pressure compressors, the paper proposes to use various instrumentation during type 1 tests to draw a surge limit line and locate the stall region in compressor map. These can be used also as a part of diagnostic tool to identify the real source of sub synchronous vibration at site

KEY WORDS: Surge ,Impeller stall , diffuser stall , dynamic pressure sensor , sub-synchronous vibration, High pass filter

INTRODUCTION:
For mission critical high-pressure compressors or for a high speed, high pressure centrifugal compressor, it sometimes becomes imperative to conduct an ASME PTC10 type 1 test along with standard API 617 compliant spin test before equipment is towed out.
For high pressure centrifugal compressors installed at site, aerodynamic induced (such as stall, surge) vibration may lead to resizing diffuser and return channel leading to enormous delay if issue is not identified at OEM test bench itself.
Vibrations in a centrifugal compressor occurs mainly due to three phenomena –

- Forced excitation
- Self-excitation due to lubrication / instability due to cross coupled forces and

• Purely due to aerodynamic instability inside compressor flow path

So far in a standard API 617 test and even at full load full speed test only the first two points are identified as influencing factors for compressor vibration . The objective of this paper is to address the third point proposing a plausible method.

A brief refresher of stall and surge is presented below for the reader before describing the proposed method.

BRIEF NOTES ON COMPRESSOR SURGE AND STALL:

Compressor surge –

Surge is defined as the operating point at which centrifugal compressor peak head capability and minimum flow limits are reached. When the plenum pressure behind the compressor is higher than the compressor outlet pressure, the gas tends to reverse or even flow back in the compressor. Consequently, the plenum pressure will decrease, inlet pressure will increase and the flow reverses again. A compressor is in "surging" condition when these sudden, repeated reversals of the fluid's flow direction occur. Flow reverses within 20-50 milliseconds and surge cycles as duration of 0.3 sec to 3 seconds

Compressor surge is the fastest phenomenon that occurs in process industries around turbomachinery, and with the most disastrous consequences. One of major factors contributing to developing surge is the rotating stall phenomenon, which can push the operating point into full surge.

Compressor stall – A centrifugal compressor normally has multiple stages. Each stage comprises of impeller, diffuser, return channel and volute as shown for a typical centrifugal compressor (Fig 2 below). Individual components such as Impeller and diffuser do have their unique stall phenomena which is described with brief note.

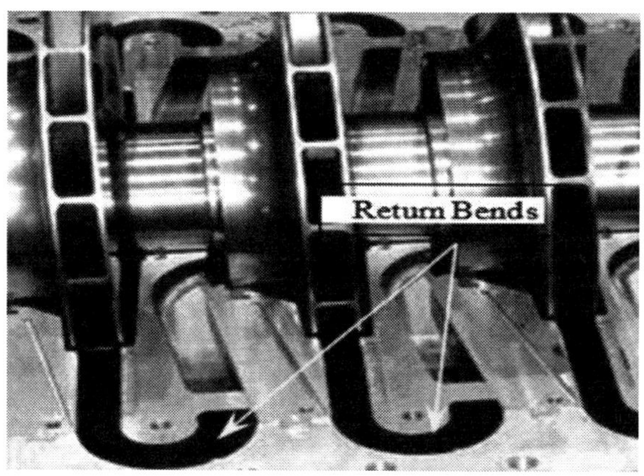

Fig 2 - a typical compressor stage - Courtesy Siemens -Dresser Rand

Impeller rotating stall –

When gas inlet angles into the impeller (measured in the axial direction) are increased, separation of the blade boundary layers can occur, as shown by the streamlines around the blade. Boundary layer separation occurs on the trailing side of the vanes. Stall cells are formed in the separation area and when they reach the impeller exit, it moves in a counter-rotation direction inside flow passage

This separation and distortion of the circumferential flow pattern cause the gas flowing through the compressor to slow down or stagnate at portion and sometimes result in reverse flow. In this situation, flow is very low, but still "going forward". If there is not enough gas flow to fully fill the blade passages in the impeller, boundary layer separation occurs on the trailing side of the blade. Stalling may occur over a few blades or a section of some stages.

These stall cells travel around the circumference of the compressor at 10–90% of the shaft speed, sometime rotating stalls occur just before the time of surge.

Rotating stall is a local pressure disturbance, in contrast to surge which causes fluctuations throughout the compressor system. If the flow is heavily interrupted i.e., there is no forward flow, pressure may surge back through the compressor. In this paper, progressive stall phenomena is considered.

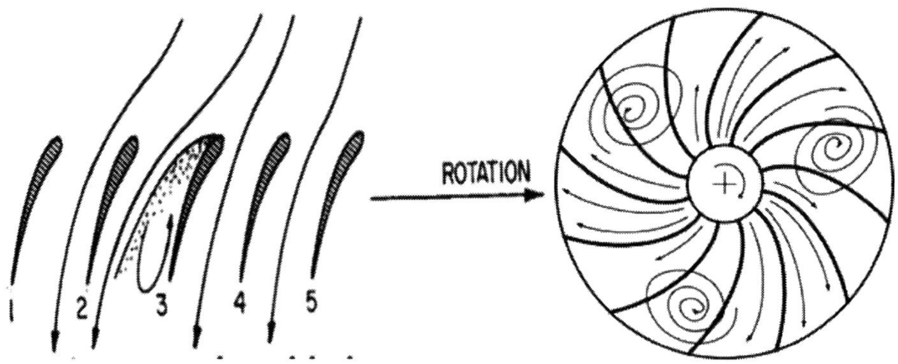

Fig 3 Progression of stall and stall cells in an impeller

Diffuser stall

There are two types of diffusers used in between bearing centrifugal compressors . The vaned diffuser types give better pressure recovery and efficiency than the vaneless diffuser, but the operation range is narrower that of the vaneless diffuser

Similar to the cause for impeller stall, when the flow angle exceeds a certain angle of incidence, the growth in boundary layer thickness results in increased blockage in the diffuser channels. The diffuser pressure recovery and stability decrease as inlet blockage increases.

Off-design operation continuously changes the diffuser flow angle and results in periodic flow unsteadiness in the diffuser inlet. As shown in Fig 4 the stall cells grow in diffuser and rotate against the direction of rotation of impeller.

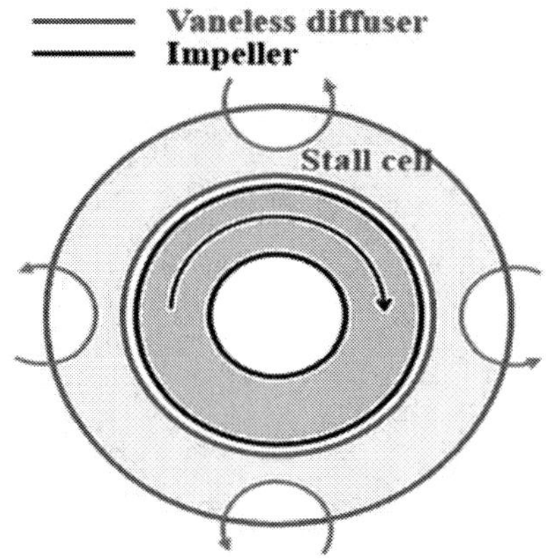

Fig 4– Diffuser stall

Each stage can experience stall and then stall encompasses the stage and finally full stage goes into stall.

Centrifugal compressors equipped with vaned diffusers are widely used in industry because of their high pressure-rise characteristics. However, they are likely to be prone for unsteady phenomena of rotating stalls, as compared with vaneless diffusers.

DISTINCT VIBRATORY PHENOMENA DURING SURGE AND STALL

In rotating stall condition, a radial pulsation is produced that acts orthogonally of the axis of rotation of compressor rotor. The pulsation across the impeller due to a rotating stall in a stage can be expressed as approximately: $\Delta P \approx \rho\ U2/20$ where ρ is the density of the gas and U2 is the impeller tip speed. The exciting frequency of the pulsation transgresses into radial asynchronous vibration

Surge propagates axially in gas flow path whereas rotating stall propagates circumferentially at some fraction of rotating speed.

The radial sub-synchronous vibrations are the effect of dynamic pressure pulsations that occur during a stall condition which are basically function of the gas density and impeller tip speed squared. The magnitude of the pressure pulsation causes vibration.

The aerodynamic instability due to impeller stall typically causes forward sub synchronous rotor vibration at a frequency of "less than 1X", typically 0.6X to 0.8X (60 to 80 percent of the rotor speed).

The diffuser stall is often accompanied by forward sub synchronous rotor vibration at a low frequency of around 0.2X to 0.4X (20 to 40 percent of the rotor speed). And the disturbances in the area between stationary and rotating channels (for instance, due to axial misalignment) can produce a forward sub synchronous component from the 0.4X to 0.6X range. Flow reversals do not occur, and axial vibration at the above frequency or significant movement is not detected. However if the process parameters are found to be within limits as compared to a bench mark data of stable operation , then reason to be explored to find possible flow obstruction or fouling.

From the diagnostic standpoint, diffuser stall differs from the other instabilities of the whirl type in its strong dependence on the compressor operating conditions; it should disappear with increased flow. It differs from surge by its frequency proportionality to the rotative speed as well as by the predominantly axial character of surge-caused vibrations.

Impeller rotating stall -

The sub-synchronous radial vibration frequency is observed in the range of from 50% to 80% of the compressor running speed.

The sub-synchronous vibration frequency tracks with running speed.

There may be a discontinuity or "droop" in the performance curve associated with the onset of the sub-synchronous vibration.

Diffuser stall –

The sub-synchronous radial vibration frequency is observed in a range of 6% to 33% running speed.

The vibration frequency is sensitive to flow rate and, after onset, will typically increase as the flow rate is reduced.

If dynamic pressure probes are installed on selected locations, their output shows a response in the 6% - 33% range and corelate with sub-synchronous vibration frequency.

After studying few inhouse testing by few OEMs and using the empirical formula it can be concluded that as we go forward to higher stages the amplitude of pulsation caused by stall and sub- synchronous vibration goes up with a good correlation as shown in figure below-.

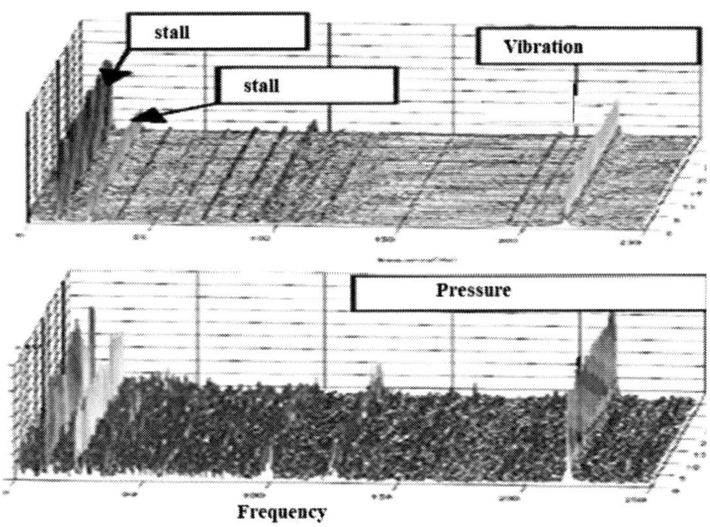

Fig 6 – Correlation between Pulsation and vibration during stall

Surge [When It Surges, It Shakes] –

The cyclic changes of the pressure difference across the compressor during surge, causes pressure pulses acting in the axial direction. These pressure pulses cause the thrust forces on the rotor to reverse. The thrust reversals, if large enough, can lead to a failure of the thrust bearings, and excessive axial displacement. The high-pressure centrifugal compressor does have comparatively lighter rotors. Repeated thrust reversals of sufficient magnitude in those rotors can damage the thrust bearing, inter-stage seals, shaft-end seals and shaft sleeves. High amplitude axial vibration – rotor shuttling, high radial vibration at quite low frequency may be considered as indication of surge.

Avoiding occurrence of compressor stall during design stage -

To mitigate the issues of aero instability and high sub synchronous vibration in rotor as a ramification, following actions are taken by compressor designer and manufacturer.

a. **Use of widely accepted critical angle criterion** – This is related to inlet flow angle and if angle of incidence crosses the critical angle, chances of occurrence of stall is high.

Senoo critical angle, as calculated below:

α= 105.91(b3/r2)1/2-117.27(b3/r2)-8.18 (1)

Nishida critical angle (known as modified Senoo criterion) as calculated below:

α*= α +(17.02-74.2 b2/r2) (1-b3/b2) (2)

With r2= impeller outlet diameter

b2= impeller exit width

b3= diffuser width

The modification in equation was carried out after multiple validation of Seeno criterion. As seen from the above equation b2 and b3 are of prime importance and ratio is normally termed as "diffuser pinch".

b. CFD analysis and its limitation –

CFD simulation of stages (particularly stages closer to discharge side / volute) are analyzed using required Navier Stokes Flow field equations with various commercially available solvers. These simulations are carried out -at the guarantee point and one operating point corresponding to surge protection line: The analysis normally proves that the compressor is free of any stall, either generated at inlet, or in the impeller, or at exit of the impeller.

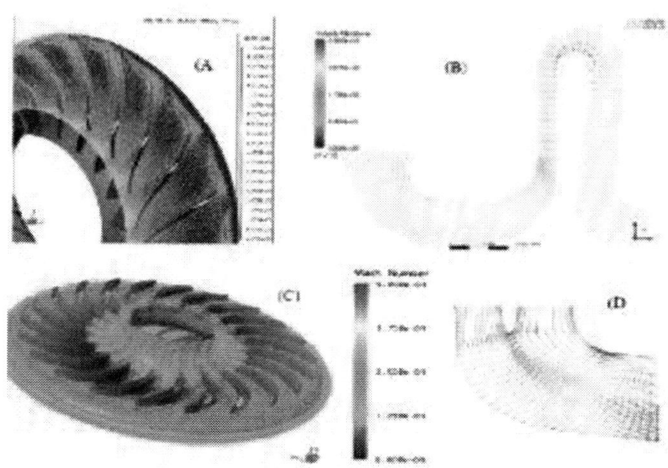

Fig 7 – CFD process during compressor flow path design

Although CFD analysis is based on correct assumptions, there are numerous reasons for CFD not matching test results. Some are limitations of CFD itself; others come from the practical limitations. Mathematical models of fluid flow are not perfect as exact modelling of gas path including surface roughness and aspects of actual fabrication cannot considered as built .

c. Low solidity vaned diffusers (LSD's). Experiments have proven that such diffusers delay the onset of rotating stall based on fine tuning the chord and pitch ratio. However, LSD is primarily used to augment static head rise in diffuser.

Fig 5 - Low solidity vaned diffuser- - Courtesy Siemens -Dresser Rand

OBSERVATIONS DURING FEW FIELD TESTS -

Noticing an audible squeal during compressor mapping at site, Machinery Vibration trends were taken from event log from diagnostic system. It was observed that time interval of squeal sound and Rotor axial movement trend had some distinct relationship. This excursion was detectable as RPM of rotor was quite low, Compressor was at recycle mode at low suction pressure. Further investigation concluded the reason of squeal noise as obstruction in gas recycle line path.

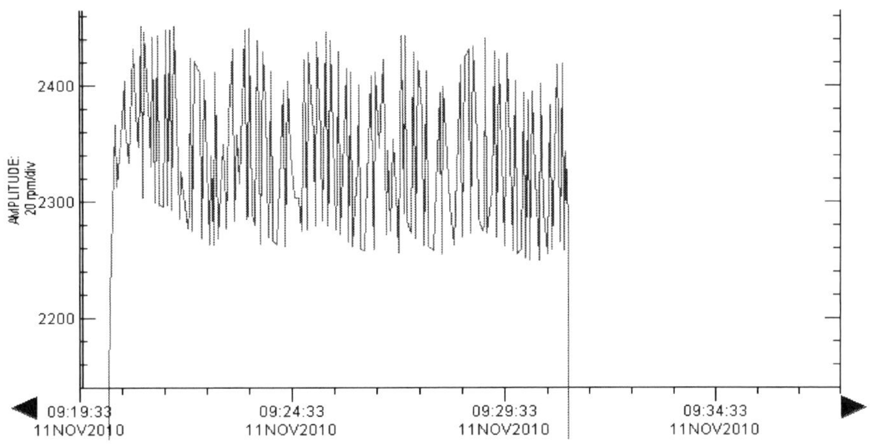

Fig-7 Axial excursion reading during low level pressure pulsation in loop

It was evident from the above observation that pressure fluctuation during surge shall cause axial excursion as above observation (Fig 7). In addition, gear box pinion radial vibration fluctuation was observed because of fluctuation of load resulting change in dynamic stiffness of pinion bearing.

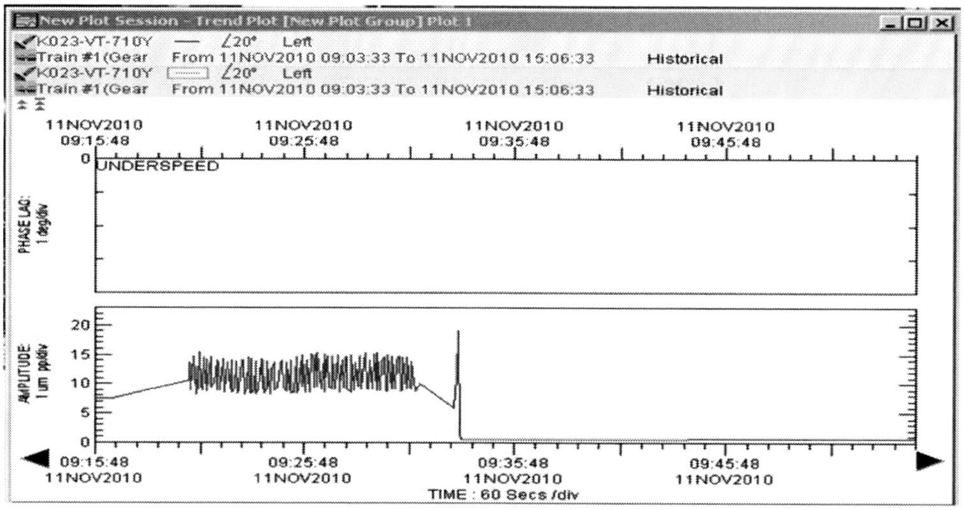

Fig 8 – Hunting in pinion vibration reading.

The onset of surge / machine stall is accompanied by a low frequency asynchronous vibration which is detected only when the surge energy is quite high. If the Compressor train is driven with a speed increasing gear box, the pinion being the lightest part of system may demonstrate some excitation in low frequency regime and a spectrum can easily reveal the symptom as witnessed during the incident (refer Fig 8).

Correlation of pressure pulses and axial excursion in another field experience

During commissioning of a turbo compressor train an incident of trip related to surge occurred with noise. The validation of capacity control algorithm with process parameters varied as mandated in data sheet, It was suspected that change in process parameter caused improper functioning of control algorithm and the surge was not real.

However, after careful analysis of time-based plot of suction pressure discharge pressure and axial shuttling of compressor rotor, all found to be complete agreement to each other, and it was concluded that a real surge event had occurred. Based on this understanding, borescope inspection was carried out and obstruction in recycle valve was found.

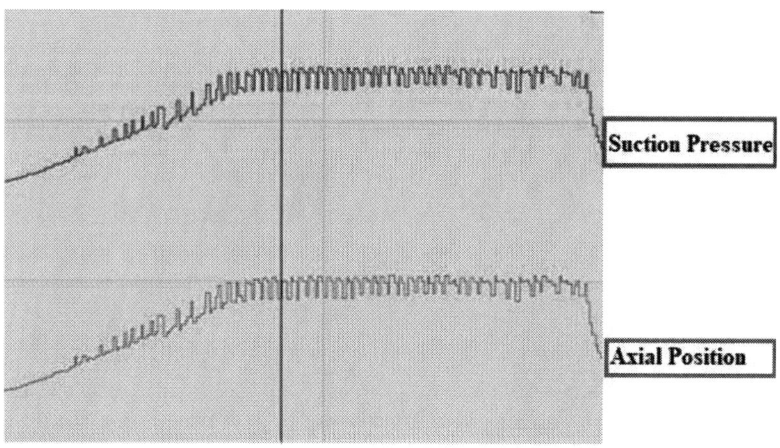

Fig 9 -Correlation between suction pressure pulsation and rotor axial shuttling

PROPOSAL TO OEM AND END USER–

Above findings have led to explore a possibility to diagnose such event during field commissioning and a road map to markup stall region based on possibility to measure the dynamic pressure fluctuations at penultimate impeller stage and prepare a proper prediction algorithm.

Detection of stall

This proposal is based on criticality of compressor unit particularly a high-pressure compressor with between bearing design and have stage drains. This write up provides a simplified approach to identify stage stalls and surge line during PTC type 1 test. This can be carried out based on measured stage pressure pulsations and sub synchronous vibration at OEM works[ ] . Refer to Figure 11

Before carrying out the ASME PTC 10 test, mechanical run test as per API 617 standard is to be carried out to identify if any sub synchronous vibration in rotor-bearing system exists . During testing of 1st unit of compressor as per PTC 10 Type 1, by throttling the flow towards the left side of curve radial and axial vibrations are measured along with FFT spectra of pressure pulsation near to return channel of impeller stages closest to discharge. If variable speed drive is used in train, the compressor shall be tested from 70% to 105% of operating speed range as shown in API 617 compressor map .

During that period, shaft radial displacement data can be filtered using a high and low pass filter. FFT spectrum of vibration can be correlated with pressure pulsation FFT. This activity can be carried out by setting up a multichannel vibration analyzer .

It is feasible to use inter-stage drains as provided by compressor OEM as per end user purchase standard as shown in Fig 10 below. Dynamic pressure transducers fixed in stage drain ports can capture the subtle pressure oscillations using a signal booster (if required) and signal conditioning system then can transform to a FFT spectrum.

It is important to note that with this set up, it is possible to differentiate the root cause of sub-synchronous vibration such as stall or low frequency instability such as bearing pad flutters and oil whirl.

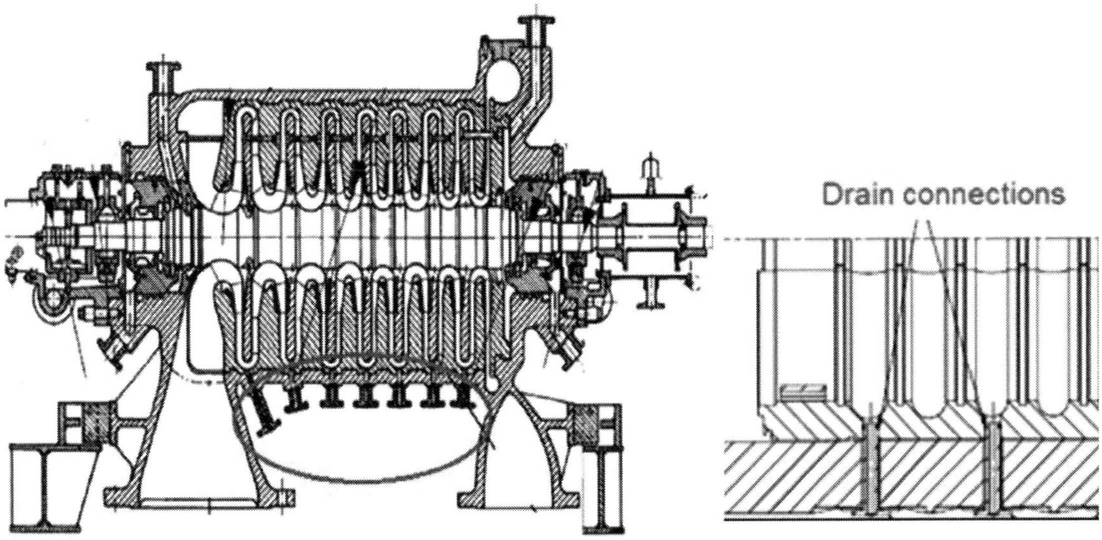

Fig 10 Utilization of stage drain nozzles.

Detection of incipient Surge

Based on few fields experience and many literatures available , vibratory responses during a mild surge , frequency of pressure pulsations and extent of compressor rotor excursions can be corelated . In addition, sub-synchronous vibration of gear box pinion also can be measured, and coherence of data can be checked.

To realize the proposed road map, it requires a signal processing system with Low and High pass filters and signal booster /charge amplifiers In addition dedicated microprocessor with a good scanning time with a logic box / comparator and a counter is also needed .. It is imperative that vibration sensors be mounted in line with the scheme shown in API 670 as per purchase order.

A schematic of compressor instrumentation is shown below in Fig 11, where vibration transducers are placed as per API 670 scheme and dynamic pressure transmitters are placed in compressor stage drain point openings.

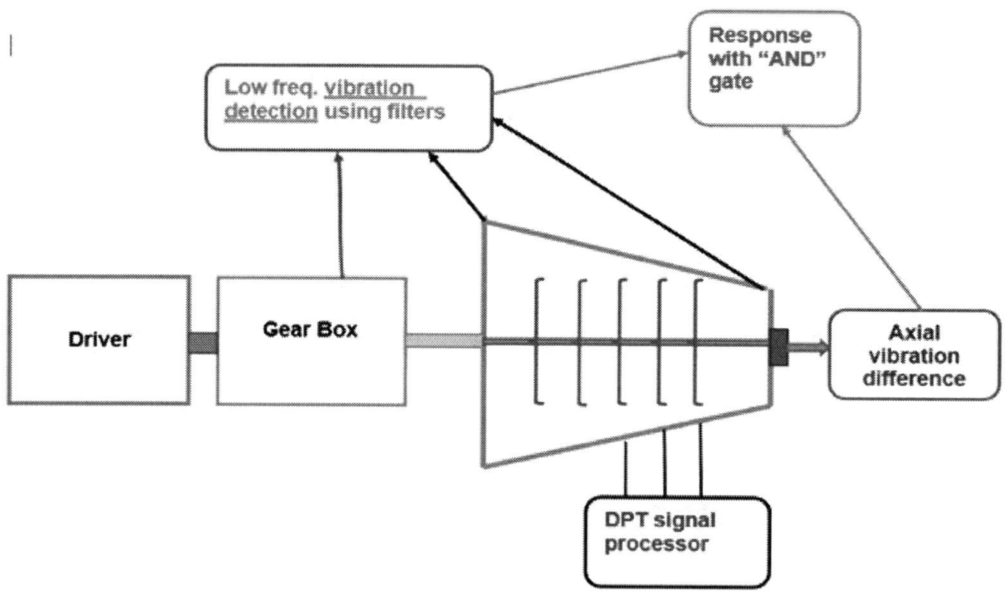

Fig 11 schematic of compressor instrumentation

To detect a mild surge the proposal is to use continuous measurement of axial excursion which can trigger with a suction pressure pulsation pulse. We can even use a comparator to find if gear box pinion low frequency vibration goes over a predetermined threshold. The detection logic can be armed as per field engineering and commissioning run requirement. This logic shall ring an alarm only and shall not override compressor capacity control and protection logic.

The logic uses real time measurement using signal splitters for getting the data to pass through a low pass filter to detect low frequency shuttling of rotor and gear box pinion vibration

Z1 = axial displacement of compressor shaft measured and saved as bench mark reference data

Z2 = axial displacement of measured at real time by proximity probes.

Z0 = difference between Z1 & Z2. This value shall be accompanied with a timer to estimate the severity of rotor excursion. The timer triggers based on value of Z0 above threshold.

Process parameter to consider – monitor Ps/Pd and compare to standard ratio from performance curve for reference and create a weightage for decision . This can be used as an "AND" logic gate .

The logic box can have block diagram in distinct colors to detect a stall and surge.

The logic box shall initiate an alarm for an incipient surge if the number of rotor axial shift is more than Z0 micron exceeds predetermined count of A in 2 seconds. Axial position transducers shall be the measuring element. The value of 2 seconds is based on numerous vendors experience that compressor can withstand a surge for 2 seconds with surge event with a counting of 2. A comparator can further be provided in software to compare the surge precursors with predetermined base line data as drawn during ASME PTC 10 test.

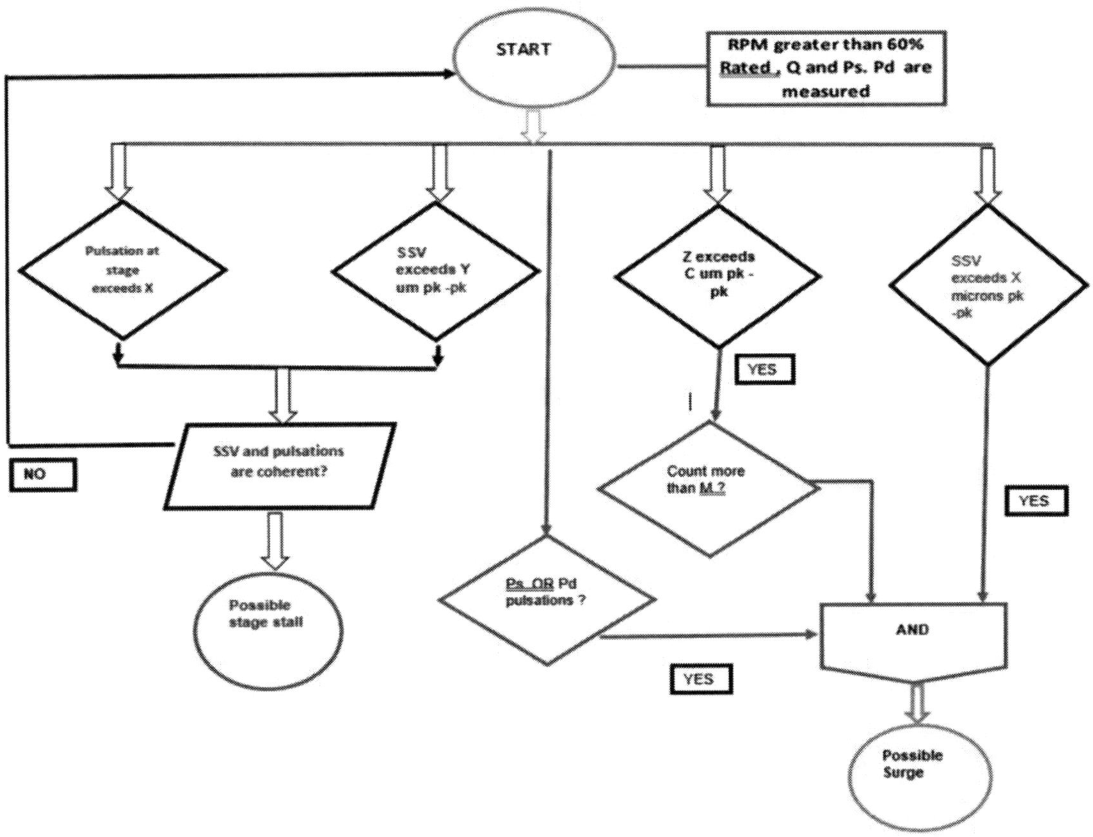

Fig 12 -Logic diagram to detect the events of stall and mild surge

Instrumentation -

Measurement and monitoring of radial vibration with low amplitude at low frequency calls for very accurate band pass filters. The proposal is to use signal splitters in the path where cables are going to vibration monitoring rack.

If the unit uses a speed increaser gear box, low frequency filtered shaft displacement probes can easily pick up the pinon vibration as pinion , normally the lightest part in transmission train.

The rotor excursion during surge is dependent on dynamic pulsation of machine and occur at low frequency hence high pass filtering may be used..

Fig 13 – Flush diaphragm type pressure transducer

Short note on required pressure sensor.

The dynamic pressure transmitter used to detect stall should use Piezoelectric transducer, with higher SNR (Signal to Noise Ratio). The proper mounting of dynamic pressure transducer with signal booster / charge amplifier can greatly mitigate attenuation of pressure pulsation signal. With integrated amplified output, the transducers are well-suited for continuous operation in hostile environments and in field test applications across long cables. Special low- noise cable and charge amplifiers may be required based recommendations by supplier of sensor. The transducer signal can be doubled and then each signal can be processed with a pre-configured with low pass and high pass filters accordingly which can detect diffuser and impeller stall properly.

Water falls spectrum of pulsation with RPM and time and water fall spectrum vibration with RPM and time can be a good comparative tool for benchmarking.

AKNOWLEDGEMENTS:

The author wishes to tender sincere thanks to Mr. MVL Narasimham and Mr. Viljoen Christo for internal review and suggestions for structuring of this paper. The author also acknowledges the support of VIATC 2023 program chair for accepting this paper for VIATC 2023 conference.

REFERENCES:

[1] A review of aerodynamically induced forces acting on centrifugal compressors and resulting vibration characteristics of rotors -James M. Sorokes, Mark J. Kuzdzal and D. Fred Marshall – 47th Turbomachinery Symposium TAMU Texas

[2] Sub synchronous Vibration of Multistage Centrifugal Compressors Forced By Rotating Stall - Fulton, J.W., Proceedings of Rotor dynamic Instability Problems In High Performance Turbomachinery, Texas A&M University1986

[3]Rotor dynamics of Centrifugal Compressor during Rotating stall – Donald Bently et al GE Orbit

[4] Vibrational Diagnostics of Rotating Stall in Centrifugal Compressors– Donald Bently et al GE Orbit

[5] Dynamic Data Analysis of Compressor Rotating Stall - Frank Kushner -25th Turbomachinery Symposium TAMU Texas

[6] A characterization of Rotating Stall behaviors in High and Low speed Centrifugal Compressors - Purdue University International Compressor Engineering Conference 1998

[7] A novel method to detect surge Mantosh Bhattacharya – Gas Machinery Journal Q2 2013

[8] System and method for detecting rotating stall in a centrifugal compressor- US patent 6857845 B2- Stabley et.al

[9] Turbo-Machinery Controls Application notes- Trisen

[10] Prediction of sub synchronous rotor vibration amplitude caused by rotating stall - Masayuki Kita et al- Proceedings of thirty seventh R=Turbomachinery symposium TAMU
[11] Stall Mode Transformation in the Wide Vaneless Diffuser of Centrifugal Compressors - Yaguang Heng et al.
[12] Centrifugal compressor diffuser rotating stall vaneless vs vaned - M. Giachi et al. Proceedings of 12th European Conference on Turbomachinery Fluid dynamics & Thermodynamics ETC12, April 3-7, 2017
[13] Axial and Centrifugal Compressors Application Guide -Bently Nevada BHGE
[14] Stall induced aerodynamic forcing and rotor vibrations in a multistage centrifugal compressor - Dr.Davide Biliotti et al -3rd Middle East Turbomachinery Symposium| ,Doha,

BIBLIOGRAPHY:

[1] API standard 670 5th Edition, November 2014
[2] Application Guideline for Centrifugal Compressor Surge – GMRC Release Version 4.3
[3] Compressor Surge and Rotating Stall: Modeling and Control (Advances in Industrial Control): Gravdahl, Jan Tommy,
[4] Centrifugal and Axial Compressor Control- Gergory McMillan

VIATC | TRAINING CONFERENCE & EXPO

NIAGARA FALLS, NEW YORK JULY 26 -28, 2023

USING TORSIONAL-LATERAL CROSS COUPLING TO IDENTIFY TORSIONAL RESONANCES

Brian F. Howard, PE
Principal Engineer, Bently Nevada
brian.howard@bakerhughes.com

Lysle Turnbeaugh
Senior Technical Product Manager, Bently Nevada
lysle.turnbeaugh@bakerhughes.com

Brian Howard, P.E. is a Principal Mechanical Engineer at Bently Nevada in Minden, Nevada. In this role, he focuses on sensor development, machinery analytics, diagnostic algorithm development, and research. Prior roles include Signal Processing Engineer and Global Technical Leader for Reciprocating Machinery Diagnostics where he provided diagnostic support to customers to solve problems with rotating and reciprocating machinery. Mr. Howard received a B.S. in Mechanical Engineering from the University of Houston and an M.S. in Electrical Engineering from the University of Washington. He is a member of ASME and IEEE.

Lysle Turnbeaugh is a Senior Technical Product manager Bently Nevada in Minden, Nevada. Over the last 15 years with Bently Nevada he has worked in product development where he has had the opportunity to develop new sensors, monitoring systems, and IEC 61508 emergency shutdown devices. Mr. Turnbeaugh received a B.S. in Electrical Engineering from the University of Nevada, Reno with recognition from Tau Beta Pi and IEEE-Eta Kappa Nu.

ABSTRACT:
The influence of shaft asymmetry due to design features (generator rotors) or machinery malfunction (the transversal crack on the lateral vibration responses of the rotor) is well documented in the literature. It is known that asymmetry of the shaft in the presence of side load can lead to high-level torsional vibrations. While the industry has a good understanding of the phenomena, the importance of cross-coupling in identifying torsional resonances under transient speed and load conditions gets less attention.

This paper defines the problem of cross-coupling and presents a mathematical model describing the coupled motion of a two-mass rotor system. It validates the model results on an experimental test rig. The paper then compares the model predictions to lateral and torsional vibration measurements, using conventional and recently developed non-contacting torsional vibration measurement systems. The case studies include large induction motors across both steady-state and transient conditions. The paper concludes with guidelines for using cross-coupling for torsional resonant identification.

Key Words: cross-coupling, torsional vibration, lateral vibration, transient response

INTRODUCTION

Condition monitoring in refineries, petrochemical plants, and power generation nearly always include measurements of lateral, or radial, and axial movement of the machine rotor, illustrated in Figure 1. Lateral vibration describes rotor movement perpendicular to the long axis of the rotor. The motion can be rigid body (translation of the rotor) or a bending motion. Excessive lateral vibration can result in contact between the rotating and stationary parts leading to poor performance or machine damage. In many cases, balancing of the rotor provide the desired control over the rotor behavior. For these reasons, lateral vibration has been monitored for many decades.

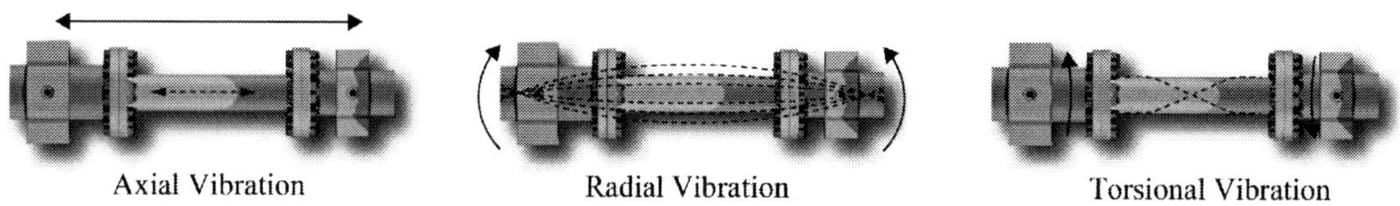

Figure 1: Types of rotor vibration

Axial vibration describes the rigid body rotor motion parallel to the long rotor axis. A number of different forces can cause this vibration. In process turbomachinery, such as compressors and pumps, fluid forces act on the rotor in the axial direction. Thermal turbomachinery, such as steam turbines or hot gas expanders, can experience differential growth between the rotor and case and the machine comes to temperature. In electric machinery improper magnetic alignment between the rotating and stationary field can result in an axial force in the direction of the rotor long axis. In any case, excessive motion can result in contact between rotating and stationary components and subsequent machine damage. For this reason, axial vibration has also been monitored for many decades. The monitoring has often focused on the slowly changing part, or static component, of the axial vibration; however, dynamic axial vibration can be rich in information.

Torsional vibration describes angular flexure about the rotor long axis. Power must flow through turbomachinery to perform work. This power is delivered as torque which causes the rotor to twist about the long axis. Torque usually changes slowly and is a static measurement. In addition to the torque on the shaft, the rotor can oscillate about the long axis. This dynamic torsional vibration indicates the rotor response to torsional forcing functions acting on the rotor. Measuring torsional vibration has been difficult to measure so it is often ignored; however, measurement technologies are advancing making the measurement more accessible for condition monitoring.

Traditional rotordynamics modeling treats lateral, axial, and torsional vibration independently. The approach is justified because lateral and axial vibration act against bearings and cause the machine case to vibrate. In contrast, torsional vibration cannot act against a bearing or other machine element so there are few cross-coupling elements. In some operating conditions and design configurations cross-coupling between lateral and torsional vibration occurs. Understanding interaction is important for turbomachinery condition monitoring.

Model Development

This section of the paper provides an overview of lateral and torsional rotor models and then explains how cross-coupling mechanisms operate in rotor systems.

Lateral View of the Rotor

Lateral rotor modelling techniques are well developed and presented in a variety of sources: API-684, etc. In this process the rotor is modelled as masses connected by lateral stiffnesses. The left pane shows a simple two-bearing, three mass system with stiffnesses and damping in the vertical and horizontal directions. For an isotropic system (no cross-coupling between the vertical and horizontal directions) the right pane of Figure 2 shows the system response. This figure represents the response, displacement of the shaft centerline at each station, from a rotating force acting on the center mass at different frequencies. The model does not assume that the force is synchronous; however, in most turbomachinery the rotating force is due to the residual unbalance which is synchronous with rotor speed. For this reason, transient analysis of lateral vibration proves essential in understanding and diagnosing rotor behavior.

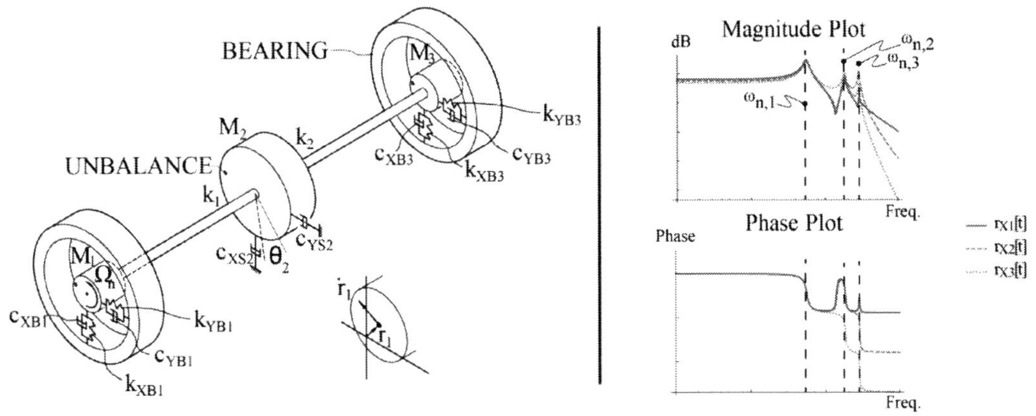

Figure 2 - Lateral System Schematic and Response.

Torsional View of the Rotor

The application of torque causes a shaft to twist about its rotational axis. Since all rotating machinery are involved in the delivery or transformation of power, all shafts experience some twisting about their rotational axis. The steady-state twisting is referred to as torque and the dynamic portion is referred to as torsional vibration. It is torsional vibration that is the focus of this discussion.

Torsional modelling follows a similar approach lateral vibration modelling. The analyst models the mechanical rotor system as inertia masses connected by torsional stiffnesses. The calculation report deflections as degrees or radians, analogous to displacement in lateral vibration. Much like the lateral system, the torsional system has stiffness and damping elements connecting the masses. Stiffness represents the resistance to an applied torque and damping indicates the resistance to the velocity at which the torque is applied.

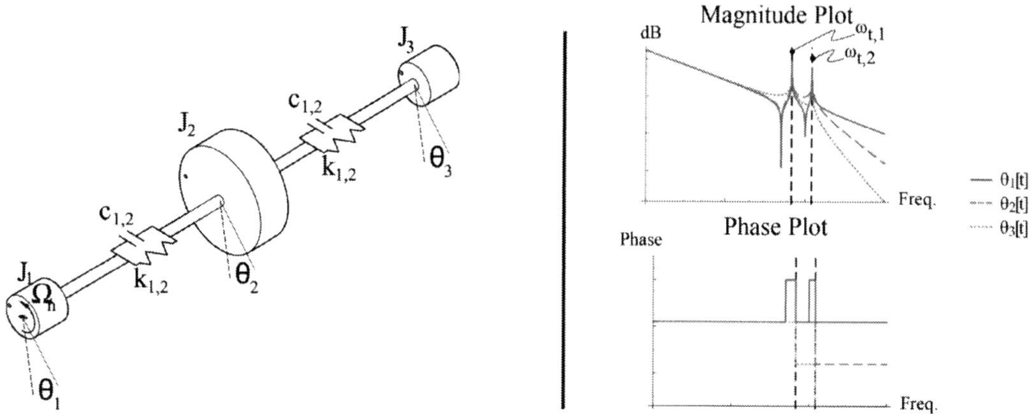

Figure 3 - Torsional System Schematic and Response.

The right pane of Figure 3 shows the torsional system response to a forcing function at the middle mass. Like the lateral vibration model, the response does not assume a cause for the forcing function. Unlike the lateral model, which can usually be connected to single forcing function related to rotor speed, torsional forcing functions can originate from multiple sources. These sources include misaligned couplings, rubs, gearbox tooth profile errors, changes in the electrical system, and cross-coupling of axial and radial vibration. These forcing functions are not always a harmonic of running speed so it can be difficult to detect them with speed-based measurements. One exception is cross-coupling of radial and torsional vibration. Since the radial vibration often results from a rotating force, when it cross-couples into torsional vibration the shaft rotation frequency will be the forcing function for torsional as well as radial. This allows many of the analysis tools developed for lateral vibration to be applied to torsional vibration measurements.

Cross-Coupling Mechanisms

One way cross-coupling can occur is if the polar mass moment of the system changes. This change could occur in a couple different modalities. For one example, assume a machine has a stiff rotor and operates well below its first balance resonance. If the shaft orbit is circular, then the polar moment of inertia is the same throughout the revolution. In contrast, an elliptical orbit indicates a change in polar moment of inertia over the revolution. Eddy-current probes are usually located near bearings (nodal points) so that changes in the orbit correlate with changes in flexure of the rotor. Flexing of the rotor causes a periodic torsional forcing function. Figure 4 shows a simple mass supported between two bearings. When the distance from the orbit to the centerline (average shaft position) is small the rotor is straighter than when the orbit is further away from the centerline. For small displacements, a mostly straight shaft, the driver sees a polar moment of inertia equal to the polar mass moment, J_p, of the rotor elements. If the shaft bends by an amount r, as shown in Figure 4, the disk center of mass processes about the shaft centerline rather than rotating about the geometric center of the mass. Since the polar mass moment is proportional to the mass times radius squared, the driver now sees a displacement equal to $J_p + m\,r^2$.

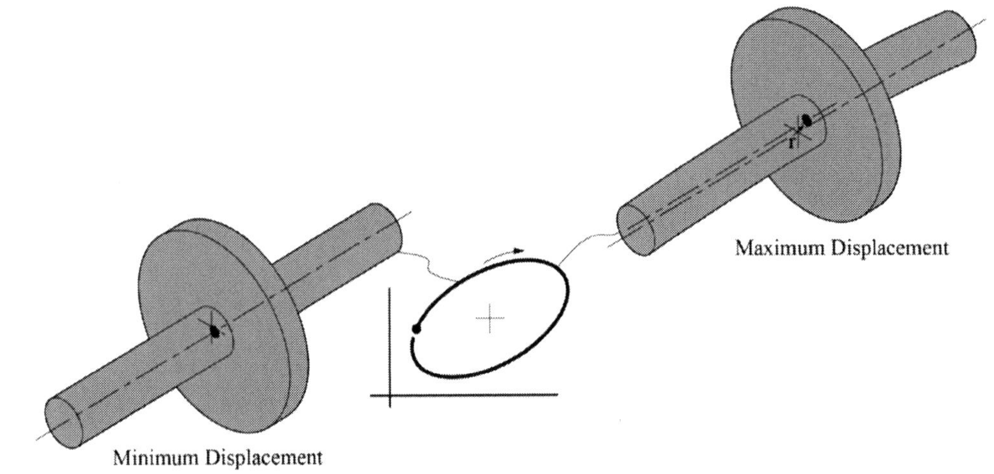

Figure 4 - Change in Bending Moment with Elliptical Orbit.

Torsional and lateral vibration can cross-couple through another mechanism. Rotors incorporating asymmetric radial stiffness (induction motors, generators, etc.) or asymmetry can develop from a shaft crack. In either case, the bending stiffness of the rotor changes as it rotates. In the presences of a static radial load the difference in stiffness causes the rotor to change shape as it rotates, illustrated in Figure 5. In this figure gravity acts on the rotor so that when the strong stiffness axis is parallel to gravity, as in the left side of Figure 5, there is little deflection. When the weak stiffness axis is parallel to the force, as shown in the right side of Figure 5, the shaft deflects more. As was the case with an elliptical orbit, the change in deflection changes polar moment of inertia. In the example shown in Figure 5 the change in stiffness repeats twice over a single revolution resulting in a 2X excitation. In this case the asymmetry is uniform and regular, repeating evenly throughout the revolution. In general, the change stiffness over the rotation is not sinusoidal so 4X, 6X, etc. harmonics may be present as well.

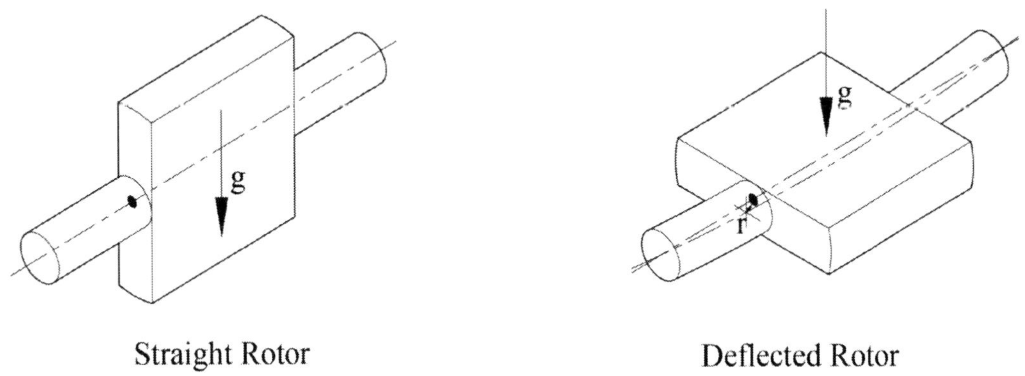

Figure 5 - Change in Bending Moment Due to Rotor Asymmetry in The Presence of a Sideload.

The cross-coupled model equates kinetic, potential, and dissipative energies in the lateral and torsional modes and use the LaGrangian to calculate the equations of motion. These models are complex and require many inputs. For example, compare the variables in the right side of Figure 6 to that in Figure 3. Some of the inputs, such as orientation and degree of asymmetry, are difficult to estimate for real rotors so predications and operating behavior can differ. Depending on the asymmetry and sideload, torsional components can appear in the lateral vibration signals, as shown in the right side of Figure 6, or lateral components may appear in the torsional vibration signal. The models show that shaft cracks can bring across different components than does the uniform asymmetry required by a design.

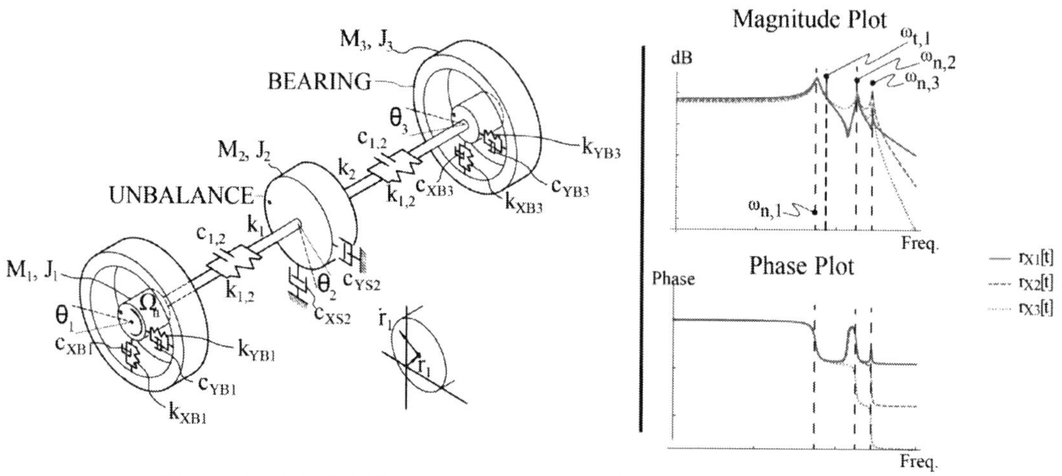

Figure 6 - Model Showing Cross-Coupling Due to Asymmetry.

Test kit application

The test kit, shown in Figure 7, consists of a drive motor driving a 3-mass rotor connected to a magnetic perturbator.The drive motor rotates the perturbator against the load motor. The perturbator has two poles so that the torsional excitation frequency is twice the difference of the load and drive motor speeds. The rotor itself has two measurement planes monitored by magnetostrictive torsional vibration sensors.

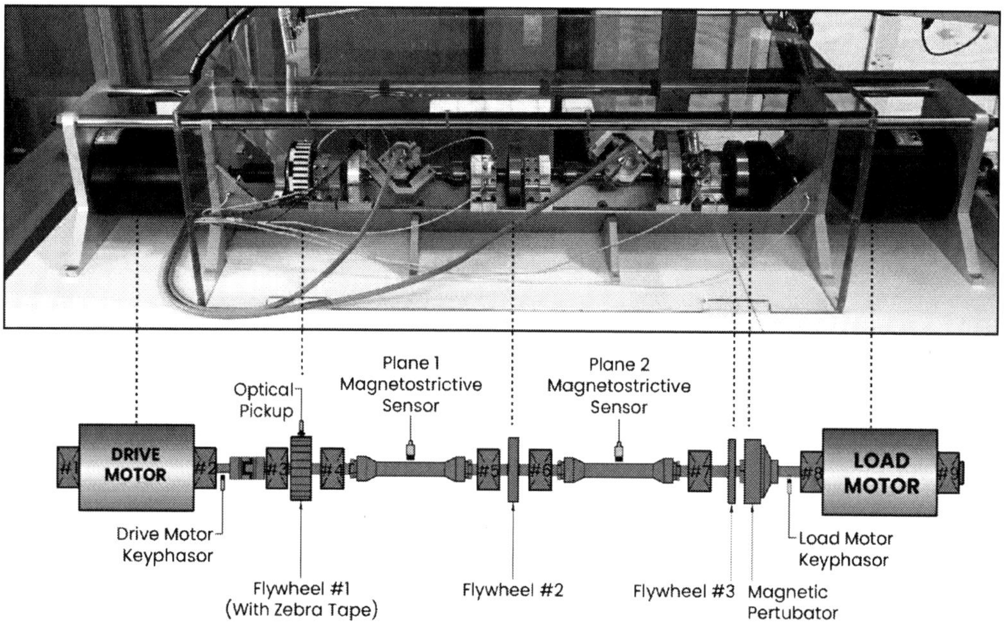

Figure 7 - Test Kit Machine Train Diagram

In this kit a cracked shaft results in shaft asymmetry. The shaft design placed the crack near a tungsten carbide flywheel. The test shaft, shown in Figure 8, is installed between mass #1 and bearing #4. The shaft had a test region machined down by 55% to get stress levels that result in a shaft life of 30-60 minutes. The test area also includes circumferential stress concentrators to accelerate the failure once cracks initiate.

Figure 8 - Shaft with Test Area

The first mode TNF response was found by varying the perturbation frequency around the natural frequency and is show in Figure 9. The bottom left pane shows the spectral response after 250 seconds of operation. The TNF is shown as 108.8 Hz, in good agreement with the model. At 600 seconds the response has shifted downward slightly to 108.5 Hz. At 850 seconds the first TNF has dropped to 107.6 Hz. Together with the 1X amplitude and phase these two indicators show the test area of the shaft has begun to crack.

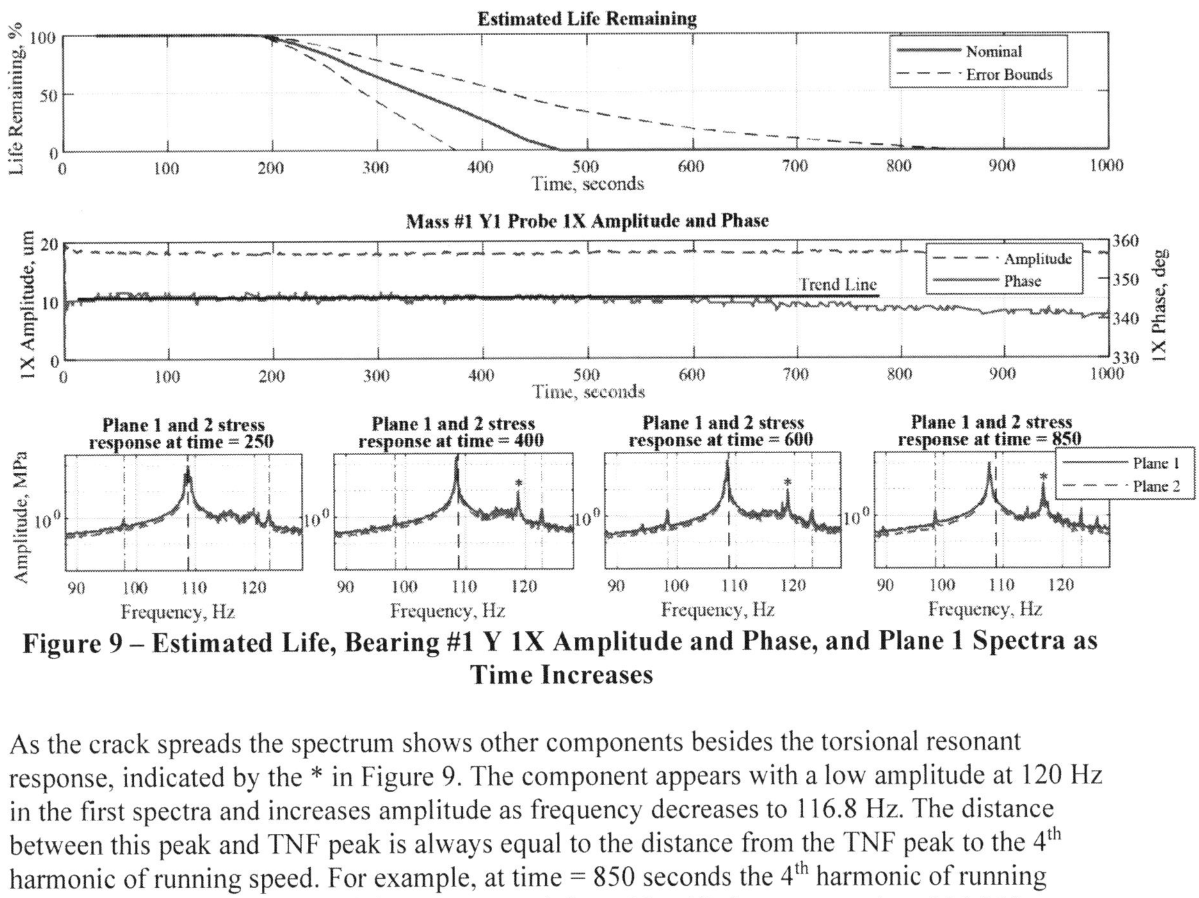

Figure 9 – Estimated Life, Bearing #1 Y 1X Amplitude and Phase, and Plane 1 Spectra as Time Increases

As the crack spreads the spectrum shows other components besides the torsional resonant response, indicated by the * in Figure 9. The component appears with a low amplitude at 120 Hz in the first spectra and increases amplitude as frequency decreases to 116.8 Hz. The distance between this peak and TNF peak is always equal to the distance from the TNF peak to the 4th harmonic of running speed. For example, at time = 850 seconds the 4th harmonic of running speed is 98.6 Hz, the TNF peak is 107.7 Hz and the unidentified component is at 116.8 Hz:

$$107.7\ Hz - 98.6\ Hz = 116.8\ Hz - 107.7\ Hz = 9.1\ Hz \mathrm{F} = \frac{\mathrm{m \cdot a}}{\mathrm{g_c}} \quad (1)$$

As discussed above 2X and 4X harmonics of running speed can cross-couple into torsional vibration if the shaft is asymmetric and a preload exists. In this case mass #1 was fabricated from tungsten and has a mass of 7.7 kg which provides a sideload to the test shaft. The asymmetry appears to worsen as the TNF drops, consistent with a developing crack.

The behavior observed in test kit as the shaft cracks can also be seen in turbomachinery. The next sections discuss case histories where the cross-coupling has been helpful in identifying torsional resonant frequencies.

Cross-coupled applications
Cross-coupling provides a useful source of excitation and in some cases can be used to detect and monitor torsional vibration. The following two sections provide application examples.

Test Stand

Figure 10 shows the test stand configuration for the back-to-back test stand. The drive motor has 2 poles and nominal power of 7174 hp (5350 kW). The load motor has 4 poles and nominal power of 6470 hp (4825 kW). Torsional models of the train predicted a first torsional natural frequency at 24.6 Hz (1476 CPM) and a second natural frequency of 48.8 Hz (2925 CPM). Two magnetostrictive sensors were mounted just outside the test motor drive end bearing, shown in Figure 10. The train also included eddy-current sensors installed near the magnetostrictive sensors and a reference strain system installed on a compliant torque cell installed near the driven generator.

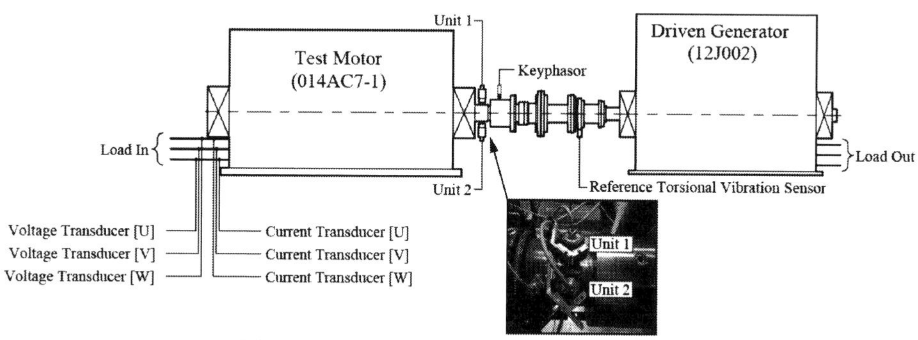

Figure 10 - Motor Generator Machine Train

The reference sensor is designed to be the most compliant member of the rotor system to enhance response to torsional vibration. Figure 11 shows the response as the machine speed ramps up. Both the test and load motor are induction designs, but the drive motor field induces more of a torsional response than does the load motor so cross-coupling tends to appear at 3X and its harmonics.

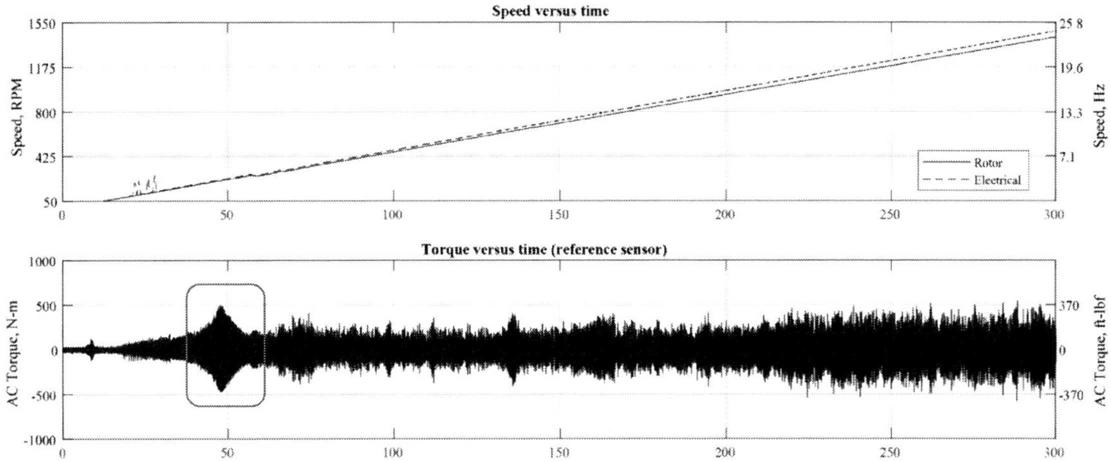

Figure 11 - Reference Torsional Vibration Sensor Response as Speed Changes.

Data is filtered at 6X for both the reference torsional and eddy-current lateral vibration sensors and plotted in Figure 12. The two left panes shown the amplitude versus shaft rotation and the right panes shown the filtered data in cascade format. The first torsional natural frequency during speed transients was previously reported as 22.5 Hz (1,356 CPM), O'Connor (2019), which agrees with the peak in the torsional vibration signal and one of the peaks in the eddy-current sensor signals. This peak is also clearly visible in the top right cascade plot. The eddy-current sensor also shows a large peak at 347 CPM implying a resonant response at 34.7 Hz. This machine has a first balance resonance at 31 Hz which, given the small amplitude of this signal and resulting uncertainty in amplitude and phase of the measurement, is the likely source of this resonance on the bode plot.

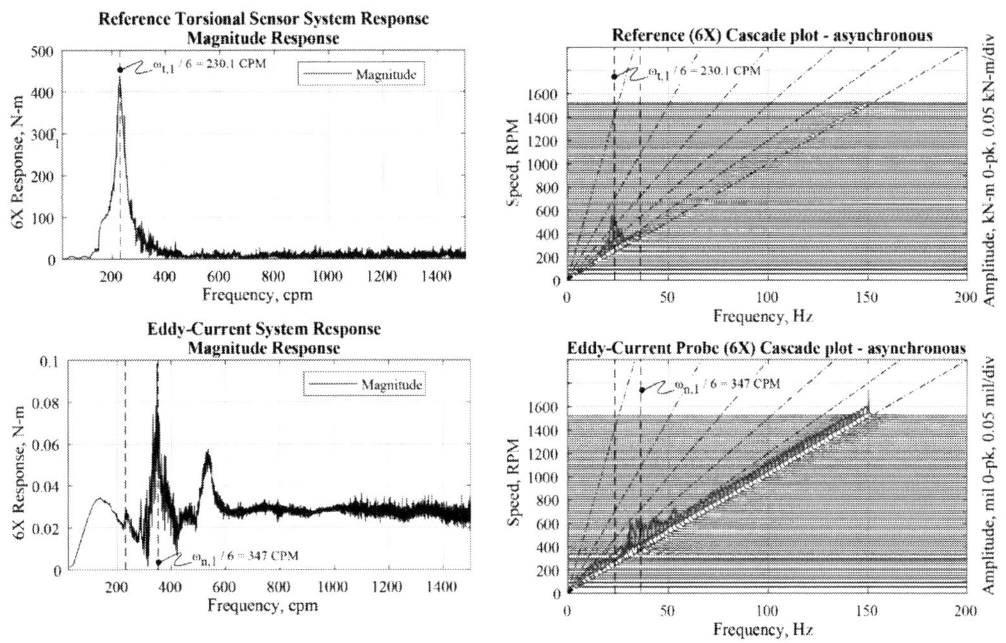

Figure 12 - Reference Torsional Vibration (Top Panes) and Eddy-Current Probes (Bottom Panes)

Another way to look at the torsional response in the eddy-current probe is to remove frequency associated with running speed by subtracting a running TSA average (a comb filter) from the signal. What is left in the signal does not relate to the running speed of the machine. Figure 13 shows the cascade plot for this signal. The torsional signal, highlighted by the rectangle, can be seen up to about 900 RPM. No other components with a vertical relationship are shown in the plot, except the components at 150.0 Hz, 3X line frequency, which is electrical noise as the load on the drive motor increases. Both the data in Figure 12 and Figure 13 are consistent with a sideloaded asymmetric rotor. This behavior provides confirmation of torsional natural frequencies.

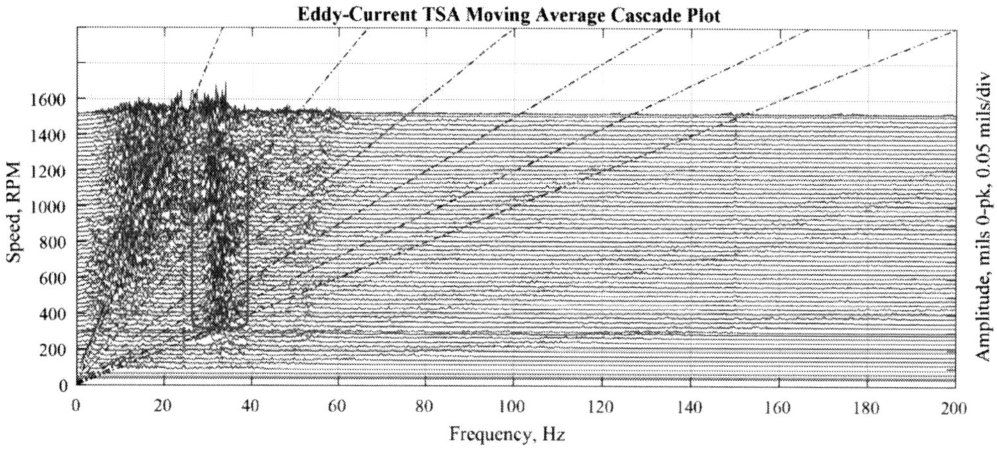

Figure 13 – Eddy-Current Probe Signal, TSA Moving Average Filtered.

Conclusion

Lateral vibration analysis has proven invaluable in turbomachinery condition monitoring. As torsional vibration sensing technology continues to gain acceptance, understanding the relationship between lateral and torsional vibration will enable detection of more machinery faults. Key to this understanding is the concept of cross-coupling in which the lateral and torsional vibrations interact. This interaction in governed by the machinery construction and operational details including shaft asymmetries in the presence of a sideload. These asymmetries can be the result of intentional design choices or developing problems in the rotor. Using models and a small test kit combined with a cyclical stress counter the resulting data showed different cross-coupling responses for uniform asymmetry, usually a design choice, and non-uniform asymmetry, possibly a cracked shaft, under a sideload.

Table 1 - Summary of data features for shaft asymmetry conditions

Data Feature	Uniformly Asymmetry	Non-Uniform Asymmetry
Lateral resonance in torsional	Not present	Can be present
Torsional harmonics	Present, depending on excitation source	
One or more torsional subharmonics	Not present	Can be present

Cross-coupling is relevant to designers and operators of turbomachinery. Where design choices permit, rotor asymmetries should be minimal to reduce cross-coupling and enhance detection of rotor problems. Original equipment manufacturers (OEM's) input on the transfer function between the measured plane of stress and the most highly stressed component in the rotor is essential to allow monitoring and detection of over-stressed components before cracks can develop. The combination of improved monitoring and continuing advancement of machine design will improve reliability and availability for equipment owners and operators.

NOMENCLATURE

AF	= Amplification factor	(-)
A_{hp}	= Half-power amplitude	(N-m)
A_r	= Amplitude at resonance	(N-m)
B_s	= Stator induction field	(T)
c	= Damping	(N-m-s/rad)
J_p	= Second polar Moment of Inertia	(m^4)
m	= Mass	(kg)
P_m	= Mechanical losses	(W)
P_{mag}	= Magnetization losses	(W)
P_{mec}	= Output power	(W)
P_n	= Nominal power	(W)
r_n	= Lateral displacement at station n	(mm)
s	= Laplace variable	(-)
S_{ut}	= Ultimate tensile strength	(MPa)
T	= Applied torque	(N-m)
ξ	= Damping ratio	(-)
τ_{tq}	= Torsional stress	(MPa)
ω	= Stator field frequency	(rad/s)
ω_{drive}	= Stator pulsation	(rad/s)
$\omega_{n,i}$	= Lateral resonance number i	(rad/s)
$\omega_{t,i}$	= Torsional resonance number i	(rad/s)
Ω_n	= Nominal rotor speed	(RPM)

REFERENCES

Adachi, Akira; Murphy, Brian (2016). Torsional Instability of Cooling Tower Fan During Induction Motor Startup. Texas A&M University. Turbomachinery Laboratories.

API Standard RP 684, 2005, "API Standard Paragraphs Rotordynamic Tutorial: Lateral Critical Speeds, Unbalance Response, Stability, Train Torsionals, and Rotor Balancing," Second Edition, American Petroleum Institute, Washington, D.C.

Bently, Donald; Hatch, Charles T. (2002). "Fundamentals of Rotating Machinery Diagnostics". First Edition.

Corcoran, J. P., Kocur, J.A. and Mitsingas, M. C., 2010, "Preventing Undetected Train Torsional Oscillations," Proceedings of the Thirty-Ninth Turbomachinery Symposium, Turbomachinery Laboratory, Texas A&M University, College Station, Texas, pp. 135-146.

Muszynska, Agnes; Goldman, Paul; Bently, Don. (1992). Torsional/Lateral Vibration Cross-Coupled Responses Due to Shaft Anisotropy: A New Tool in Shaft Crack Detection. IMechE.

O'Connor, David; et. al. (2019). Motor/Generator set non-contacting torsional vibration measurements. Turbomachinery Laboratory, Texas A&M Engineering Experiment Station.

O'Connor, David; Gutierrez, Jose. "Reveal the Umonitored Risks in Your Rotating Equipment-Part 1". Hydrocarbon Processing, January 20202. Pp 71-73.

Schijve, J. (2003). "Fatigue of structures and materials in the 20th century and the state of the art". International Journal of Fatigue. 25 (8): 679–702. doi:10.1016/S0142-1123(03)00051-3.

Shigley, Joseph Edward., et al. (2006) Shigley's Mechanical Engineering Design. McGraw Hill.

Smith, Natalie; Wilkes, Jason; Allison, Timothy; Moore, J. Jeffrey; Kulhanek (2018). Troubleshooting of Sub-synchronous Torsional Interaction Phenomena on an Electric Motor-Driven Centrifugal Compressor. Texas A&M University. Turbomachinery Laboratories.

Wu, Xi; Meagher, Jim (2008). A Two-Disk Extended Jeffcott Rotor Model Distinguishing a Shaft Crack from Other Rotating Asymmetries. Hindawi Publishing Corporation. International Journal of Rotating Machinery. Available electronically from https://doi.org/10.1155/2008/846365.

ACKNOWLEDGEMENTS

Special thanks to Olga Malakhova her help in developing and applying the cross-coupling models to the induction motor-generator set.

VIATC | TRAINING CONFERENCE & EXPO

NIAGARA FALLS, NEW YORK JULY 26 -28, 2023

OIL WHIRL INSTABILITY ON A GEARBOX AND GAS TURBINE UNBALANCE CASE STUDY

Sergey Drygin, PhD
Bently Nevada Baker Hughes
Machinery Diagnostic Technical Leader,
Dubai, UAE,
Sergey.Drygin@bakerhughes.com

Sergey Drygin, PhD joined Bently Nevada in 2006 in the Machinery Diagnostic Services group and has 25+ years of experience in condition monitoring and machinery diagnostics on various types of industrial machinery as Senior Machinery Diagnostics Engineer. Sergey has a PhD. degree in Machinery and Technology, Engineer's and bachelor's degrees from Faculty of Electro-mechanical Engineering, at Kuzbass State Technical University, Russia. Sergey is certified trainer for Bently Nevada products for more than 15 years. He is CAT III certified expert from Russian Society of NDT Certification, as well as ISO CAT III certified expert.

Nicolas Péton, PhD
Bently Nevada Baker Hughes
Nantes, France,
Nicolas.Peton@bakerhughes.com

Nicolas Peton, PhD joined Bently Nevada in 2006 in the Machinery Diagnostic Services (MDS) group. Previously he worked for two different manufacturers, Alstom Steam turbine and Cryostar expander/compressor. He worked in the chemical industry for PPG industry in USA and as an independent consultant for start-up activities worldwide. He has also worked as a mechanical/acoustical research engineer in research institutes (Technion, Haifa and TU Berlin). He is currently MDS Global Director for Machinery Diagnostic Services. He has a Diplôme d'ingénieur and a PhD degree from the Université de Technologie de Compiègne, France.

Michael J. Titone, PE
Machinery Diagnostic Services Technical Leader, Bently Nevada
Michael.Titone@bakerhughes.com

Michael J. Titone, P.E. is presently a Machinery Diagnostic Services Technical Leader for Bently Nevada. He joined the company in 2011. In prior job assignments, he worked for a power generation company and a vibration consulting firm in various roles. Since 1989, Michael has been a practicing engineer in the areas of vibration analysis, balancing, predictive maintenance/monitoring, mechanical system design/analysis, and I&C system design.

Michael earned a BS in mechanical engineering from Louisiana Tech University and an MBA from Mississippi College. He also earned 18 hours of graduate credit in mechanical engineering classes. He is a licensed professional engineer in the state of Mississippi and certified as a Category III Vibration Analyst through the Vibration Institute.

ABSTRACT:

The subject Gas Turbine Generator was a new unit that had experienced unexpected impacts due to the breaker closing and opening spuriously twice on 20th October 2017. After the incident, gas turbine and gearbox radial vibration increased and tripped the unit due to high vibration on the Drive End Gearbox High Speed Shaft, DE GB HSS (bearing #3). High vibration on the turbine side which exceeded the OEM Alert limit had been observed since the first startup.

A subsynchronous vibration (SSV) problem at 0.41X was detected on DE GB HSS (bearing #3). The problem occurred when the generator was synchronized, and the vibration increased and decreased with the megawatt load. The SSV was likely the result of oil whirl on the DE GB HSS (bearing #3). While the SSV was present, the direct vibration amplitude was not stable. On the turbine bearings, the dominant frequency of the vibration was 1X.

The Machinery Diagnostic Services (MDS) team suspected that a misalignment between the turbine and gearbox could be unloading the gearbox bearing. Other possible causes of the issue were excessive clearance in the gearbox bearings and oil seals, gear pairs misalignment, and possible twist on the gear casing. It was strongly recommended to perform balancing on turbine side. After one and a half months of investigation work, findings included (1) alignment out of tolerance between turbine to HSS GB, (2) HSS GB oil seals rubbing, (3) LSS GB bearings clearance not acceptable due to Babbitt damage, (4) GB casing twisting, and (5) imperfection on gear tooth contact.

After correction of the gearbox problems, the subsynchronous vibration on DE GB HSS (bearing #3) was eliminated, and the machine successfully reached base load (22MW). However, the turbine imbalance remained an issue. MDS proposed single plane balancing on the non-drive end of the turbine side in the accessories coupling. The first balance shot decreased the vibration to acceptable levels.

KEY WORDS: Gearbox; Fluid induced instability; Gear forces; Gas turbine; Balancing

NOMENCLATURE:

DE Brg– Drive End bearing
NDE Brg – Non-Drive End bearing
HSS – High Speed Shaft
LSS – Low speed shaft
GB – Gearbox
FSNL – Full Speed No Load condition

INTRODUCTION:

Bently Nevada's Machinery Diagnostic Services (MDS) was invited to perform vibration data acquisition and subsequent analysis during startup of newly commissioned 22 MW gas turbine generator. The train consisted of a gas turbine driving a generator through a gearbox. The generator was a two-pole air-cooled unit operating at 50 Hz, 3000 RPM. The rotor was supported by elliptical bearings. The coupling between the generator rotor and gearbox rotor was a rigid coupling.

Each of the bearings on the machine train was monitored by Bently Nevada XY Proximity probe pairs connected to the control panel. The gas turbine rotor rotated from Y to X, or clockwise, when viewed from gas turbine to generator (driver to driven). A machine train diagram is provided as Figure 1.

The gas turbine generator was a new unit that had experienced unexpected transients due to the generator breaker closing and opening spuriously twice on 20 October 2017. After the incident, gas turbine and gearbox radial vibration increased and tripped the unit due to high vibration on the Drive End Gearbox High Speed Shaft, DE GB HSS (bearing #3). High vibration on the turbine bearings exceeding the OEM Alert limit had been present since the first startup.

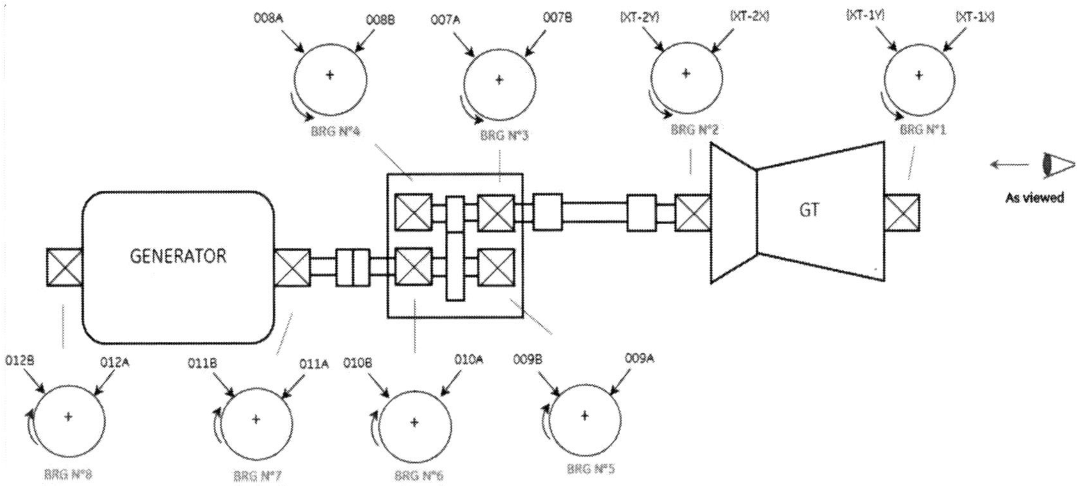

Figure 1: Machine Train Diagram with transducer location

DATA COLLECTION:

Initially, the vibration data in the control system was evaluated against the OEM limits. The Generator unit when synchronized, but before reaching 2MW load, indicated vibration amplitudes exceeding the OEM limits. A screenshot from the HMI is provided as Figure 2. The vibration reading that caused the trip was the 007A probe on Bearing #3.

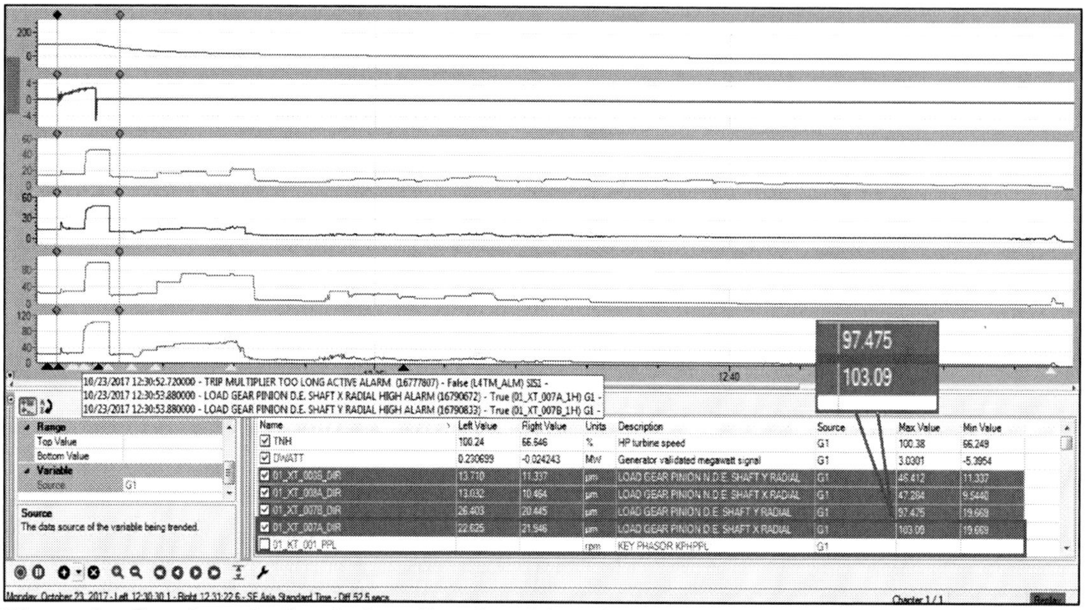

Figure 2: Gearbox shaft relative vibration, historical trends from HMI, µm pp

After this unit trip, a multi-channel data acquisition system (ADRE 408) was used for on-site vibration data collection. Vibration signals were connected from the control panel using BNC connectors. During the megawatt loading process, the vibration on bearing #3 was shown to increase and fluctuate. Note that the 1X vibration amplitude component remained low while the direct vibration amplitude increased and fluctuated as shown in Figure 3. Although not shown in the trend plot below, when the load decreased, the vibration decreased.

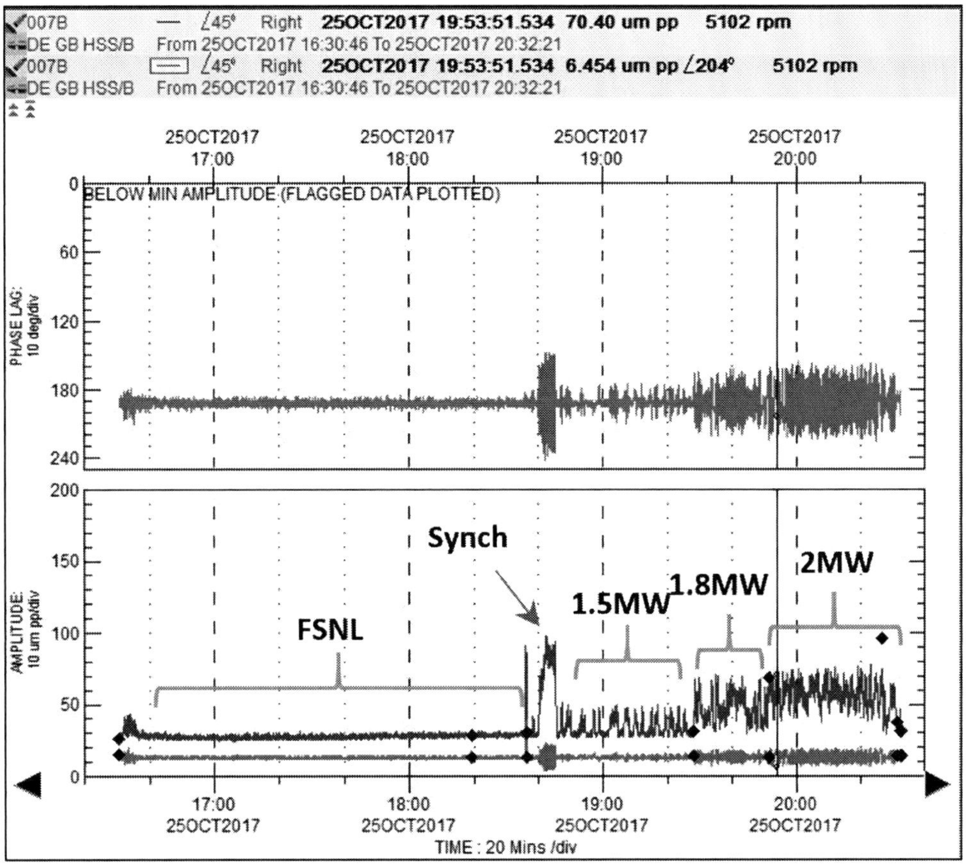

Figure 3: Gearbox shaft relative vibration FSNL, loading up to 2 MW

Based on ISO 7919-4 (Mechanical vibration of non-reciprocating machines – Measurements on rotating shafts and evaluation criteria – Part 4: Gas turbine sets) standards, the generator bearing vibration levels for the relative vibration transducers were in zone B – machines with vibration within this zone normally considered acceptable for unrestricted long-term operation. However, for newly commissioned machines vibration within zone A is expected.

DATA ANALYSIS AND MALFUNCTION INVESTIGATION:

The vibration response at full speed no load (FSNL) conditions was normal. When the vibration reached the OEM trip levels, the unit was operating at 2 MW. The direct (compensated) orbit/time base plots of bearing #3 provided in Figure 4 and Figure 5 show the response at 2 MW and at FSNL.

At FSNL conditions the orbits were elliptical with the dominant frequency of 1X. At 2 MW, the orbits showed a dominant subsynchronous vibration (SSV). The motion was mostly circular motion with forward precession. The dominant vibration frequency was less than 1/2X as indicated by eight Keyphasor dots per 3.5 revolutions.

Figure 6 provides a full spectrum waterfall plot of GB HSS (bearing #3 and #4) during steady state conditions from FSNL to 2 MW. The plot shows the initiation of the dominant subsynchronous vibration component at 0.41X when the unit was synchronized.

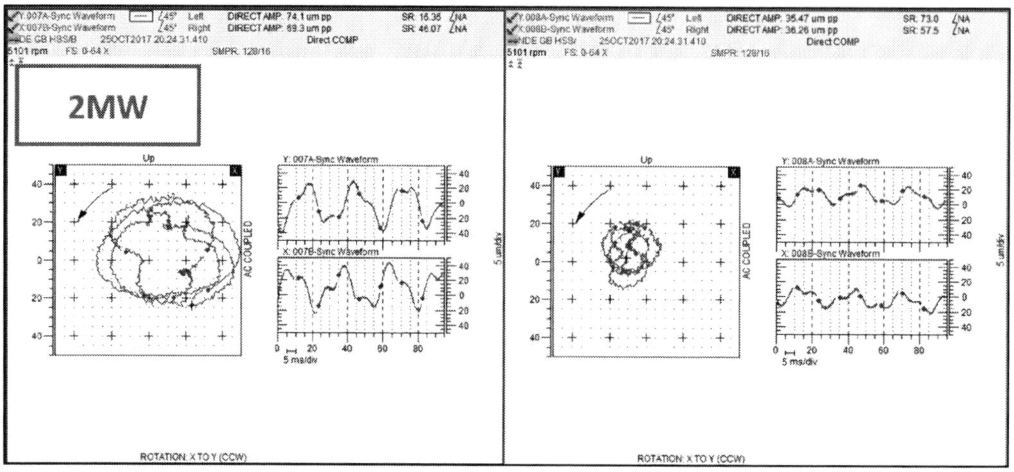

Figure 4: Direct orbit timebase plots of Bearing 3 and 4 steady state at 2 MW load

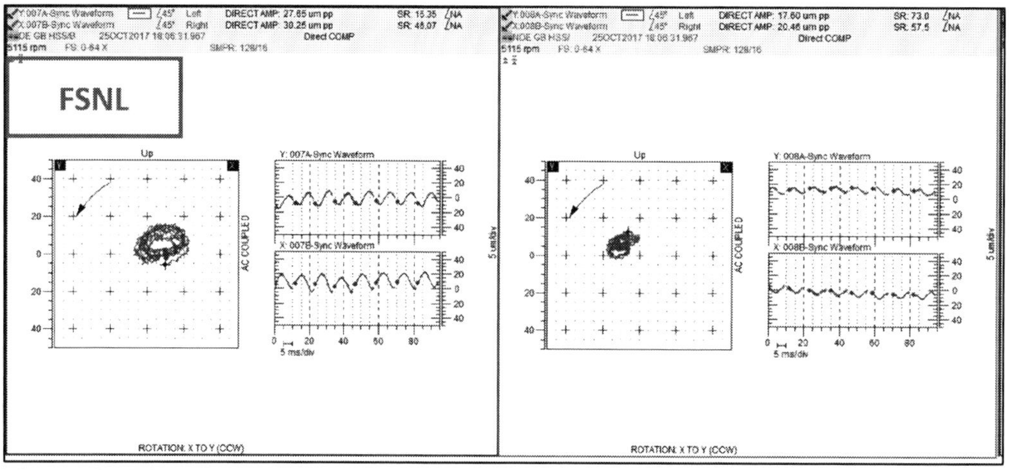

Figure 5: Direct orbit timebase plots of Bearing 3 and 4, steady state at FSNL

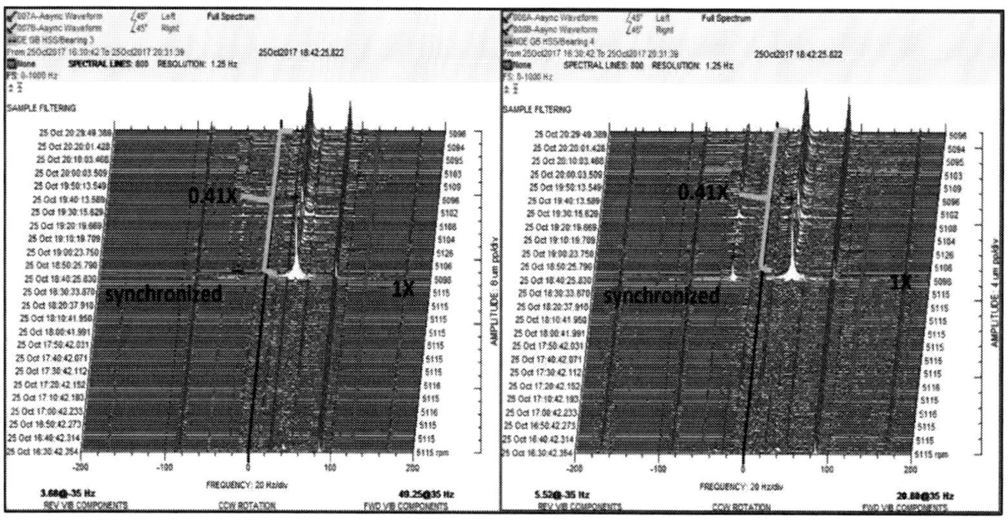

Figure 6: Full spectrum waterfall plot of bearing #3 and #4. Subsynchronous spike appeared at the Unit synchronization.

MDS suspected that a fluid induced instability, oil whirl, was causing the SSV on the Gearbox high speed shaft bearings. However, oil whirl is only likely on an unloaded bearing. The reader should refer to Bently, Chapter 22, [1] and Eisenmann, pp. 422-435 [2] for additional information on oil whirl.

MDS calculated the gear forces to estimate the resultant forces, as shown in Figure 7 and Equations 1 – 4. The following parameters were used in the calculations:

- The weight of the high speed shaft (HSS) was 769.6 kg (1697 lbs.)
- The weight of the low speed shaft (LSS) was 6371.7 kg (14025 lbs.)
- Rated power was 20 MW
- Pinion speed was 5102 RPM
- Pinion pitch diameter was 230.084 mm

$$Tangential\ force = \frac{2*torque}{pinion\ pitch\ diameter} = \frac{2*\left(\frac{9549*power\ [KW]}{pinion\ speed\ [rpm]}\right)}{pinion\ pitch\ diameter[m]} = 324311\ N \quad (1)$$

$$Separation\ force = tangential\ force * \tan(20*\pi)/180 = 118038\ N \quad (2)$$

$$Resultant\ force\ on\ HSS = \sqrt{(324311-7548.6)^2 + 118038^2} = 338040.5\ N \quad (3)$$

$$Resultant\ force\ on\ LSS = \sqrt{(324311+62386.3)^2 + 118038^2} = 404311\ N \quad (4)$$

The resultant force on the HSS at rated load was 45 times the rotor weight, definitely putting the pinion rotor into the upper half of the bearing.

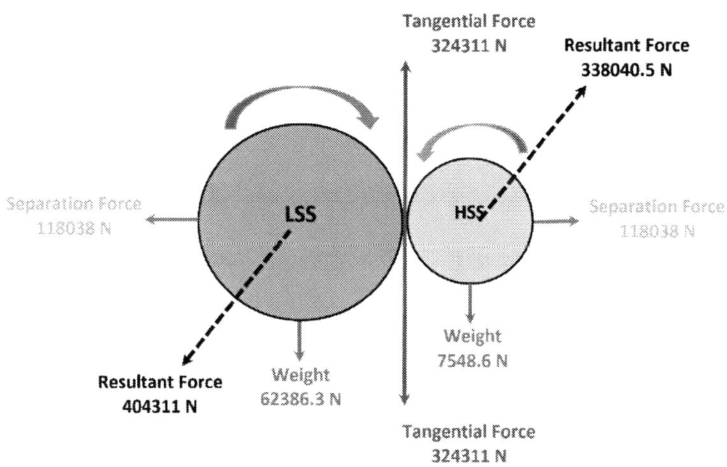

Figure 7: Estimation of active gear elements forces at steady state condition.

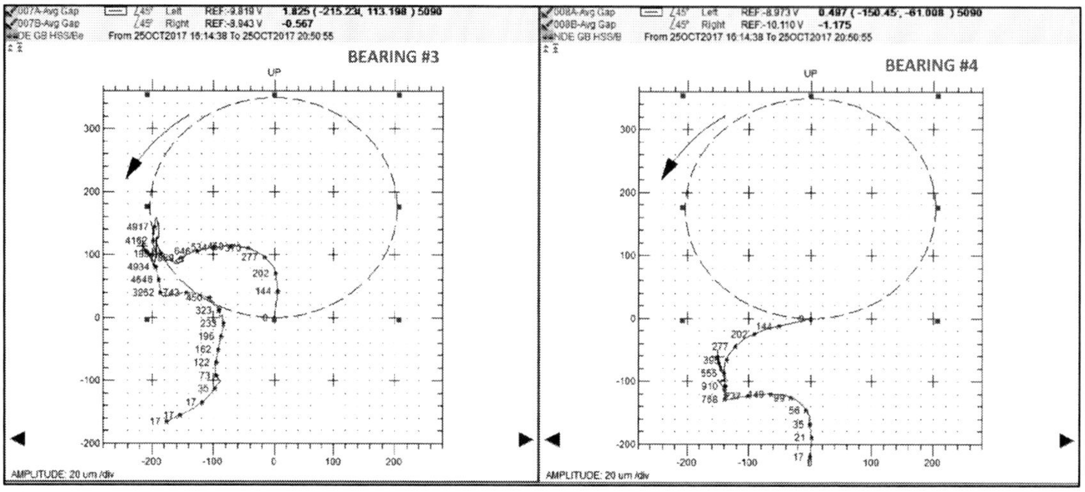

Figure 8: SCL plot (compensated for SU position) of bearing #3 and #4 during startup-steady state-shutdown condition

Based on the calculated resultant forces, the HSS should have moved up and to the right during operation. However, the shaft centerline (SCL) plot provided in Figure 8 shows movement to the left. Note that the plots were referenced assuming the rotor was in the bottom of the bearing at startup. For gearboxes, this assumption may not be correct. Therefore, the clearance circle shown on the plot should be used as a reference only. The plot indicates an improper load condition affecting the movement of the rotor.

MDS suspected that the gear box casing had experienced some deformation, followed by strong forces during the abnormal synchronization event and provided the following recommendations to the customer:

- Strongly recommended – check the alignment condition between gas turbine rotor and the GB HSS
- Strongly recommended – check the bearing #3 clearances and inspect the bearing for damage
- It is recommended to check the lube oil path through the bearing #3.
- Check gear pairs alignment.
- Perform blue contact test on gear pairs and check for possible twisting on the gear casing.

During the inspection, the following malfunctions were found:

- Out of tolerance alignment between GT and HSS GB.
- HSS GB bearings (bearing #3 and #4) clearances still acceptable
- HSS GB oil seals rubbed the shaft
- LSS GB bearings (bearing #5 and #6) clearances not acceptable (wipe white metal bearing)
- GB casing twisting
- Gear tooth contact was not correct.

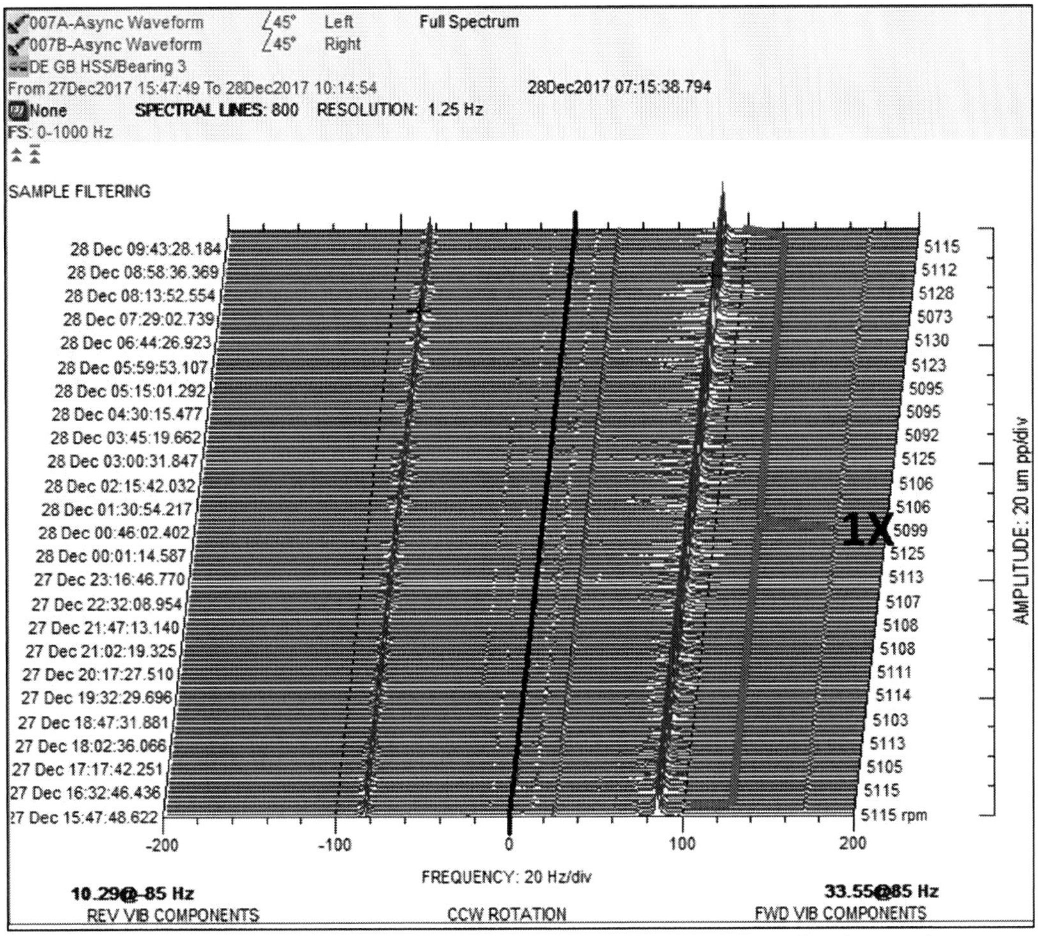

Figure 9: Full spectrum waterfall plot of Bearing #3 after GB major inspection

During the work resulting from the inspection, the following items were addressed:

- The gearbox casing twisting was corrected
- The gear pair was aligned
- The gearbox to GT and generator were aligned
- The gearbox HSS oil seals and gearbox LSS oil seals and bearings were replaced.

As result of these corrective actions, the gearbox HSS bearings were properly loaded. The fluid induced instability problem was resolved as shown in Figure 9 and Figure 10.

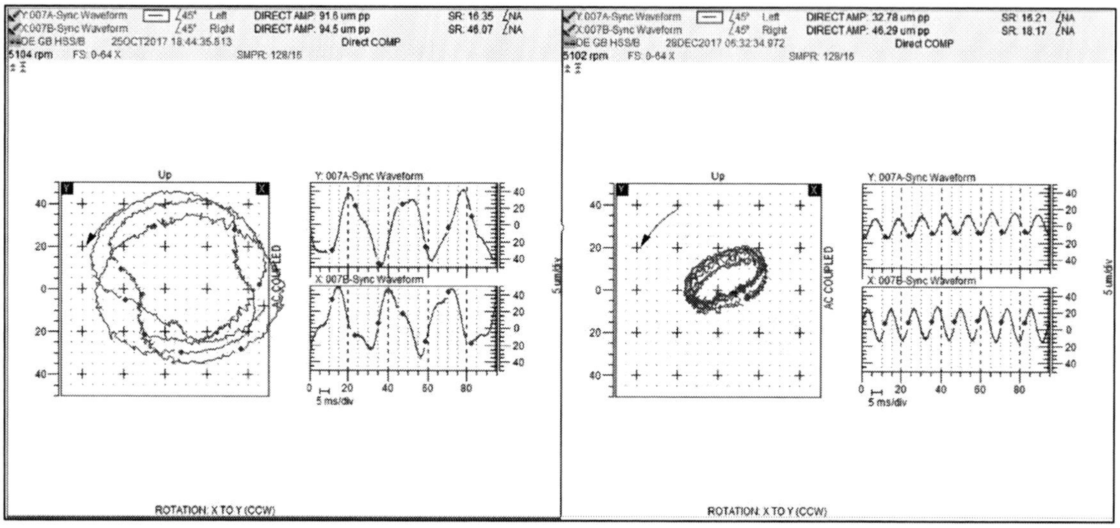

Figure 10: Direct orbit/time base plots (compensated) of bearing #3 and #4 comparison before GB major inspection (left) and after GB major inspection (right) during steady state condition

ON-SITE BALANCING:

To address the remaining machine train issue, MDS recommended on-site field balancing. Figure 11 provides 1X compensated polar plots from the NDE turbine bearing. Figure 12 and Figure 13 provided orbit time-based plots of the turbine at steady state conditions. The dominant frequency of the vibration is 1X, and there are no obvious indications of any issues except for some imbalance.

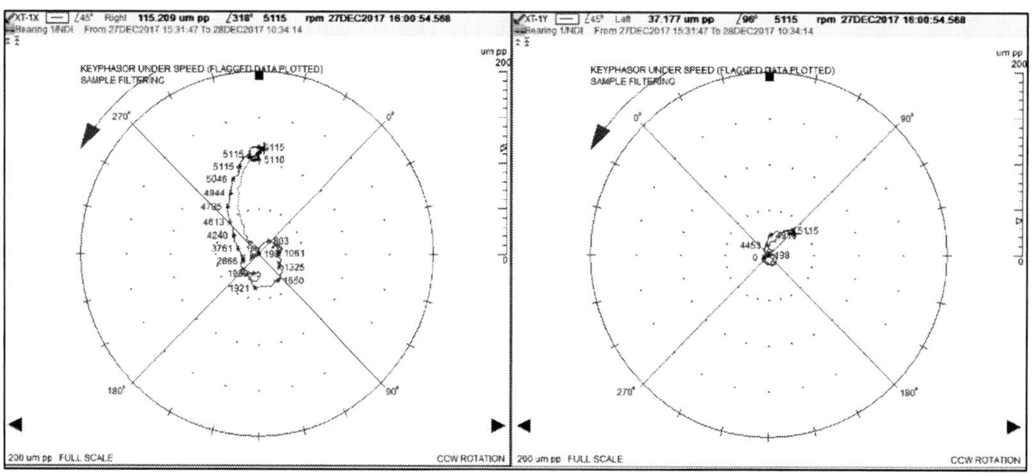

Figure 11: 1X compensated polar plots, Turbine bearings

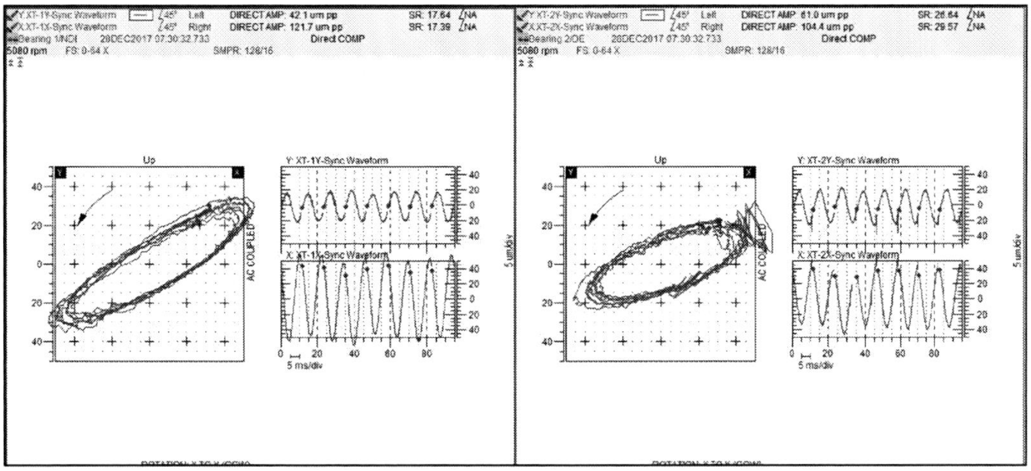

Figure 12: Compensated direct orbit plots, turbine bearings

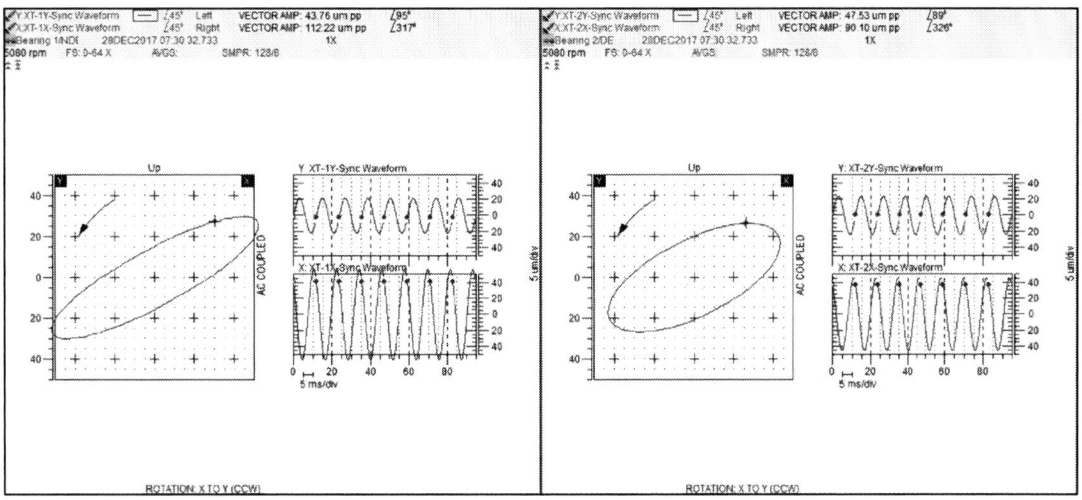

Figure 13: Filtered 1X orbit timebase plots, turbine bearings

Balance weights were installed on the accessories coupling at the NDE end of the turbine. The vibration decreased significantly as a result. The polar plot provided in Figure 14 and the 1X filtered orbit plot provided in Figure 15 show a comparison between before and after balancing. The plots show that the imbalance force decreased significantly during the first trial run, indicating that the calculated heavy spot location was correct, and the weights were installed opposite the actual heavy spot.

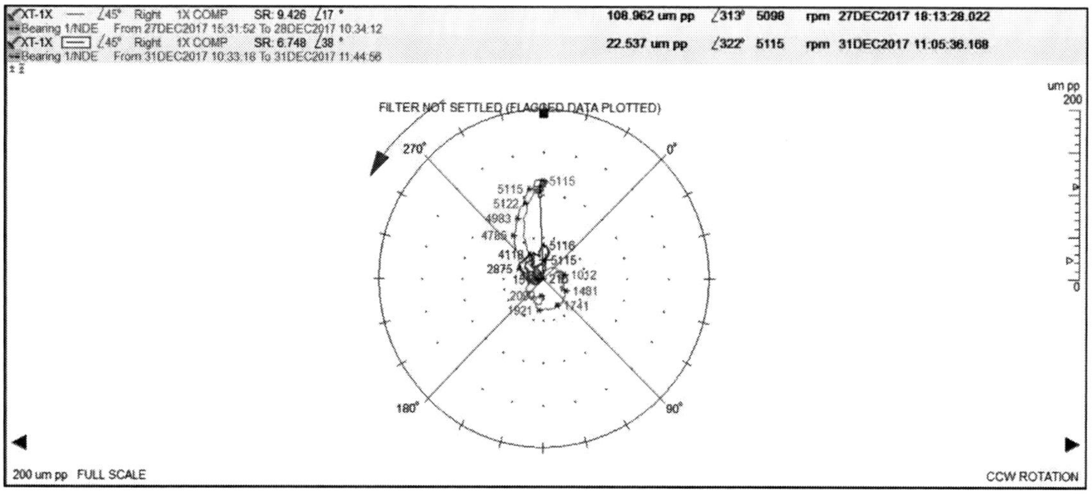

Figure 14: Polar plot comparision before (red) and after (blue) balancing

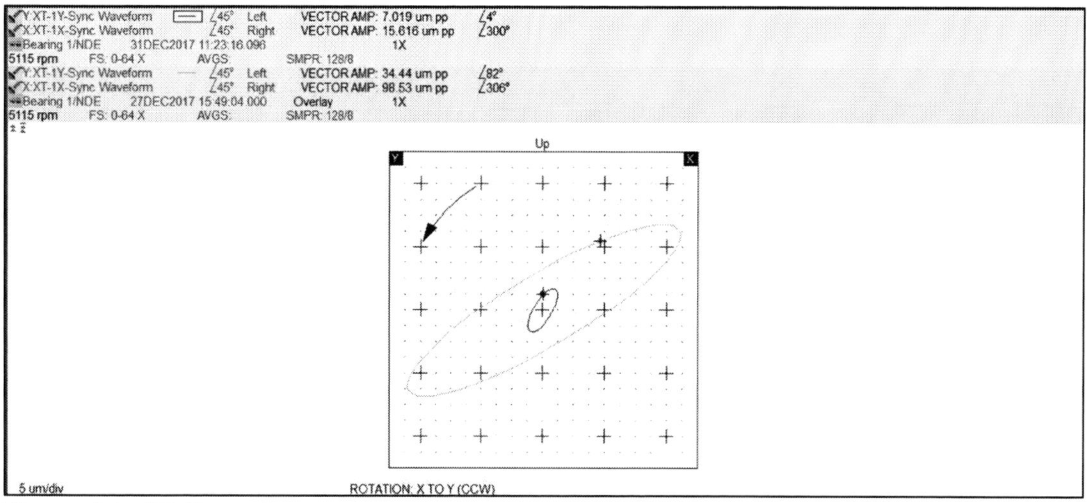

Figure 15: 1X filtered compensated Orbit plots comparison before (orange) and after (blue) balancing.

CONCLUSION:

This case is good example of the use of proper machinery diagnostic skills to inform corrective actions. As a result, the work performed to address the issues was minimized based on the machinery diagnostic services engineer's recommendations.

REFERENCES:

[1] D. E. Bently und C. T. Hatch, Fundamentals of Rotating Machinery Diagnostics, Minden: Bently Pressurized Bearing Press, 2002.

[2] R. S. Eisenmann, Sr. und R. S. Eisenmann, Jr., Machinery Malfunction Diagnolsis and Correction: Vibration Analysis and Troubleshooting for the Process Industries, Upper Saddle River, New Jersey: Prentice Hall PTR, 1997.

VIATC TRAINING CONFERENCE & EXPO

NIAGARA FALLS, NEW YORK JULY 26 -28, 2023

Developing Instincts for Better Data Collection and Analysis

Shawn Covington
Cat III Vibration Specialist
Horner Industrial Group
Shawn.Covington@hornerindustrial.com

Shawn Covington, is a vibration specialist with Horner Industrial Group (Indianapolis, Indiana). He has over 16 years of experience in the field as a vibration analyst working in a broad range of industries including steel manufacturing, food processing, auto manufacturing, paper mills, chemical/pharmaceutical plants, commercial residence and farming applications. Shawn is certified and experienced in field balancing as well as a Cat III Vibration analyst with the Vibration Institute.

ABSTRACT:
We have all seen or purchased those valuable charts and read as many books regarding vibration analysis, but the reality is that when we get into the field the machines do not always read the books. Sometimes the sounds of hooves are actually a herd of Zebras. One of the things it takes to be a better analyst is getting to know the machines you are analyzing. This requires holding lightly everything you think you know and learning to ask better questions. Calling a machine out too early can be just as harmful to your credibility as missing a serious problem. As analyst, we often must make judgment calls; our customers do not just want to know if the machine is bad (they can hear the problem), they want to know "How long it will last". There is no crystal ball for predictive maintenance, so it is up to the vibration analyst to develop good instincts in the field and at the computer to provide quality analysis and reliable feedback to the customers. This paper is an adaptation of a presentation about lessons learned over the years with case studies to show how things are not always what they seem and how asking good questions and understanding the history of a machine can save the analyst from looking like a fool.

KEY WORDS: Vibration trends, ISO Standards, Twice-line frequency, SCR Firing Frequency, VFD Drives, AC & DC Motors, Peak-Vue.

A problem is not a problem, unless it is a problem:

Baseline data was gathered for a new customer on a belt driven rotary lobe style blower at a food processing plant. Two jackshaft bearings help support the load on the driven shaft. Vibration data gathered on the outboard jackshaft showed 1.8 in/sec peak.

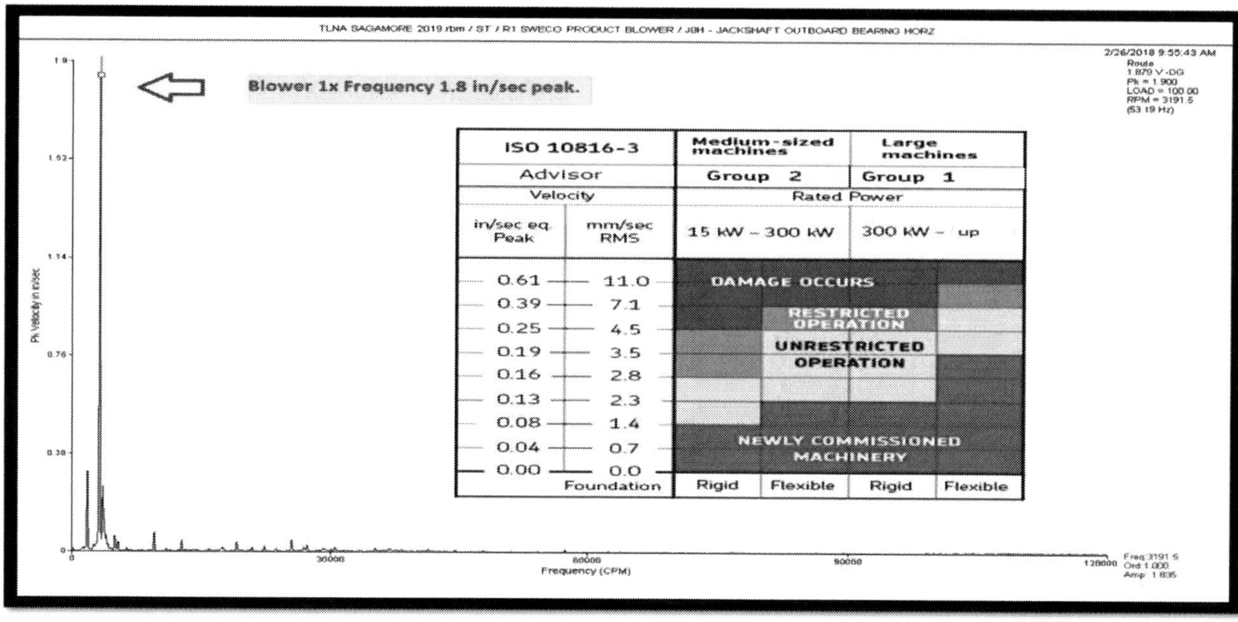

Figure 1: Jackshaft vibration spectrum w/ ISO 10816-3 Standard

According to the ISO 10816-3 vibration standard, this machine is well above the place where damage will occur. As analysts in the field, most of us would be very quick to call out this machine and maybe even have the customer shut it down so no further damage would occur. According to the standard we have a serious problem. Fortunately, an experienced analyst was making the call that day and decided to investigate further. After speaking with the operators and plant maintenance it was learned that this blower had been historically high for a long time. It was decided to monitor the vibration before making any further recommendations. The blower in question ran for over 8 years with the same vibration and no components were damaged. The books say that this kind of vibration is a problem, this machine apparently never read those books and an analyst without good instincts would likely make a premature call if they did not learn to develop a better habit of knowing who they are working for, what they are working on, as well as understanding the data they are analyzing and understanding themselves.

This paper was created to discuss four principles I have developed to aid in developing better instincts as a vibration analyst. Those four principles are as follows

1. Know the Who's
2. Know the What's
3. Know the Data
4. Know Yourself.

Principle #1 Know the Who's:
Who you are working for will not affect the amount of vibration you collect in the field, but it does matter when you are determining whether you should report a problem or not. Remember a problem is not a problem unless it is a problem. Some facilities such as a pharmaceutical manufacturer require all their equipment to meet some industrial standard such as the ISO 10816-3. If you are called out to a farming operation they may be pleased so long as their teeth are no longer chattering. The example below seen in figure 2 is a newly installed fan at a pharmaceutical manufacturer.

Figure 2: Photo of air handler

Data gathered on this air handling unit showed very low vibration at any mechanical fault frequency. As an analyst, I saw no reason why this fan should not be commissioned and put into service. The fan's 1x turning speed frequency at 830 CPM was only showing a max amplitude of less than 0.15 in/sec and no blade-pass or bearing faults were present in the data. Although the vibration data showed no mechanical faults present in the machine, the overall vibration levels observed on the fan did not meet the plants standard and the fan was not allowed to be commissioned. Had this problem occurred at a steel mill, the fan would have likely been accepted and put into production.

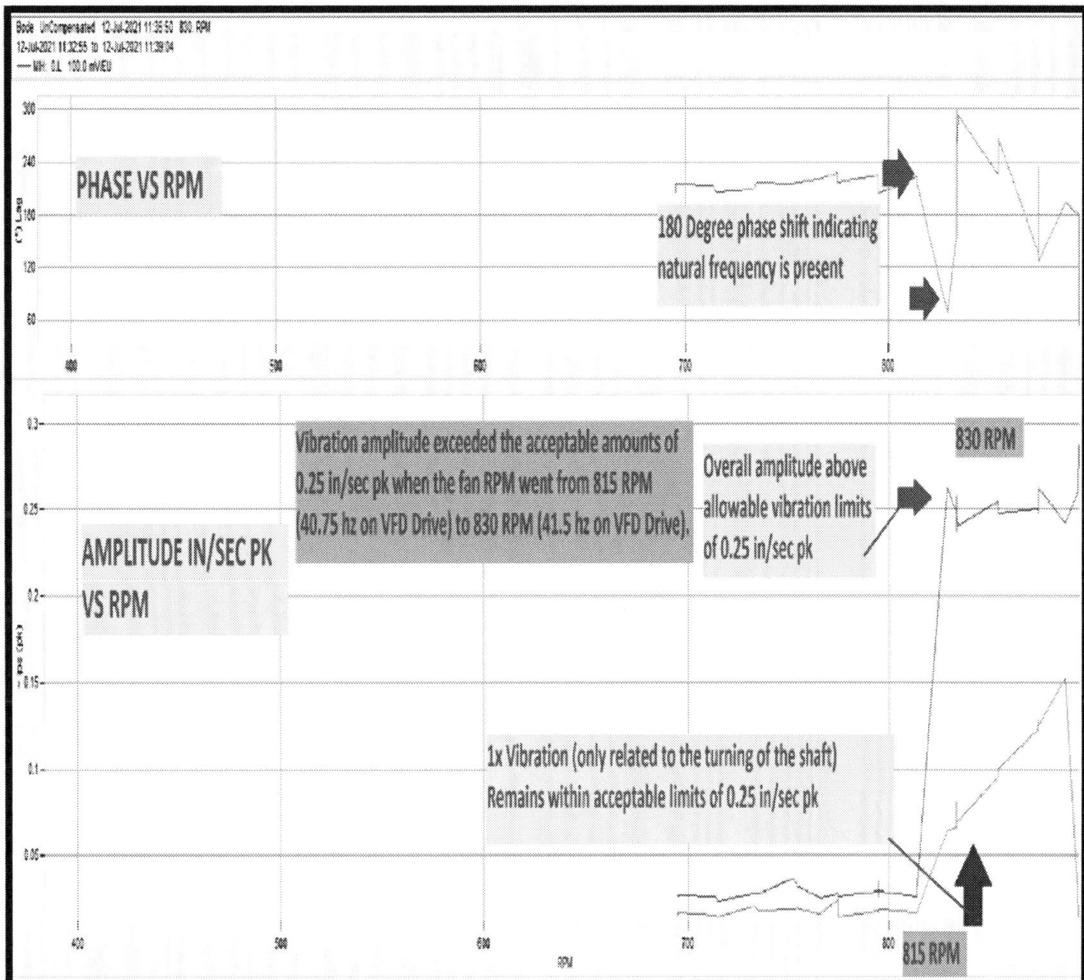

Figure 3: Bode plot of vibration data on the air handler.

The data seen in Figure 3 shows a bode plot of the vibration observed on the air handler. The 1x vibration is seen to be well within the ISO 10816-3 standard of less than 0.25 in/sec, but the overall vibration crossed the acceptable threshold and the fan was not accepted since this particular facility would not accept any equipment that did not meet the industry standard.

But was there a problem with fan? Mechanically no, but the data in Figure 4 shows that there was a vibration showing up at 600 cpm or 0.72x the turning speed of the fan. This frequency was not related to any mechanical faults within the machine. It was suspected that there was some natural frequency within the system being excited by broadband energy created by the airflow of the fan when operating at high speeds. The fan was never to operate at these high speeds but the standard required the machine to be tested at 100% on the variable frequency drive.

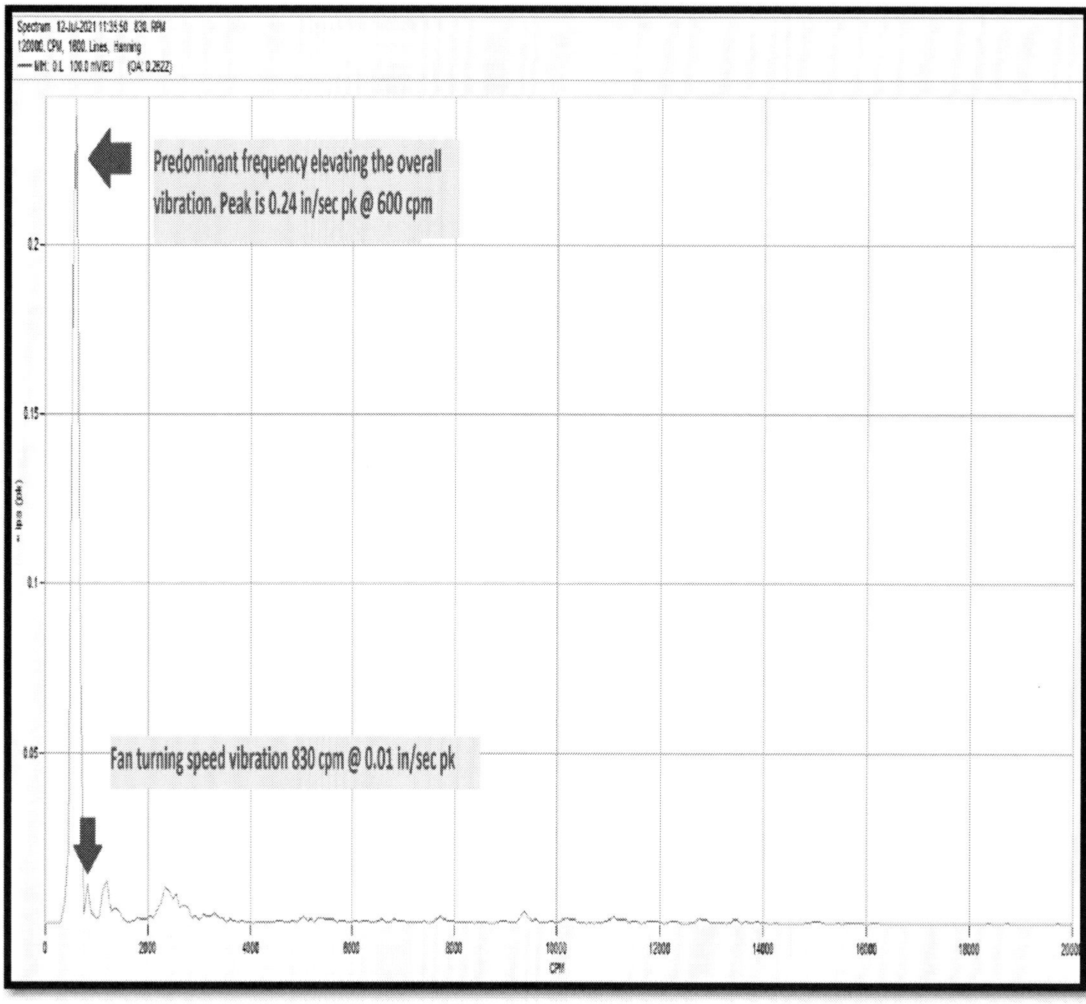

Figure 4: Vibration spectrum of air handler.

Since the fan would never operate near the RPM where the higher vibration occurred and the vibration present was not related to the mechanics of the fan, I did not think there was a problem with the fan. But knowing who I was working for and what priorities they put on newly installed equipment, this problem which may not be a problem elsewhere, was a problem here and I condemned the fan.

Principle #2 Know the What's:

Knowing what kind of equipment that you are analyzing is an obvious fact. but often we group every type of machine into neat little groupings so that a machine at one facility should act the same at another facility. Making such assumptions may get you into trouble. Therefore, it is not only important that you know what you are working on but also that you know the particular machine. What kind of work is it performing, how important is this machine to plant operations, are great questions to consider. Knowing the temperament of your equipment is equally important when considering whether or not to tell your customer about a problem. Remember a problem is not a problem unless it is a problem. The more you understand each individual piece of equipment the better the analyst will be at determining whether the vibration you are observing is a problem and how important it is.

Figure 5: Five-stage Centrifugal Blower

Data is gathered weekly on the centrifugal blowers seen in Figure 5. This facility asks that we monitor the blower bearings closely due their habit of failing very quickly.

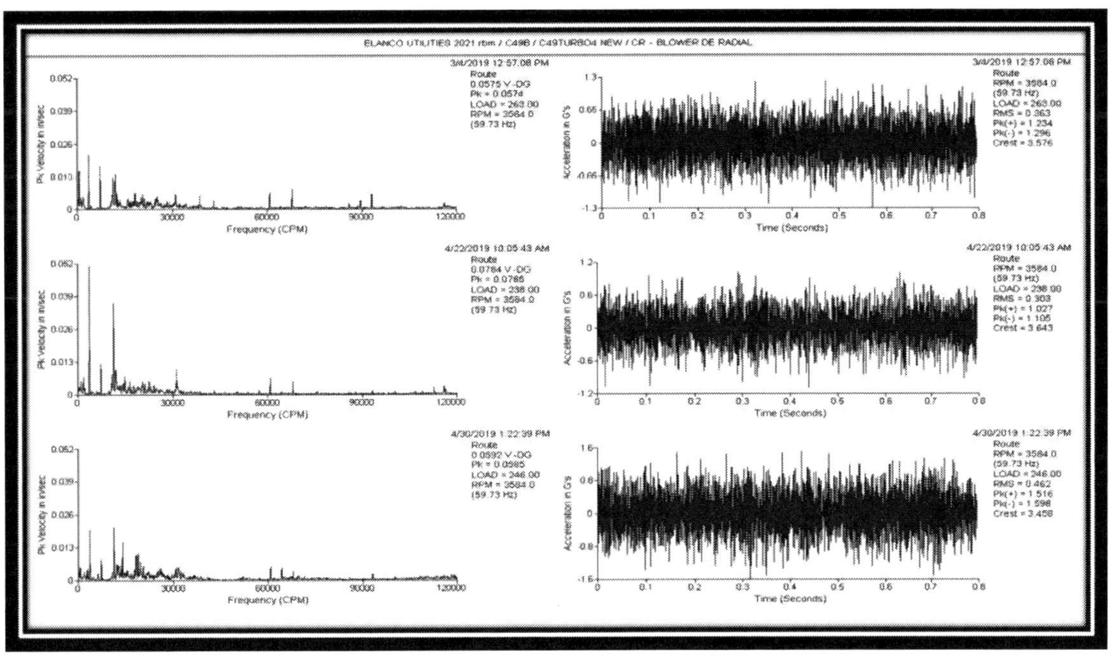

Figure 6: Blower velocity spectrum and acceleration waveform.

Figure 6 includes three sets of data gathered on the blower inboard bearing. The velocity spectrum and waveform data indicated no change in amplitude between March 4th 2019 and April 30th 2019. Yet the bearing had a fault on April 22nd and was replaced before the April 30th.

We often rely on the velocity spectrum and acceleration waveform to determine severity and employ Peak-Vue readings to look for early signs of bearing damage in the rolling element bearings. Both the velocity data as well as the peakvue readings are recorded and we are able to observe their trends as each data collection performed. You can see the Peak-Vue data in Figure 7.

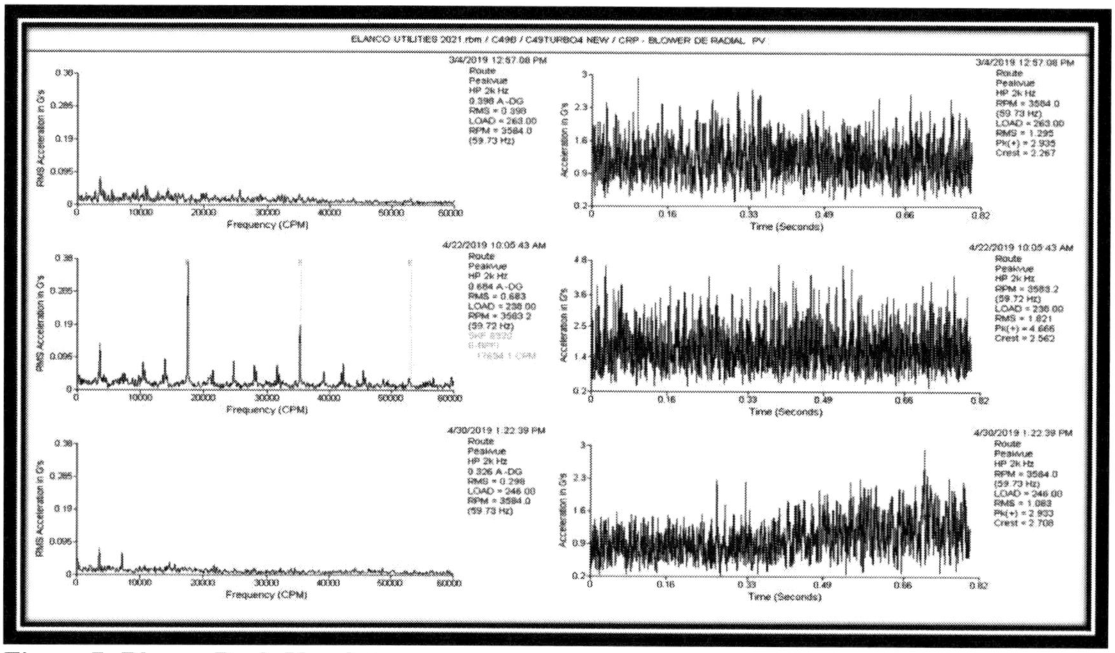

Figure 7: Blower Peak-Vue data.

As you can see in the data, April 22 2019 showed a fault frequency in the spectrum that matched the inner race defect for the blower's inboard bearing. Because the levels were low and nothing was showing up in the velocity data, multiple experienced analysts recommended that the bearing be monitored and replaced once the vibration levels increased to higher levels. Because we knew the history of this machine and its tendency to lose bearings very quickly, we called the customer and told them to shut the machine down and replace the bearing. By the time the machine was shut down and the bearing removed, a large spall had already started to appear on the surface of the inner race. Baseline data was gathered on the new bearing on April 30th 2019. The reason why the seasoned analyst thought this problem was a minor issue was because he did not have any personal knowledge of this particular machine. It may not always be easy to know the machines you are working on, but asking good questions and keeping a detailed history of failures will often make all the difference between a bad call and a right one.

Principle #3 Know the Data:

The data that we gather in the field is very important. Making sure you acquire good data when you are taking your readings will save you from making a bad call or at least a second trip. Keeping historical vibration data and trends will also give you better insight into the way your equipment is operating. Equally important to good data is the analyst's ability to understand what they are looking at. While this paper is not meant to teach one how to analyze vibration data, the example below will illustrate how important it is to understand the type of signatures you see in your vibration spectrum and waveforms.

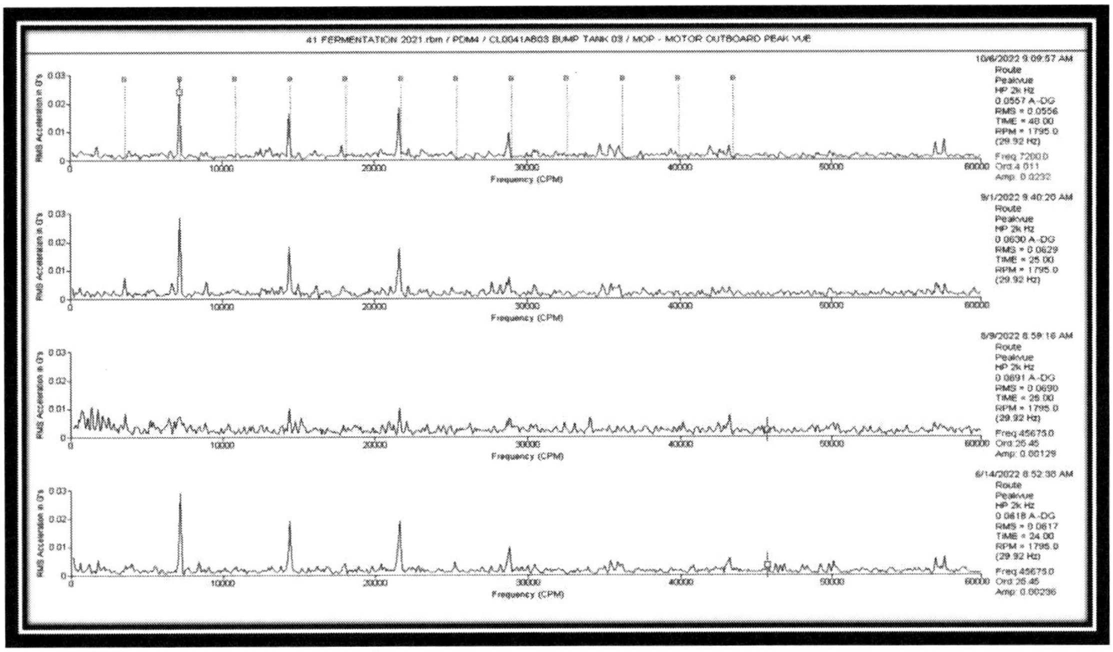

Figure 8: Peak-Vue data on AC motor.

The data seen in Figure 8 shows a Peak-Vue reading on a 75hp AC motor. The spectrum shows several harmonics of a peak which seems to match up with 2 times ball-spin fault frequency (BSF) of the bearing installed on this motor. 2x BSF is a common fault seen on rolling element bearings and at first glance it would appear that we are seeing the early stages of a bearing failure. An inexperienced or impetuous analyst may be quick to write this up on a report and warn their customer of a problem. The problem is not with the bearing but with the analyst's diagnosis. Although the 2x BSF seems to line up with the first peak in the spectrum it begins to miss the mark on the 2nd and 3rd harmonic. Upon closer observation, it can see that the first peak in the spectrum is dead center on 7200 CPM (120Hz). This is not a bearing fault, but 2x line frequency which notoriously shows up in Peak-Vue data when taken on AC motors. Taken a second look and understanding the data saved this analyst from getting egg all over his face.

Frequencies such as 2x Line Frequency on AC motors and the SCR frequency on DC motors often show in your vibration signatures. It is rare that these signatures indicate a problem, but it is important to be aware that they will show up in your spectrum. (And it is important to know when they are indeed indicating a problem). It is also important to note that AC motors driven by a VFD will show 2x line frequency, but it will not be at 120 Hz unless the motor is being driven at 100% on the drive. A six-pole motor being driven at 75% will show a 2x line frequency around 5,451 CPM and not 7200.

Principle #4 Know Yourself:
It always feels good to be called an expert at something but in reality, my title was obtained by passing an exam and does not necessarily reflect my experience in the field. Even with many years in the field and experience in numerous industries every seasoned analyst will run into problems that are over their head. It is important to know yourself well and to know what you do not know. You may find yourself needing to admit to your customer that you do not know what the problem is nor how to help them. This may cost you a job but your credibility will be damaged more if you provide poor analysis and are found to be mistaken. Your willingness to admit what you don't know and your willingness to learn more will make you a better choice for your customer than an analyst who doesn't know and chooses not to admit it.

In my 16 years of experience in the field, I have made a lot of mistakes, those mistakes have been some of my best teachers and although I wish at the moment that I hadn't made them, I know I am a better analyst now because I was willing to learn from them.

The vast field of vibration analysis will never be mastered. Sometimes the problems you run into will be complex and simple answers to your problem cannot be found no matter how much experience you have. These 4 principles are simple guides to help you develop better instincts in the field but nothing will replace time and experience; therefore be willing to admit you do not know everything but also be willing to try your darndest to figure it out.

4 Applications to implement to aid in developing better instincts as a vibration analyst:

The four principles above are meant to aid in developing better instincts as vibration analyst, but how does one go about implementing these principles? I suggest these four applications to get you on the right path.

Ask lots of questions: Nothing will save time as much as asking questions. When you are called out to diagnose a problem or see an issue in your data, start talking to everyone involved. What is the problem? How important is this machine to the production and safety of the facility? How much time is available for analysis and repairs? The mechanics that work on the equipment have a good idea what kind of faults are common in their machines, likewise operators are often valuable resources. They may not know anything about vibration but they are often around the equipment more than anyone else. Ask the reliability group or the coordinators about previous vibration data that may have been taken by them or another company. Try and find out as much about the machine's operation and its maintenance history as possible.

Investigate: How is the problem manifesting? What does this machine do? Is it a motor, pump, fan etc.? What are the drive and driven components? Often understanding more about the specific machine can eliminate a bunch of faults and bring you closer to understanding the problem. For example, you will not find bearing fault frequencies on sleave bearings. Sometimes asking questions is not enough, you may need to get on your hands and knees and look around for cracks in the foundation or chunks of coupling lying on the floor. What does the machine look like, sound like? How does it feel, is it hotter than normal? Even smell can help especially on belt driven machines. I ask these questions before I arrive at a facility but I never take anyone's word for it if I can help it. Trust but verify! If you are able to get to the machine, a few extra minutes investigating with your 4 senses (don't taste the machine!) and this extra time will often save you time at the computer and save face with your customer.

Educate yourself: Never ignore any historical data you have on the machine you are testing. If you do not personally have vibration trends on this machine, ask the customer if they can provide past vibration reports or at least someone who is familiar with the history of the equipment. Compare data on one machine with others like it and get to know the machine and what it does. There is a vast resource available online, typing your problem into a search engine will often yield hundreds of articles written by other experts in the field. Join a group like the Vibration Institute to take advantage of their articles and training opportunities. Lastly get a network of analysts who you can call and brainstorm while you try and troubleshoot the problem.

Be Humble: It is important to know what you do not know so that you do not find yourself drowning in a mess created by your own unwillingness to admit you didn't know something. Know your common mistakes and watch out for them, have a team to rely on and discuss the problem. Lastly be honest! Admitting a mistake or owning up to not fully understanding something goes a long way in establishing a good reputation. Your credibility matters in this field and I have often found that my customers appreciate honesty and are more apt to forgive a bad call when an analyst doesn't come across as a know it all.

Summary:

If you want to know if a problem is a problem remember the four principles to building instincts as an analyst:

1. Know **WHO** you are working for.
2. **WHAT** you are working on.
3. Know your **DATA.**
4. Know **YOURSELF**

Therefore:

- **Ask good questions (Lots of them)**
- **Investigate**
- **Educate yourself**
- **Be Humble**

VIATC | TRAINING CONFERENCE & EXPO

NIAGARA FALLS, NEW YORK JULY 26 -28, 2023

BACK TO THE BASICS, MASSES AND SPRINGS AND WHAT THEY TEACH US ABOUT VIBRATION

M. Austin Creasy, Ph.D.
Associate Professor
Purdue University
mcreasy@purdue.edu

M. Austin Creasy, Ph.D. is an associate professor in the School of Engineering Technology at Purdue University. He is the Mechanical Engineering Technology (MET) program lead where he leads the curriculum and assessment committees that define the content of the MET program. He has taught 20 different courses during the 12 years as a faculty member of Purdue that include content related to mechanics, materials, and programming. Dr. Creasy has over 35 published articles in journal and conference proceedings. He is currently a member of the Board of Directors of the Vibration Institute (VI) and is leading the efforts in providing VI training online.

ABSTRACT:
A mass and a spring are the fundamental unit that allows us to understand a lot about vibrating systems. A vibrating system contains oscillating force, mass, stiffness, and damping. A mass and a spring make a system where these four items can be easily manipulated to understand their individual relationship on a vibrating system. That understanding allows us to develop equations that we can use to model a system and predict how that system will react to changes. This paper will discuss the vibration of a mass and spring, show how math models are derived, and show how models can be manipulated to understand vibration.

KEY WORDS: Damping, Derivation, Derivative, Integral, Natural Frequency, Single Degree of Freedom System, Undamped

INTRODUCTION:

A lot of vibration understanding is derived from a single degree of freedom (SDOF) system that consists of a mass and a spring (which includes a small amount of damping). Displacing the mass of the system from the equilibrium point and releasing it will cause the mass to oscillate. This system has been used to derive and understand equations that explain to a vibration analyst how mass, stiffness, and damping affect the vibration of a system. These equations are based on Calculus principles that are introduced here for the vibration analyst that may not have an extensive math background. This paper will show how a SDOF system is modeled and how these math principles are used to obtain the natural frequency equation and understand underdamped, critically damped, and over damped systems. An assumption is made that a vibration analyst reader will be somewhat familiar with vibrations in general (position, velocity, and acceleration) and the math concepts of basic trigonometry operations.

MATH REVIEW:

Calculus principles can be used to show the relationship between position, velocity, and acceleration of a SDOF system. The two principles that need to be understood are the derivative and the antiderivative. The mention of calculus, derivatives, and antiderivatives may cause anxiety for some vibration analysts that do not have an extensive background in math. This paper looks to introduce these principles and simplify their definition for the slightly experienced vibration analyst. To start reviewing these calculus principles, the definition of a function first needs to be provided. A textbook definition [1] is as follows:

> *A function from a set domain to a set range is a rule that assigns a unique element f(x) in the range to each element x in the domain.*

In simpler terms, you have an input (the domain: typically, time in vibration analysis) that causes an output (the range: typically position, velocity, or acceleration in vibration analysis) that is obtained through some type of math procedure. For a SDOF system, vibration is modeled as an oscillating system where the input (domain) will be time, the output (range) could be position, and a trigonometry function is the math procedure. So, the function could look like:

$$f(t) = \sin(\pi t) \quad (1)$$

where f is the math symbol of the function, t is the domain (time in this case), and *sin* is the trigonometry function that defines the oscillating range. The vibration analyst reader should note at this point that the amplitude (the constant value that is multiplied by a function to define the maximum range value) is unity (the value of "1" and does not need to be included in the function when the amplitude is "1") and that the frequency (the constant value that is multiplied by the time) is set to the value of π with units of rad/s. Figure 1 shows the function of Eq. (1) plotted over a domain of 0-s to 2-s with the range of the function shown with a maximum value of 1 and minimum value of -1. The frequency was set to the value of π in order for the output to complete a cycle in 2-s, which is the frequency equivalent of ½-Hz. The conversion factor of the frequency between rad/s and Hertz is 1-Hz is the equal to 2π-rad/s and a slightly experienced vibration analyst should be able to show that a frequency of π-rad/s is equal to ½-Hz.

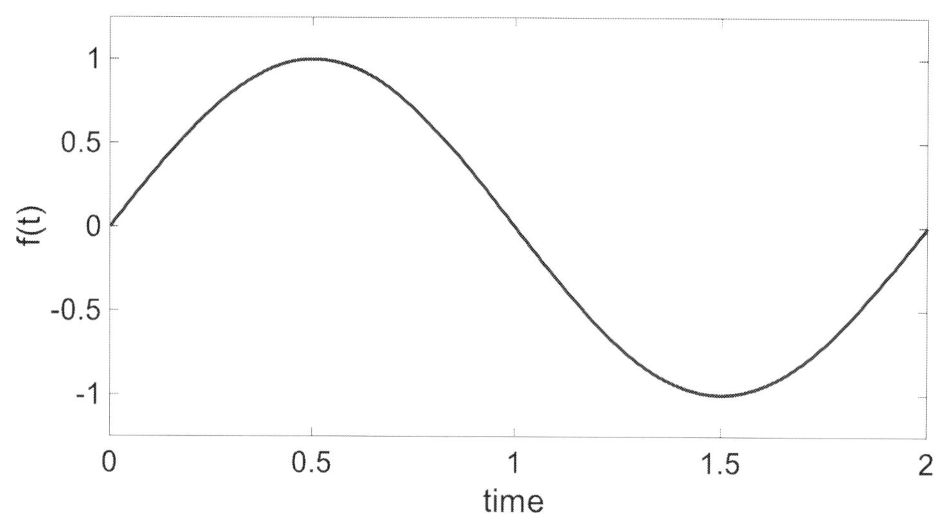

Figure 1. Sine function with a domain of 0-s to 2-s, an amplitude of unity, and a frequency of π-rad/s.

A derivative of a function written as a limit is defined as follows:

$$f'(t) = \lim_{\Delta \to 0} \frac{f(t+\Delta) - f(t)}{\Delta} \tag{2}$$

where the ′ (apostrophe) symbol indicates the first derivative of the function, Δ is the increment to the next point of the function, and *lim* is the limit. Again, this equation probably causes anxiety for the typical vibration analyst and this paper is not intended to discuss limits in detail (any reader interested in limits or limit theory should review an introductory calculus text). A simple way of stating the derivative definition, is the derivative at a specific time is the instantaneous slope of the function. The slope of a line can be calculated by two points on the line where the slope is the rise over the run (the change in the y-axis values divided by the change in the x-axis values). Removing the limit part of the equation provides an approximation of the derivative when you have two points of your function (or two points of measured data if data is obtained instead of a math model). This change is also called the instantaneous rate of change,

$$f'(t) = \frac{f(t+\Delta) - f(t)}{\Delta} = \frac{\Delta y}{\Delta x} \tag{3}$$

Figure 2 shows an example of this calculation for the function, *sin(πt)*, at the point t = 0.2-s and Δ = 0.3-s. This value is the slope of the line shown in red. The actual derivative of the stated function at t = 0.2-s is the value shown by the green dashed line and this figure graphically shows a large variation in the calculated derivative compared to the actual derivative.

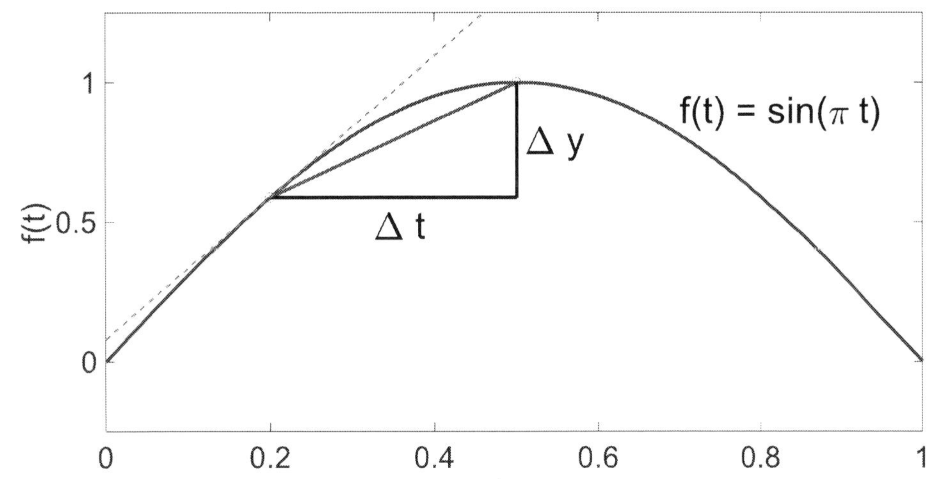

Figure 2. Derivative approximation (slope of red line) at t = 0.2-s and Δ = 0.3-s compared to the actual derivative (slope of the green dashed line).

Now if the value of Δ is made smaller as indicated by Eq. (2) (the approximation is better as Δ gets closer to zero), the value gets closer to the actual derivative shown by the green dashed line. Again, with t = 0.2-s and Δ reduced to 0.1-s, Figure 3 shows the calculated derivative (slope in red) results of Eq. (3) compared to the actual derivative (dashed green line). The red line's slope is much closer to the actual slope shown by the green dashed line. Shrinking that Δ value closer and closer to zero, will cause the slope of the red line to approach the slope of the green dashed line.

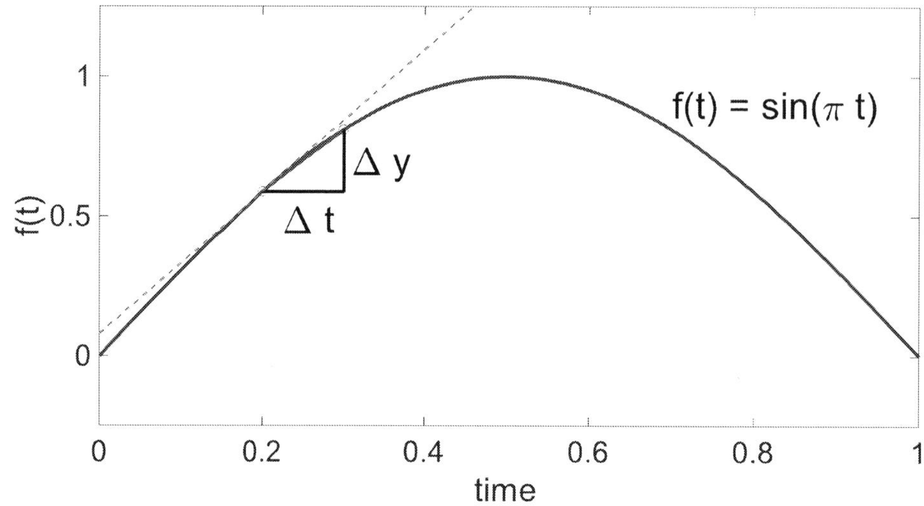

Figure 3. Derivative approximation (slope of red line) at t = 0.2-s and Δ = 0.1-s compared to the actual derivative (slope of the green dashed line).

Performing this procedure at multiple points and plotting the points at each time provides an approximation of the derivative function. Figure 4 shows a plotted example of this calculation in Figure 4B where the derivative is found at each 0.25-s as shown by the red dots connected by straight lines. Figure 4A is the function, *sin(πt)*, that has been used in this discussion. Several slopes are also plotted on the function of Figure 4A at points t = 0.25-s, 1.0-s, and 2.0-s for reference to the plotted values in Figure 4B. The reader should note that the slopes at t = 0.25-s and 2.0-s are positive, the slope at t = 1.0-s is negative, and the rise over the run provides the approximation of the graphed values. Reducing the interval step at which the derivative is found will produce a smoother function for the derivative curve.

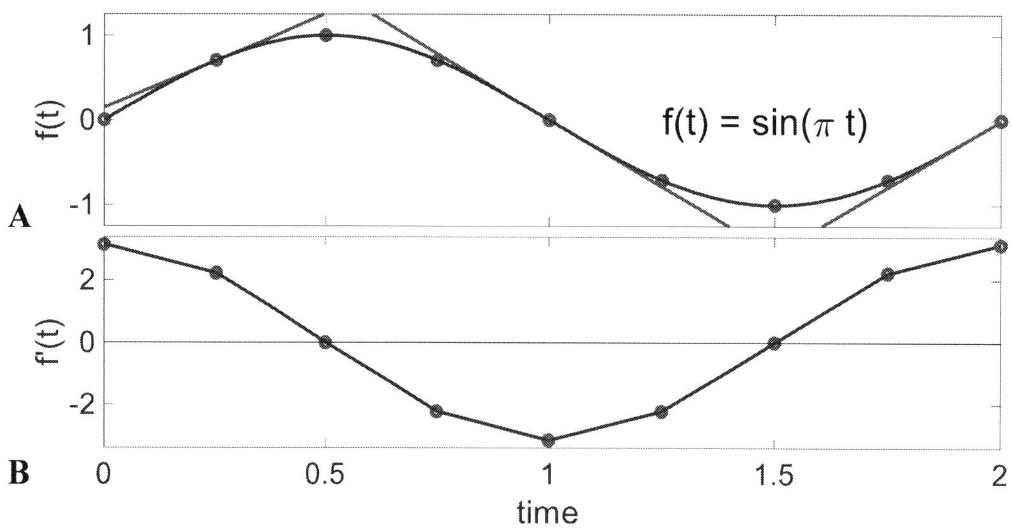

Figure 4. A) Sine function of Figure 1 with slopes indicated at t = 0.25-s, 1.0-s, and 2.0-s. and B) Derivative plot of Figure 4A with derivative values calculated at every t = 0.25-s.

If the function is mathematically known or modeled, there are tables provided in calculus references that show the exact function of the derivative without needing to use the approximation shown above. For the function used in this discussion, the derivative will be

$$\frac{d}{dt}\sin(\pi t) = \pi\cos(\pi t) \tag{4}$$

where the *d/dt* symbology is a mathematic definition of the derivative.

Why is this math necessary to understand a SDOF system? Because the velocity of a SDOF system is the time derivative of the position function. Therefore, from Eq. (2), the velocity is

$$v(t) = \lim_{\Delta t \to 0} \frac{\Delta s}{\Delta t} \tag{5}$$

and in a similar fashion, the acceleration of a SDOF system is the time derivative of the velocity. Again, from Eq. (2), the acceleration is

$$a(t) = \lim_{\Delta t \to 0} \frac{\Delta v}{\Delta t} \tag{6}$$

As a vibration analyst, any time that position is measured, this calculus principle called the derivative can be used to approximate the velocity and acceleration. One item that the vibration analyst should also note is that there are an infinite number of functions that will have the same derivative function. If any constant is added to Eq. (1),

$$f(t) = \sin(\pi t) \pm constant \tag{7}$$

the slopes at the same domain values between Eq. (1) and Eq. (7) will be identical. Figure 5A shows instances of Eq. (7) plotted over the domain of 0-s to 2-s. For each plot, the slope at t = 1.0-s is shown to understand that the derivative is identical for each plot. The red lines are parallel to each other indicating the same slope. The derivative is plotted for all three curves in Figure 5B. Each derivative function overlays the derivative function of the other curves and appears as the single curve, $\pi\cos(\pi t)$.

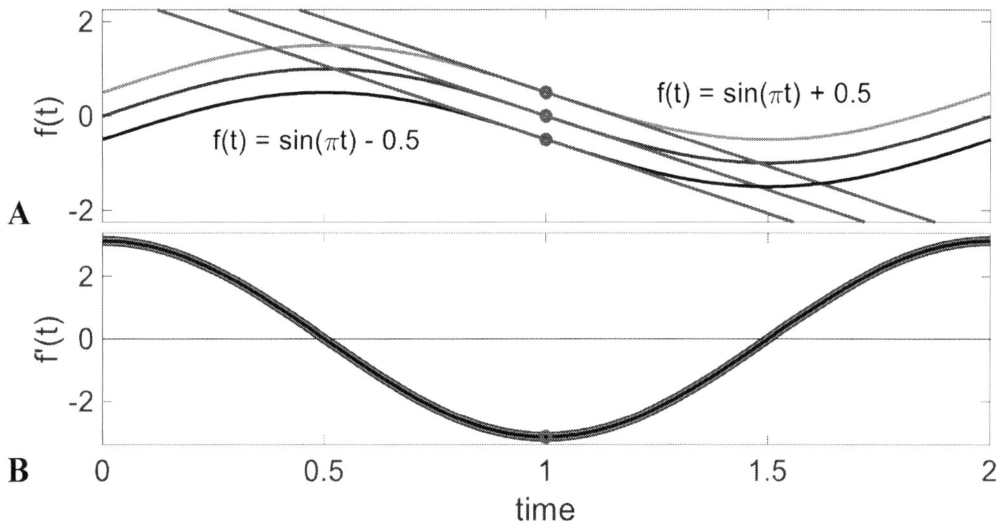

Figure 5. A) Sine functions plotted with a domain of 0-s to 2-s with the indicated constants added to the different functions with the derivative slope at *t* = 1.0-s plotted for each function and B) Derivative function plotted for all functions of Figure 5A.

Since a derivative can be the derivative of multiple functions (as shown in Figure 5), the antiderivative (also called the integral) of a function has numerous potential functions as well. This type of integral is called an indefinite integral and mathematically is written as follows:

$$\int f(t)dt = F(t) + C \qquad (8)$$

where the function, *f(t)*, is the function to be integrated, *dt* indicates the variable for the integration (which is time in this case and will not be discussed in more detail here), *F(t)* is the antiderivative function, and *C* is a constant that can be any value (usually found by knowing a point of the curve, typically called a boundary condition).

Another type of integral is called the definite integral. The definite integral can be approximated by calculating the area underneath the function being integrated. With this integration, the constant from Eq. (8) is ignored and you are calculating the integral that starts at zero on the y-axis. This process is call Riemann Sums where the function can be approximated with small rectangles or trapezoids (to get a more precise solution with less error) in order to find the area. Figure 6 shows an example of this integration. Figure 6A is the function from Figure 5B where small rectangles are taken at each 0.1-s and plotted in Figure 6B. The area above the x-axis is considered positive area (red rectangles) and the area below the x-axis is considered negative area (cyan rectangles). When those areas are summed together over time, starting at zero, the graph from Figure 5A is recreated without the offset constants. Reducing the size of the rectangles will increase the accuracy of the definite integral and more closely approximate the desired curve. A careful analysis of Figure 6B shows small corners at each of the calculated values that will become smoother as the size of the rectangles is decreased.

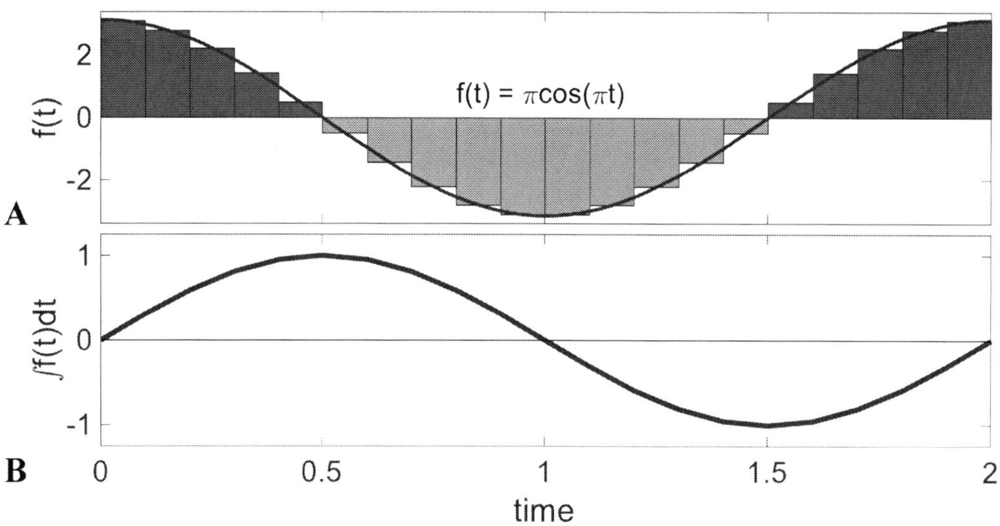

Figure 6. A) Definite integral estimation using Riemann Sums at every 0.1-s and B) Plotted definite integral estimation.

Why is this integration necessary to understand a SDOF system? Because the velocity of a SDOF system is the antiderivative (integral) of the acceleration function. Therefore, from Eq. (8), the velocity is

$$v(t) = \int a(t)dt \tag{9}$$

and in a similar fashion, the position of a SDOF system is the antiderivative (integration) of the velocity. Again, from Eq. (8), the position is

$$p(t) = \int v(t)dt \tag{10}$$

As a vibration analyst, any time that you have the measured acceleration, the integral can be used to approximate the velocity and position. The analyst should be aware that numerical issues can arise and cause errors in the calculations, these numerical issues will not be discussed here.

UNDAMPLED SINGLE DEGREE OF FREEDOM:

A spring/mass system is a SDOF system that allows the vibration analyst to start understanding the concepts of vibration. Figure 7 shows a schematic of a spring/mass system. Here there is a rest position where if no motion is present, the mass will stay at rest with the spring stretched from the unstretched length by a distance (*s*), where

$$s = \frac{mg}{k} \tag{11}$$

m is the mass, *g* is the gravitational constant, and *k* is the spring stiffness. If the mass has a force applied to it or is displaced to either position A or position -A and released, the mass will oscillate indefinitely until frictional forces remove the energy from the system or another force stops the motion.

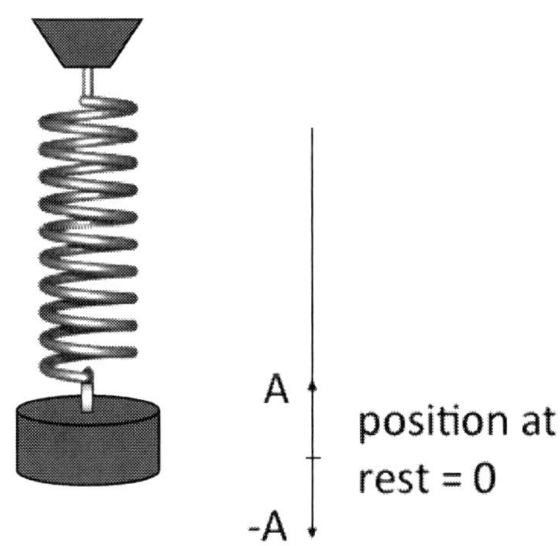

Figure 7. Schematic of a spring/mass system.

This spring/mass system can be analyzed using Newton's Second Law of motion to model the position, velocity, and acceleration of the mass when a force or displacement is applied to give the mass motion. This discussion will ultimately focus on the mass being displaced for motion. Newton's Second Law states that forces on a mass is equal to the time rate of change of the momentum of that mass or stated as an equation

$$\sum F(t) = \frac{d}{dt}(mv) \tag{12}$$

where m is the system mass, v is the system velocity, F is the applied force, and $\sum$ is the mathematical symbol for summation. This equation shows that a time derivative of the mass and velocity (momentum) product is equal to the applied forces on the mass. Since the mass does not change with time for our typical spring/mass system, the time derivative is only applied to the velocity. Equation (6) indicates that the derivative of the velocity is the acceleration and therefore provides the reader with the typically referenced Newton's Second Law of

$$\sum F(t) = ma(t) \tag{13}$$

In order to apply Eq. (13) to our spring/mass system, all of the forces on the mass have to be summed together. Therefore, a free-body-diagram (FBD) should be drawn of the forces applied to the mass of Figure 7. An FBD is a drawing that shows all of the forces that are acting on a mass or structure. The only forces are the weight of the mass and the force of the spring. The FBD is shown in Figure 8 where the blue arrows indicate forces. The spring force may oscillate with time, but the mass force will stay constant assuming that gravity is pulling down. Implementing Eq. (13) and assuming that down is positive (for later simplification purposes) produces the following equation

$$-ks(t) + mg = ma(t) \tag{14}$$

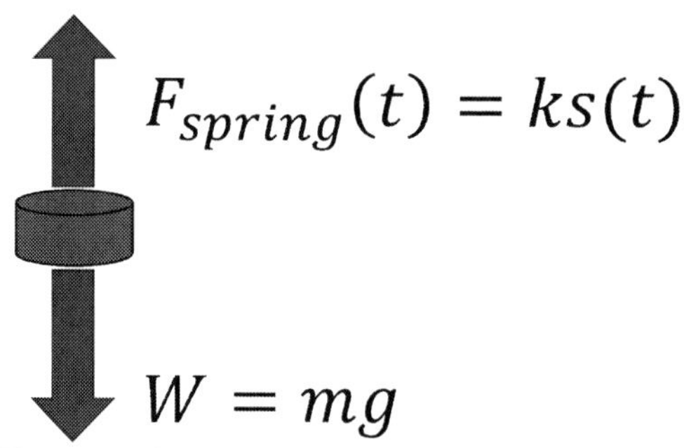

Figure 8. FBD of a SDOF mass with an attached spring.

Now, applying superposition, the force due to gravity can be removed because the position of that force causes the mass to move to the rest position shown in Figure 7. Superposition is the process of separating an equation into multiple parts and solving the individual parts separately. Once the individual parts are solved, they can be added together to obtain a total solution of the equation. But, this process is only valid for linear systems (linear and non-linear equations will not be discussed here in detail). If the vibration analyst desires to model the motion from the unstretched length of the spring shown in Figure 7, the gravity force would need to stay in the equation, but is not necessary for this discussion. Now, the spring force can be added to both sides of the equation to produce the differential equation

$$ma(t) + ks(t) = 0 \tag{15}$$

A differential equation is a math model that relates a function and the function's derivatives. This equation is a function of both the position and the acceleration (the second derivative of the position) of the mass. A solution to a differential equation may be any solution that makes the differential equation true, but it may not be the solution for the physical phenomena modeled by the differential equation. Since this system will be vibrating, the solution is assumed to have some sinusoidal trigonometry function as the solution. If the mass is displaced the distance A from Figure 7 and released, the vibration analyst would therefore assume the position solution to be

$$s(t) = Acos(\omega t) \tag{16}$$

where ω is the natural frequency of the system and A is the initial displaced distance. Recall that from Eq. (5), the velocity is the derivative of position and from Eq. (6), the acceleration is the derivative of position. Therefore, reviewing function derivatives from any calculus textbook, the velocity will be

$$v(t) = \frac{d}{dt}(s(t)) = \frac{d}{dt}\big(Acos(\omega t)\big) = -A\omega sin(\omega t) \tag{17}$$

and the acceleration will be

$$a(t) = \frac{d}{dt}(v(t)) = \frac{d}{dt}\big(-A\omega sin(\omega t)\big) = -A\omega^2 cos(\omega t) \tag{18}$$

Substituting the solution for position from Eq. (16) and the solution for acceleration from Eq. (18) into Eq. (15) produces the following:

$$m\big(-A\omega^2 cos(\omega t)\big) + k\big(Acos(\omega t)\big) = 0 \tag{19}$$

The amplitude, A, and the trigonometry function, $cos(\omega t)$, are found in both terms of Eq. (19) and factoring these terms will produce the following:

$$Acos(\omega t)(-m\omega^2 + k) = 0 \tag{20}$$

If the system is in motion, neither the amplitude nor the trigonometry function can be zero over all time. Therefore, the system frequency has to be a specific value for the equation to be true. Dividing both sides by the amplitude and the trigonometric function removes those values from Eq. (20) and leaves the vibration analyst with the following equation, called the characteristic equation

$$(-m\omega^2 + k) = 0 \tag{21}$$

which can be solved for the system natural frequency

$$\omega = \sqrt{\frac{k}{m}} \tag{22}$$

Equation (22) is a well know estimation for the natural frequency of a system and is used to understand how the natural frequency of a system will change when the stiffness and/or mass of a system is changed. This equality is one of the main equations seen in vibration analysis and is found through analyzing the motion of a spring/mass SDOF system as shown in this derivation.

DAMPED SINGLE DEGREE OF FREEDOM:

A spring/mass/damper system is still a SDOF system that allows the vibration analyst to understand the concepts of vibration with the inclusion of damping. The schematic of Figure 7 is still valid where the damping is energy absorption of some form and is assumed to be viscous (a function of the mass's velocity). Other forms of damping could be present, but viscous damping will allow the differential equation to be linear. As with the undamped SDOF system, an FBD is drawn to apply Newton's Second Law to the system. Figure 9 shows this FBD with the addition of viscous damping. Implementing Eq. (13) and assuming that down is positive (for later simplification purposes) produces the following equation

$$-ks(t) - cv(t) + mg = ma(t) \tag{23}$$

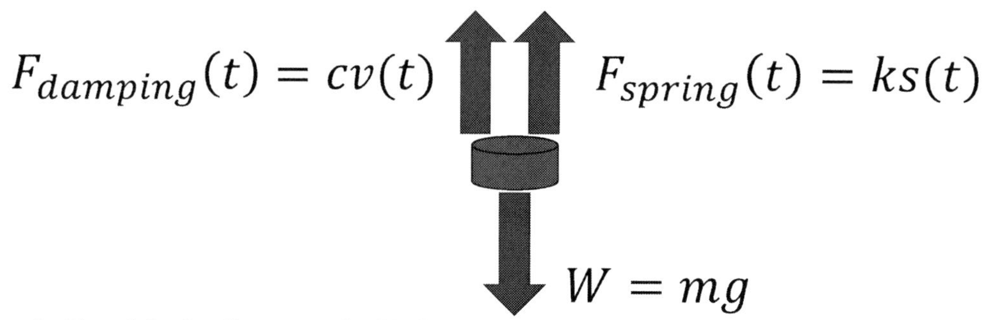

Figure 9. Freed-body-diagram of a SDOF mass with an attached spring and viscous damping.

Again, applying superposition, the force due to gravity can be removed because that force causes the mass to move to the rest position shown in Figure 7. Now, the spring force and damping force can be added to both sides of the equation and you arrive at the differential equation

$$ma(t) + mv(t) + ks(t) = 0 \tag{24}$$

Since this system will be vibrating, the solution is assumed to have a sinusoidal trigonometry function similar to the undamped solution and some type of decay due to the damping. There will be issues just assuming a *sin* or *cos* solution because of the damping term. One solution that works here is to assume an exponential solution that may have imaginary parts. Therefore, the assumed solution could be

$$s(t) = Ae^{\lambda t} \tag{25}$$

where e is Euler's number that has some interesting properties and λ is related to the natural frequency of the system and may be imaginary. Allowing λ to be imaginary provides the oscillating trigonometry functions in the solutions because of Euler's formula

$$e^{it} = \cos(t) + i\,sin(t) \tag{26}$$

where i is the imaginary number $\sqrt{-1}$.

Avoiding a long introduction to Euler's formula and imaginary numbers, this paper will assume that Eq. (25) is a solution and follow the same steps to a solution shown with the undamped SDOF system. Recall that from Eq. (5), the velocity is the derivative of position and from Eq. (6), the acceleration is the derivative of position. Therefore, reviewing function derivatives from any calculus textbook, the velocity will be

$$v(t) = \frac{d}{dt}(s(t)) = \frac{d}{dt}\left(Ae^{\lambda t}\right) = A\lambda e^{\lambda t} \tag{27}$$

and the acceleration will be

$$a(t) = \frac{d}{dt}(v(t)) = \frac{d}{dt}\left(A\lambda e^{\lambda t}\right) = A\lambda^2 e^{\lambda t} \tag{28}$$

Substituting the solution for position from Eq. (25), the solution for velocity from Eq. (27), and the solution for acceleration from Eq. (28) into Eq. (24) produces the following:

$$m\left(Ae^{\lambda t}\right) + c\left(A\lambda e^{\lambda t}\right) + k\left(A\lambda^2 e^{\lambda t}\right) = 0 \tag{29}$$

The amplitude, A, and the exponential function, $e^{\lambda t}$, are found in all terms of Eq. (29) and factoring these terms produces the following:

$$(m\lambda^2 + c\lambda + k)Ae^{\lambda t} = 0 \tag{30}$$

where if there is motion, neither the amplitude nor the exponential function, $e^{\lambda t}$, can be zero over all time. Therefore, the variable λ has to be a certain value for the equation to be true. Dividing both sides by the amplitude and the function removes those values from Eq. (30) and leaves the vibration analyst with the following characteristic equation

$$(m\lambda^2 + c\lambda + k) = 0 \tag{31}$$

This parabolic equation will provide information about how the modeled SDOF system will vibrate. The damping value may cause the system to be underdamped, critically damped, or over damped. In the majority of cases that a vibration analyst will encounter, the vibrating system will be underdamped. The solution to Eq. (31) will be briefly explored to show where these three different types of solutions are obtained.

The solution to a parabolic equation can be obtained from the quadratic equation. Therefore, the variable λ of Eq. (31) could have two real solutions, one real solution, or two imaginary solutions, depending on the values of the mass, stiffness, and damping:

$$\lambda = \frac{-c \pm \sqrt{c^2 - 4mk}}{2m} \tag{32}$$

Which of the three solutions is obtained will depend on the solution to the radical (the square root value). If $c^2 = 4mk$, the value of the radical is zero and the system will have one real solution and is considered critically damped (the solution will not oscillate due to a displacement and the system will return to the equilibrium position as quickly as possible without any overshoot). If $c^2 > 4mk$, the value of the radical is real and the system will have two real solutions and is considered over-damped (the solution will not oscillate due to a displacement and will return to the equilibrium position). If $c^2 < 4mk$, the value of the radical is imaginary and the system will have two imaginary solutions and is considered under-damped (the solution will oscillate due to a displacement and decay toward the equilibrium position over time with some amount of overshoot). Figure 10 shows examples of these three conditions where the mass and stiffness of the system is held constant and the damping value is varied. The displacement here is assumed to be an initial mass displacement that is released to allow the mass to move and ultimately return to equilibrium.

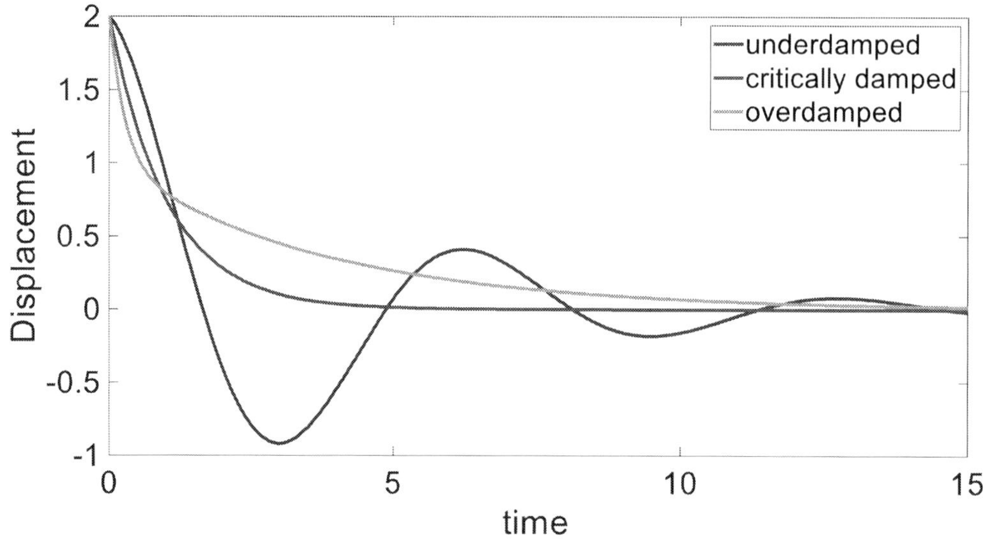

Figure 10. Modeled displacement of a SDOF system with an initial displacement and various damping values applied.

DISCUSSION/CONCLUSION:

Single degree of freedom systems, consisting of a mass and spring, provide the fundamental unit for understanding a lot about vibrating systems. This system can be mathematically modeled to provide equations that estimate how a system will reaction and what to expect when a system is excited. All of the equations and work presented here have been focused on the "free" response (where an initial displacement sets the system into motion). The derivative and antiderivative were also introduced. These calculus principles can be used to help the vibration analyst understand the relationship between position, velocity, and acceleration. In many instances, the vibration analyst will have data of either the position, velocity, and acceleration instead of a mathematical function. The principles shown here can be used with the data to approximate the other derivatives or antiderivatives of the data. Performing these calculations does not require an extensive knowledge and background in complex math. But understanding the principles of the calculations will assist the vibration analyst in understand the relationship of the equations and what to expect from a vibrating system.

REFERENCES:

[1] Thomas, G., and Finney, R., *Calculus and Analytic Geometry, 9th ed.*, Addison-Wesley Publishing Company, Reading, Massachusetts, 1996.

BIBLIOGRAPHY:

[1] Boyce, W.E., DiPrima, R.C., and Meade, D.B., *Elementary Differential Equations and Boundary Value Problems 7th ed.*, John Wiley & Sons, New York, New York, 2001.

[2] Hibbeler, R., *Engineering Mechanics: Statics and Dynamics, 14th ed.*, Pearson Prentice Hall., Hoboken, New Jersey, 2016.

[3] Inman, D.J., *Engineering Vibration 2nd ed.*, Prentice Hall, Upper Saddle River, New Jersey, 2001.

An experimental verification on inducing instability in a tall structure under seismic load using a single axis shake table.

Kameron Nez, *Undergraduate/Graduate Student*, Daniel Latimer, *Undergraduate Student*

Mosfequr Rahman, *Professor*, Aniruddha Mitra, *Professor*

Department of Mechanical Engineering, Georgia Southern University

Kameron Nez is a graduate student attending Georgia Southern University. Kameron is continuing research under Dr Mitra with vibrations and seismic loading and plans to complete his master's degree before Spring 2025.

Daniel Latimer is an Undergraduate Student Graduating Spring 2023. Daniel has accepted a job with Gulfstream and plans to continue to learn and gain experience to move up within the company.

Mosfequr Rahman, Ph.D.
Professor Of Mechanical Engineering
Georgia Southern University

Aniruddha Mitra, PhD, PE
Professor of Mechanical Engineering
Georgia Southern University

ABSTRACT:

Instability of structures, such as large vertical pressure vessel, is always an issue under the seismic load. In the current work, a lab setup has been created to experimentally verify the critical acceleration/deceleration level that would trigger the instability in the structure. An educational R-191 single axis shake table with maximum displacement of ±15 cm is used for this purpose. This shake table is programmed to simulate sinusoidal waves at different frequencies using FOXTROT program. Different frequencies in sinusoidal waves generate different levels of maximum accelerations and minimum decelerations. A 3D printed 1:50 scaled down model is used for vertical pressure vessel which is supported by its three legs at the ground. Forces transmitted through these supports are measured using loadcells and recorded. The critical levels of accelerations (and decelerations) are estimated at the instants when the transmitted load though the joints are measured zero or "negative". This situation triggers instability in the structure. The experimental results are compared with the theoretical calculation. The future work will involve the prediction of level of acceleration/deceleration that will initiate the failure of the actual structure in the field. It is also important to verify from the available earthquake data whether that level of acceleration is generated in a particular earthquake event.

KEY WORDS: Load Cells, Seismic Loading, Stability, Static structural analysis, Vibrational analysis, Impact load.

NOMENCLATURE:

A	Amplitude, m
a_{max}	Maximum Acceleration, m/s^2
G	Gravity, m/s^2
f	Frequency, Hz
F_I	Force (Inertia), N
F_n	Force (Normal), N
m	Mass, kg
t	Time, sec
T_I	Torque (Inertia), Nm
T_n	Torque (Normal), Nm
x	Displacement, m
x_1	Vertical Distance, m
y_1	Horizontal Distance, m
ADC	Analogue-to-digital converter
DAQ	Data acquisition device

INTRODUCTION:

Instability is defined as the state of being unstable; for the vertical pressure vessel this occurs when the force on any load cells supporting the structures reads zero or a negative number. With the failure; the maximum acceleration and deceleration can be determined. A model with a 1:50 ratio was used to carry out experimentation. Testing was conducted to find the minimum acceleration causing the instability in the small model, the minimum acceleration can be scaled up for the actual vertical pressure vessel.

Acceleration is the rate of change of velocity within a body of motion. Acceleration can be derived from position by taking a double derivative of a displacement function. Using a sine function to model the position of a seismic event the maximum acceleration is a function of amplitude and frequency. The maximum acceleration can be determined by testing different frequencies until the vertical tank is dislodged from the load cell.

Amplitude and frequency are changed to determine the effects of seismic force upon a vertical pressure vessel. The maximum amplitude is constant while being dependent upon the stoke length of the shake table. Frequency is an independent variable for a sinusoidal wave, used to quantify the position. By changing the amplitude and frequency of a sinusoidal wave, the position, velocity, and acceleration can be varied, and the instability of a tall structure can be induced.

Provided is the displacement sinusoidal wave formula which determined the output of the shake table.

$$x = A * sin\,(2\pi f t)$$

$$\frac{d^2}{dt^2}(A * sin(2\pi f t)) = -A * (2\pi f)^2 * sin(2\pi f t)$$

$$a_{max} = A * (2\pi f)^2 \qquad \text{Eq.(1)}$$

LITERARY SEARCH:

Testing of scaled models is commonplace within Civil Engineering. A majority of the results gathered from seismic testing are for bracing and the effect of concrete when subjected to seismic loading. Most seismic testing occurs using simple shake tables to replicate the effects of a seismic event.

When seismic testing is performed a scaled model is used in order to scale results toward the actual design specifications used in the field. When the scaled model is constructed, the material used is similar to the materials of the actual structure. For the vertical pressure vessel, the material chosen had set properties that can be then scaled to the corrected values using a scale factor.

Different structures being created are using unique bracing techniques to support buildings. The different method of bracing yields an increase in stability during a seismic event. The bracing is normally on the steel frame that supports the structure. The bracing of the structures is tested to prove or disprove the validity of incorporating additional bracing for the purpose of reducing seismic effects. Most structures within a high-risk area, already have seismic bracing or seismic damping systems built within the design. With the seismic preventive system further testing is needed to improve and optimize designs [1-6].

The simulation of seismic events uses sinusoidal waves to depict the resulting displacement. Sinusoidal waves are the simplest form of creating a displacement and an acceleration that is easily measurable. Sinusoidal waves are affected by two variables, amplitude, and frequency. The sinusoidal waves are changed through different methods to graph actual seismic events. The different methods that are used is a Fourier Analysis and Endurance Time Methodology. With a Fourier analysis a seismic event can be broken down into multiple sinusoidal waves that are upon a wave train. Different sinusoidal waves with different amplitudes and frequencies are used to create estimation of actual seismic performances[7-8]. Endurance Time methodology can be used to create seismic activity rather than a Fourier analysis. The Fourier analysis uses the displacement (amplitude) to create different peaks of the wave train. Endurance Time methodology uses different accelerations to create a special intensity that increases with time [8-9].

The problem that arises from seismic testing is the amount of noise within the testing processes. With using a Fourier transform the noise of a seismic event can be reduced by protecting the equipment being used. Sinusoid waves are broken down with a Fourier transform to an acceleration and frequency dependent graphs. Sinusoid waves, when converted, limit the amount of noise attributed to the wave form. The noise is detrimental when taking a sample of the wave form because the sample does not show the true nature of the seismic activity and the response of seismic sensitive equipment [10-11]. The noise limiting allows for a smoother transition for the shake table. The smother transition allows for less fatigue on the testing components. This limiting of noise allows for testing to be accurately measured without the significant increase of outliers.

Materials & Methods:

The hardware used within the experiment is listed. ADC, DAQ, ANCO R-191 Uniaxial Shake Table, Assorted Wires, Breadboard, ElEGOO MEGA 2560 R3, S-Type Load Cell (0-5kg), Scaled Vertical Pressure Vessel (1100 f^3). The software used is listed below. Arduino, Excel, FOXTROT, SPECTIME, MATLAB, PuTTY.

To determine the instability of a vertical pressure vessel through seismic loading, the experiment is measuring the mass of three legs of the vertical pressure vessel. The experiment uses S-Type load cells to measure the mass of the vertical pressure vessel. With the mass being measured, when subjected to a seismic load, the max acceleration can be determined. The maximum acceleration will be scaled to the actual acceleration that the vertical pressure vessel is subjected to.

The S-Type load cell singles are transmitted in analog and need to be converted to digital. The use of Analogue-to-digital converter allows the signal to be collected within the ElEGOO MEGA 2560 R3. The wiring for the load cells is displayed below.

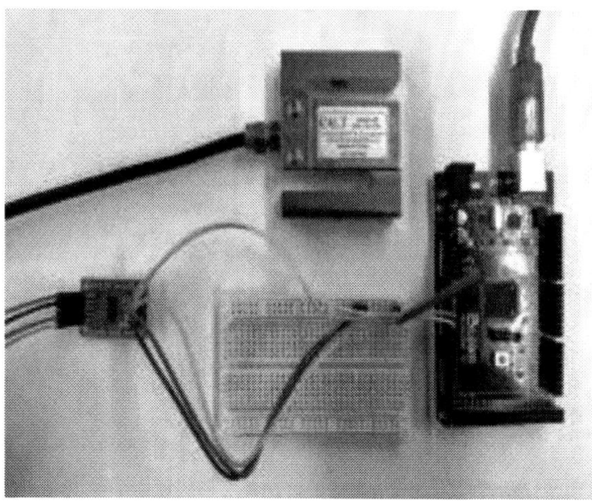

Figure 1: Displays the wiring diagram for the load cell to ElEGOO MEGA 2560 R3 [12].

To attach the load cell to the shake table, a 3D printed holder had to be designed and attached to the shake table. This assembly can be seen in *Figure 2*.

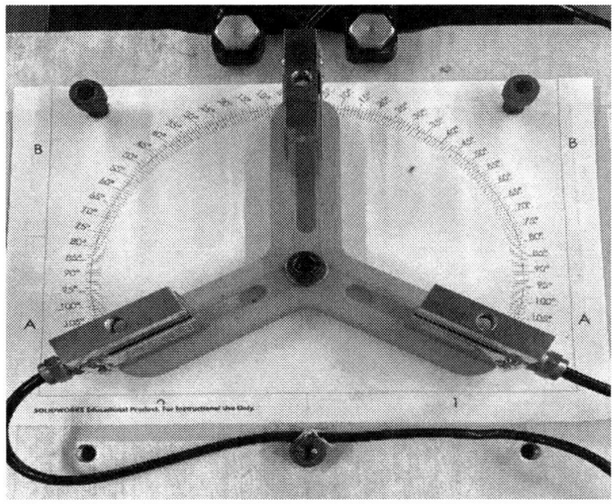

Figure 2: 3D Printed load cell holder.

The vertical pressure vessel was created within SOLIDWORKS to be a scaled down model of a practical vertical pressure vessel. The practical vertical pressure has a volume of 1100 ft^3 .With the scaled down model, the acceleration can be calculated for the larger structure.

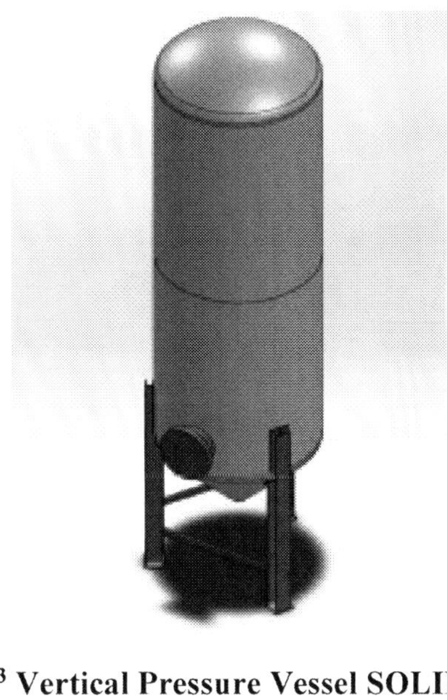

Figure 3: 1100 ft^3 Vertical Pressure Vessel SOLIDWORKS Model [12]

The vertical pressure vessel is connected to the load cells by an adhesive device securing the tank from becoming disconnected from the load cell in the lateral direction. This allows movement along the vertical axis that is related to the maximum acceleration that is applied to the tank.

The position of the vertical pressure vessel is set with the nozzle of the vessel positioned in the direction of the seismic force. The nozzle of the vertical pressure vessel contains the additional mass that will be impacted more by the seismic load. The position of the pressure vessel is set where the two legs of the pressure vessel receive the seismic load first. Figure 4 shows the two loads' cells within the front of the seismic load.

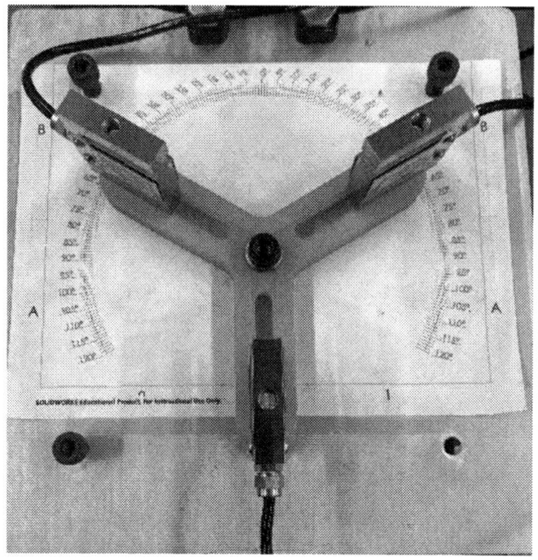

Figure 4: Position of Vertical Pressure Vessel (Nozzle Pointed Upwards)

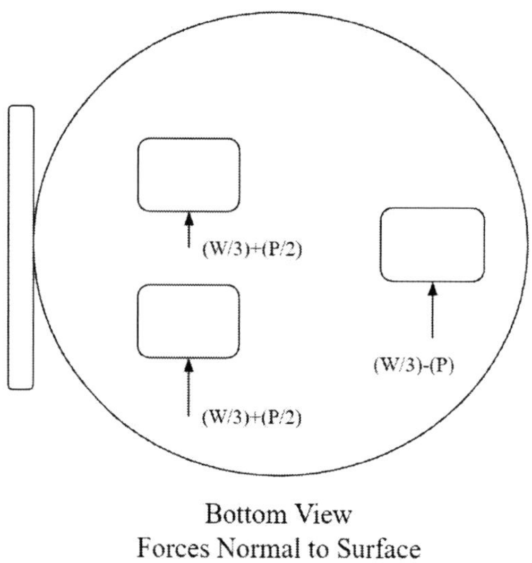

Figure 5: Position of the Vertical Pressure Vessel (Skematic)

The seismic load is applied to the left of the vertical pressure vessel. Figure 5 has the positions of the legs marked and the forces that are being applied to the Vertical Pressure Vessel.

The Shake Table is an ANCO R-191 Uniaxial Shake Table that uses FOXTROT as the software that is used to actuate the shake table using .wav files. Within the software the experiment required a differentiable wave, so a sinusoidal wave was chosen for the position file for the shake table to be recreated. A sine wave was chosen for this experiment because it is dependent on the frequency and the amplitude, which can be set at certain values so the maximum acceleration can be calculated based on those variables. The frequency used within testing ranged from 1Hz to 2Hz with a step size of 0.1 Hz. The amplitude of the experiment is 20% of the maximum displacement of 0.127 meters. At keeping the amplitude at a constant value, the maximum acceleration is only dependent on the frequency of displacement sinusoidal wave. The amplitude and the frequency can be set easily, the amplitude has a sliding scale, and the frequency is dependent on the .wav file selected.

S-type load cells are used to collect mass from the vertical pressure vessel. The signal transmitted is converted using an Analog-to-Digital converter to then be interpreted by the ElEGOO MEGA 2560 R3 and the code "data_collection". Using the "data_collection" code displays the time and the mass of the three load cells. The units of time and mass are in milliseconds and grams respectively. The values received by the Arduino software are collected using PuTTY; the software is used to write a text file containing the values of time and the mass values of the three load cells. The time and mass values are processed in MATLAB to graph the mass readings versus time. The graphs give an indication of when the mass falls to zero; when that happens the vertical pressure vessel has failed or fallen off the load cell.

THEORETICAL RESULTS:

Different seismic loads are dependent on the frequency of the seismic event. For the experiment a sinusoidal wave is used; and the frequency can be changed. Now working backwards and using torsional moments, the acceleration needed to displace the vertical pressure vessel can be determined. The acceleration can be determined by Eq.(2). Solving for the inertia force within these theoretical results are shown below.

$$F_I = a_{max}(m) \qquad \text{Eq.(2)}$$

The Free Body Diagram is provided within Figure 6. The Free Body Diagram has the lengths and forces acting on the Vertical Pressure Vessel.

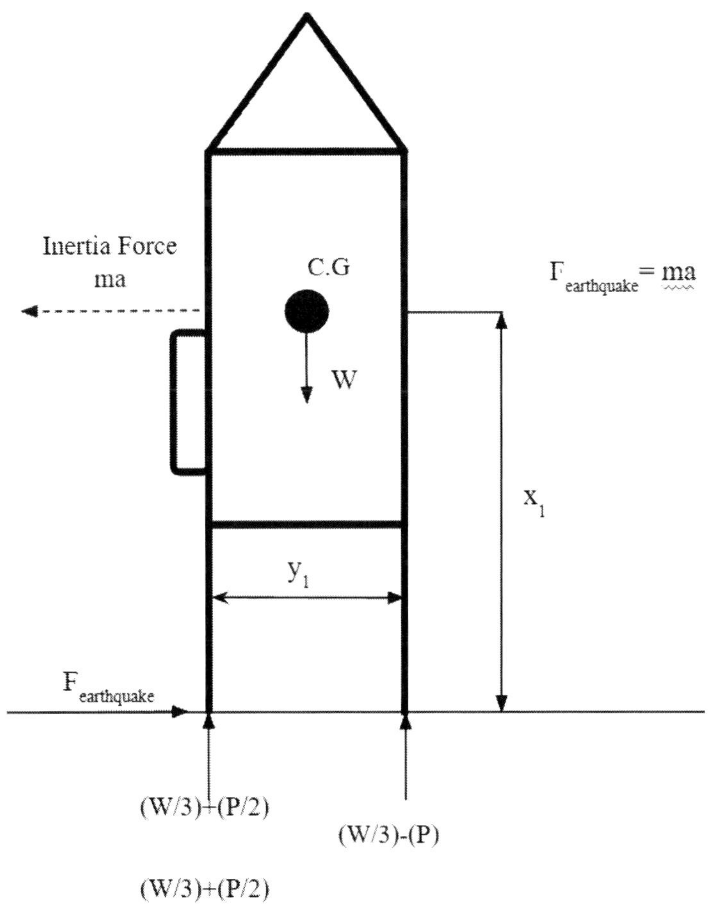

Figure 6. Free Body Diagram to Determine the Acceleration of F_I

Torque inertia is the force inertia multiplied by the vertical distance of the centroid to the ground. The vertical distance from centroid to the grounds is 0.324 m. Within Eq.(3) the torque inertia can be evaluated.

$$T_I = F_I * (x_1) \qquad \text{Eq.(3)}$$

$$T_I = F_I * (0.324m)$$

$$T_I = 0.324F_I \; Nm$$

The forces applied due to the weight of the pressure vessel can be calculated using the Eq.(4). The force normal is calculated by the mass of the vertical pressure vessel multiplied by gravity. The mass is divided by three to calculate the force applied to the critical load cell. The mass of the vertical pressure vessel is measured at 0.361 kg and gravity is 9.81 m/s^2

$$F_n = \frac{m}{3}(g) \qquad \text{Eq.(4)}$$

$$\frac{0.361kg}{3}\left(9.81\frac{m}{s^2}\right) = 1.18\ N$$

The force applied to the critical load cell due to weight is 1.18 N. The distance between the legs is calculated to be 0.17m. Using the distance the torque resulted can then be calculated by multiplying the normal force by the distance calculated above using Eq.(5).

$$T_n = F_n * (y_1) \qquad \text{Eq.(5)}$$

$$1.18\ N * (0.17m) = 0.201\ Nm$$

The torque normal is calculated to be 0.201 Nm. Using the different torques calculated the force required to topple the vertical pressure vessel can be determined using Eq.(6).

$$T_N = T_I \qquad \text{Eq.(6)}$$

$$0.201\ Nm = 0.324F_R\ Nm$$

$$\frac{0.201Nm}{0.324m} = 0.620N = F_I$$

Using the Eq.(2) the theoretical acceleration can then be determined using the mass of the vertical pressure vessel.

$$F_I = a_{max}(m) \qquad \text{Eq.(2)}$$

$$\frac{0.620\ N}{0.361\ kg} = 1.72\frac{m}{s^2} = a_{max}$$

EXPERIMENTAL RESULTS:

The acceleration of the experiment can be determined using the Eq. (1). Different frequencies that were tested ranged from 1.25 Hz to 1.30 Hz. The range of frequencies was determined by testing different frequencies ranging from 1.0 Hz to 2.0 Hz in the preliminary experiment. This range was determined from physical testing of the shake table. The shake table supports 0 Hz to 20Hz , however with the frequency higher than 2 Hz, the safety of the operator and the shake table had to be considered. The use of Eq. (1) is displayed below with a frequency of 1.25 Hz.

$$A_{max} = a * (2\pi f)^2 \quad \text{Eq.(1)}$$

$$A_{max} = a * (2\pi f)^2 \Rightarrow (0.127m) * (0.20) * (2\pi * (1.25Hz))^2 = 1.57\, m/s^2$$

The amplitude of the maximum acceleration is dependent upon the maximum displacement of the shake table. For an ANCO R-191 Low-Cost Educational Shake Table the maximum displacement is 0.127 meters. The value of 0.20 is the displacement factor used for the experiment. The shake table is not using the maximum displacement; however, it is using 20% of the maximum. The acceleration of the experiment changes with the frequency of the initial displacement sinusoidal wave. The experimental values of acceleration are recorded within *Table 1*.

Table 1. Frequency and Maximum Acceleration Calculated

Frequency Hz	Maximum Acceleration m/s^2
1.25	1.57
1.26	1.59
1.27	1.62
1.28	1.64
1.29	1.67
1.30	1.69

The acceleration for testing is based on the position of the tank. With the set position the maximum frequency that can be applied to a sinusoidal displacement is 1.25 Hz. The acceleration for the frequency of 1.25 Hz is 1.57 m/s^2. Shown in the following figure is the mass response at which the vertical pressure vessel is losing contact with the load cells.

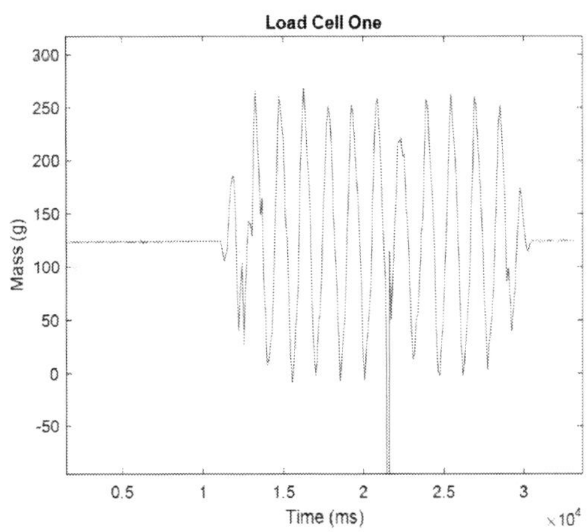

Figure 7: Mass response for Load Cell One for frequency 1.25.

With the seismic force being applied to the vertical pressure vessel the resultant inertia force has an acceleration of 1.57 m/s^2 that is compared to the theoretical results of the 1.72 m/s^2. The percentage error can be calculated with the expected value and the actual value, calculated to be 7.64%. The frequency required to displace the vertical pressure vessel is determined to be 1.25 Hz.

CONCLUSION:

A vertical pressure vessel was applied to a simulated seismic load to obtain the different mass values for a set time period The one configuration of the vertical pressure vessel is tested. For the set position the frequency required is 1.25 Hz. At this frequency the vertical pressure vessel is dislodged from the load cell and the toppling condition is met. The resulting acceleration required to cause toppling is 1.57 m/s^2. The theoretical acceleration that was calculated was 1.72 m/s^2. The percentage error of the experimental results compared to the theoretical results was 7.64%. With the results from this manuscript, values can be scaled up to the actual pressure vessel and the acceleration determined. With acceleration now the tank can now be tested with similar seismic activity. The stability of the actual pressure vessel can then be determined and verified.

ACKNOWLEDGEMENTS:
The Authors would like to thank Mechanical Engineering department of Georgia Southern University for the Equipment used in this research project. Authors would also like to extend thanks to Valesia Davis and Evan Hayes for their editorial help.

REFERENCES:

[1] Khan Fasih Ahmed, Khan Sajjad Wali, and Khan Wajid. "Seismic Performance of RC-ECC Composite Frame by Eliminating Shear Reinforcement in Beam-Column Joint: Shake Table Tests." *Advances in Structural Engineering* VOL. 25 No. 14 (2022): pp. 2966-2980. doi:10.1177/13694332221113044.

[2] Leelataviwat Sutat, Suksan Bunyarit, Srechai Jarun, and Warnitchai Pennung. "Seismic Design and Behavior of Ductile Knee-Braced Moment Frames." *Journal of Structural Engineering* VOL. 137 No. 5 (2011): pp. 579–588. doi:10.1061/(ASCE)ST.1943-541X.0000301.

[3] Moghaddam H., and Mohammadi R. Karami. "More Efficient Seismic Loading for Multidegrees of Freedom Structures." *Journal of Structural Engineering* VOL. 132 No. 10 (2006): pp.1673–1677. doi:10.1061/(ASCE)0733-9445(2006)132:10(1673).

[4] Suduo Xue, Mingyue Shan, Xiongyan Li, Shuanzhu Liang, Fuyun Hyang, and Yi Liu. "Shaking Table Test and Numerical Simulation of an Isolated Cylindrical Latticed Shell under Multiple-Support Excitations." *Earthquake Engineering & Engineering Vibration* VOL. 18 No. 3 (2019): pp. 611–630. doi:10.1007/s11803-019-0525-6.

[5] Yang, Rongqian, and Zhou Xuejun. "Experimental Research and Theoretical Analysis of the Seismic Behavior of Prefabricated Semirigid Steel Frame with X-Shaped Braces." *Advances in Civil Engineering* (2019): pp. 1–14. doi:10.1155/2019/1967253.

[6] Yu Mingming, Xie Xu, and Li Shuailing. "A Simplified Ductile Fracture Model for Predicting Ultra-Low Cycle Fatigue of Structural Steels." *Materials (1996-1944)* VOL. 15 No. 5 (2022): pp.1663. doi:10.3390/ma15051663.

[7] Wang Jiun-Jr., Shrive Nigel G., Parker Kim H., and Tyberg John V. "'Wave' as Defined by Wave Intensity Analysis." *Medical & Biological Engineering & Computing* VOL. 47 No. 2 (2009): pp. 189–195. doi:10.1007/s11517-008 0403 2.

[8] Damerji H., Yadav S., Sieffert Y., Debove L., Vieux-Champagne F., and Malecot Y. "Design of a Shake Table with Moderate Cost." *Experimental Techniques* VOL. 46 No. 3 (2022): pp. 365–383. doi:10.1007/s40799-021-00482-0.

[9] Valamanesh, V., and Estekanchi H. E. "Nonlinear Seismic Assessment of Steel Moment Frames under Bidirectional Loading via Endurance Time Method." *Structural Design of Tall & Special Buildings* VOL. 23 No. 6 (2014): pp. 442–462. doi:10.1002/tal.1054.

[10] Chen Binqiang, Lan Qixin, Li Yang, Zhuang Shiqiang, and Cao Xincheng. "Enhancement of Fault Feature Extraction from Displacement Signals by Suppressing Severe End Distortions via Sinusoidal Wave Reduction." *Energies (19961073)* VOL.12 No. 18 (2019): pp.3536. doi:10.3390/en12183536.

[11] Wang, Tao, Shang Qingxue, and Li Jichao."Seismic Force Demands on Acceleration-Sensitive Nonstructural Components: A State-of-the-Art Review." *Earthquake Engineering & Engineering Vibration* VOL. 20 No. 1 (2021): pp. 39–62. doi:10.1007/s11803-021-2004-0.

[12] DeMond Matthew, and Cathers Evan. "A Study of Seismic Loading on Vertical Pressure Vessels." *MENG 7891 Final Report* (2021): pp. 1-10.

Exploration of Characteristic curve in FOX Float 3 shock dampers to expedite shock damp tuning.

Joshua Moore, *Undergraduate/Graduate Student*, Mechanical Engineering

Jordan Denney, *Undergraduate Student*, Mechanical Engineering

Aditi Mitra, Lecturer, Mathematical Sciences

Aniruddha Mitra, *Professor, Mechanical Engineering*

Joshua Moore
Undergraduate/Graduate Student
Department of Mechanical Engineering
Georgia Southern University

Jordan Denney
Undergraduate Student
Department of Mechanical Engineering
Georgia Southern University

Aditi Mitra, M.S
Lecturer, Department of Mathematical Sciences
Georgia Southern University

Aniruddha Mitra, PhD, PE
Department of Mechanical Engineering
Georgia Southern University

ABSTRACT:

The shock absorber is an integral part of a vehicle suspension system and has a strong influence on its performance, especially in the case of motorsports. It is important to study the force versus velocity relationship, commonly known as the characteristic curve of the shock absorber both during compression and rebound. Vendor-supplied characteristics often reflect the behavior of the shock absorber in a particular setting. However, during the installation, the settings of the shock absorber are adjusted to increase the human comfort level and performance of the vehicle. This may change the characteristic curve of the shock. The available data and direct comparison of different tune-up settings are limited. In an ideal model, the force is directly proportional to the velocity. However, in literature, except for the steep linearity at the relatively small section closer to the origin, the characteristic curve is often found to follow a power regression model with an offset in the form of $F= a + bv^c$, where F is the force generated at the shock and v, the velocity, a, b, c are the regression parameters. In the current research, three Fox Float 3 are tested at their factory conditions to assess the relationships between the force and the velocity. Also, several settings of shock positions of two of those shocks are tested to develop a mathematical model. A predominant linear trend has been observed for all the cases. Future work will involve tracking these parameters throughout their operational life cycle.

KEY WORDS: Shock Absorber, Characteristic curve, Damping.

INTRODUCTION:

Shock absorbers are an integral part of any vehicles suspension system. In performance applications these absorbers' main goal is to increase tractive effort between the vehicles ground and the racing surface. In military and heavy vehicle applications the purpose is to provide damping effort against extreme conditions allowing the vehicles to traverse the ground plane at a quicker and safer mode. In passenger vehicles the shock absorber's purpose is to provide a balance between ride comfort and performance by damping road irregularities and normal operating conditions. These absorbers even have applications of non-vehicle type such as heavy industrial machinery which needs damped vibration for fatigue or more consistent operating conditions. In all these applications the shock absorber must provide a median between the sprung and unsprung weight of the vehicle or machine. The Fox Float 3 is an air over hydraulic shock absorber that is used in a variety of vehicle applications. The shock comes standard on a sever snowmobile models and was recently incorporated into Fox's BAJA SAE program where it grew in popularity as the shock is easy to use in theory, light weight, and easy to package. The damping, however, is not a user-friendly adjustment available on these shocks as they are monotube shocks. A twin tube shock or a monotube shock encompassing a bypass valve would be easier to tune due to external adjustment making it so that there is no need for the user to discharge the gas and disassemble the shock to adjust. The Fox Float 3 does not possess these features presumably because of its original equipment manufacturer (OEM) application on snowmobiles, one tune typically covers its original intent. If, however, a model could be made from experimental data to tune this shock then the adjustment could be done in a few, if not a single, adjustment saving time and money. Because this is a monotube shock, the main passage for hydraulic mineral oil to flow and for damping force to be developed exists inside the main orifices of the piston (main valve) these are tuned by shims that are seated on the top and bottom of the piston perpendicular to the theoretical flow of the oil and are secured with a nut attached to the end of the piston rod. By removing, adding, changing thickness, or changing diameter, these shims affect the damping characteristic curve (Force vs. velocity). Several different models have been made for shock absorbers and there are many starting points to this research. Some absorbers can be modeled solely off theoretical relationships between interactions inside the absorber. Many of these methods are valid to certain limitations whether it be speed, frequency, or displacement of the shock absorber.

LITERARY SEARCH:

The very basis of shock absorbers and their broad level purpose is to dissipate energy in an efficient way and disconnect a mass, that is meant to be protected from sometimes harsh input, from a mass that is in contact with that input. *D. Narkhede et al* [1] explore the nonlinear behavior of these dampers. The authors set up a loading test using custom shock testing machine that encompassed a weight of 98.1kN mounted on rollers to strike a damper to generate a half cycle sine wave signal. One of the conclusions they came to during testing was that the damper exponent α and coefficient Cw remain considerably constant through the shock tests.

Modeling of a shock absorber for confidence in predictability beyond what is experimentally measured is an important part of shock absorbers. By developing models for specific shock absorbers, adjustments can be made with reduced testing time. *Y. Cui, et al* [2] discuss testing on a Mazda CX-7 shock absorber to develop a nonparametric model for the shock and discuss the outcome. The authors used a single post shaker table for testing which produced a standard looking damping graph (force v. velocity) even at high speed. The authors tested shocks at up to 1.3m/s (51.18in/s) which they claim to represent road irregularities.

The main damping adjustment for most shock absorbers is done through values with various methods of providing resistance to fluid motion. The main piston value is a characteristic almost all shock absorbers share and is adjustable most times through thin plates stacked on top of the value called shims. *D. Buczkowski et al.* [3] perform CFD analysis using Ansys Fluent v 19.0 to explore different shock settings in hopes to reduce tuning timeline hold up by optimizing flow though main valve design in the software before building a physical model. The authors concluded that the next step would be to test an additional piston design.

There are several ways to adjust damping force through the main valve of the shock absorber. Sprung valves that move out of the way with increased flow due to velocity are one way, the other is through shims used as mini cantilever beams. These shims have an associated spring rate that move and return at a linear rate. The shims are stacked on the top and bottom of the main valve in various thicknesses and diameters to develop a desirable damping rate for that side of the main valve. *P. Skačkauska et al.* [4]investigate and verify a shim stack model. The group used experimental testing on a special electromechanical stand and obtained a peak velocity of 0.4m/s (15in/s). The model developed by the authors includes several different equations derived from literature. An equation to note is the area through the main valve before and after the initial deflection of the stack occurs. The authors found that higher frequency inputs and low amplitude inputs were not very accurate due to difficulty in reproducing those conditions, but most of the other scenarios were able to be predicted well.

MATERIALS & METHODS:

Rebuilding the shock absorber is a complex process that must be done repeatedly for this research. This research uses Fox Float 3s which are monotube air over hydraulic shocks. Monotube meaning the main valve is the only passage for air flow inside the absorber (figure 1)

Upon the arrival of the Fox Float 3s, the shock air cylinder was removed as it is undesirable to test on the damping dynamometer. Three different shocks were tested on the test rig with factory tune-up inside to gain a baseline characteristic curve (figures 3, 4 and 5). A test plan was then written to achieve the data necessary to begin physical analysis of the dampers to achieve a model. The initial test plan was to remove shims of equal diameter and thickness in even intervals to observe the severity that the shims had on the damping curve, testing on the dynamometer after each adjustment. After experimentation, however, it was observed that the diameter and thickness could not be held as constant as there are not enough of the same shims in the factory tune to accomplish this. This research utilizes a CTW RD3 shock dynamometer (figure 2). This testing rig encompasses a linear potentiometer for displacement and velocity derivation, a load cell to measure shock force, and a thermocouple mounted on the linear rail to capture shock temperature. The dynamometer uses a scotch yoke style (figure 2) method for power delivery from the unit's AC motor to the position that the bottom of the shock is hooked to. To test the shock absorber on the dynamometer a test plan was created using CTW Probe software. The test plan entails a 10in/s warmup of the absorber to get the oil moving and slightly increase the temperature above its static temperature. The next part of the test is a rod force test. This test is primarily for if the spring is equipped onto the shock, for this test the spring was effectively removed from the shock by disconnecting the air chamber, however; this portion of the test program was left in to increase data accuracy by removing a small amount of the seal drag or any other position-dependent forces. The dynamometer does this by moving the shock to the 0-stroke position (perpendicular position of the shock absorber) which is the theoretically highest point of force developed on the dynamometer. The rig is programmed to read these forces statically so there are no position dependent forces to influence of damping present. The shock dynamometer then creates a force V. displacement diagram and effectively removes it from the data present posttest. Next, the dynamometer completes three cycles at three velocities and records the middle cycle of each one to minimize the effect of the dynamometer accelerating or decelerating to the desired velocity abnormally.

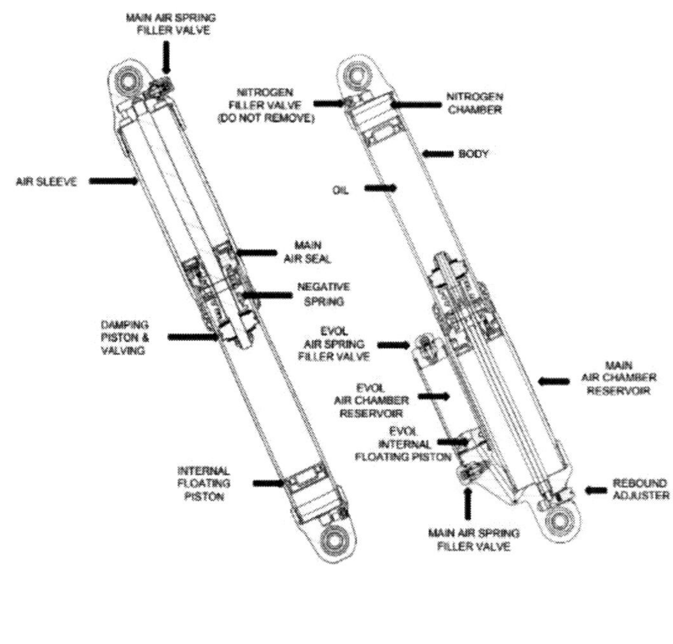

Figure 1: Fox Float 3 Diagram

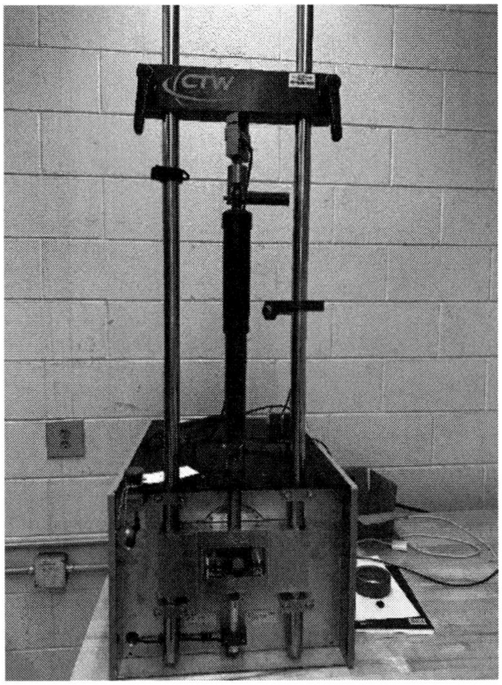

Figure 2: CTW RD3 Dynamometer

RESULTS:

All results are presented in graphical form. Figures 3, 4 and 5 show the characteristic curve at the factory setting for three different shock absorbers. Figure 6 indicates the data for different settings at the compression side including the original factory setting for one of the shocks. Similarly, figure 7 shows the data for different settings at the rebound side including the original factory setting for a different shock. Linear trends are predominant for all the settings. Finally, figure 8 and 9 indicate the change in damping coefficient with respect to number of sims for both compression and rebound. Even there the overall linear trend is quite obvious from the R squared value.

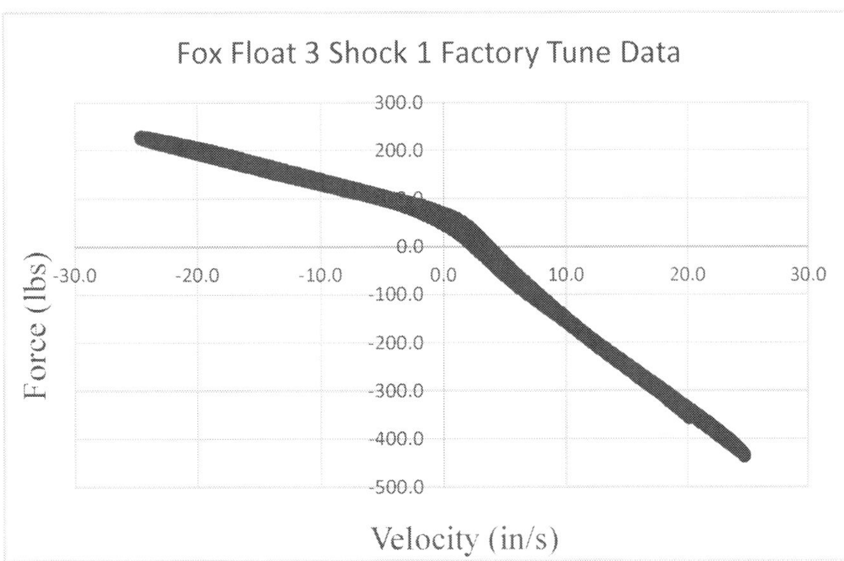

Figure 3: Fox Float 3 Shock 1 Factory Tune Data

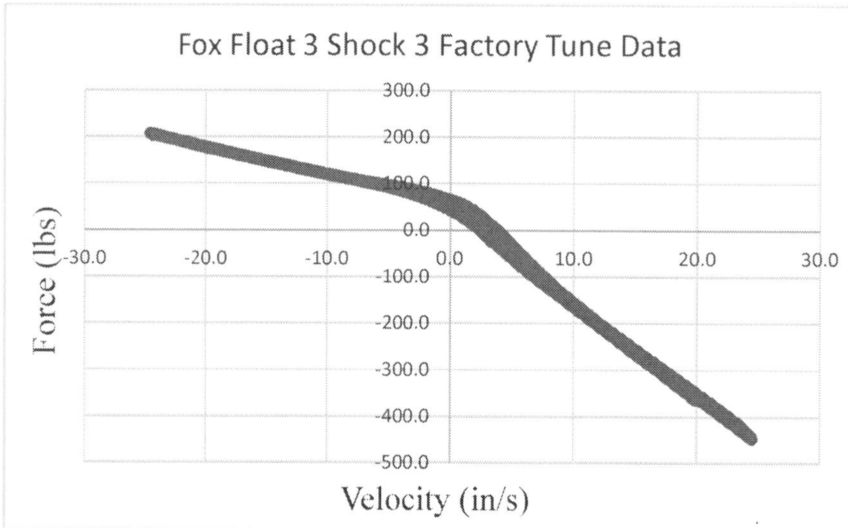

Figure 4: Fox Float 3 Shock 1 Factory Tune Data

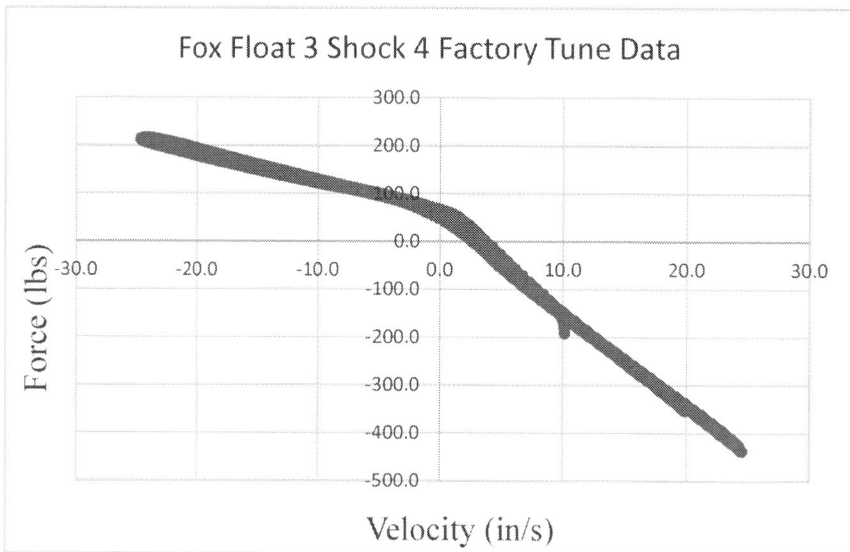

Figure 5: Fox Float 3 Shock 4 Factory Tune Data

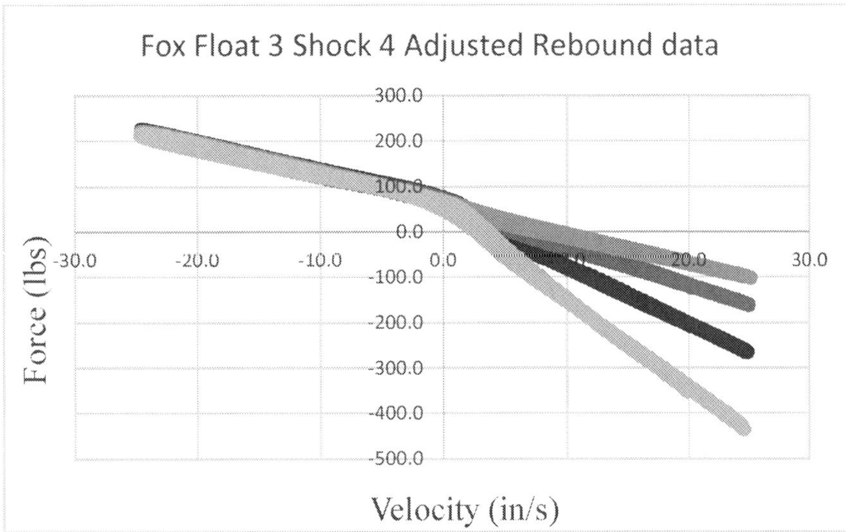

Figure 6: Fox Float 3 Shock 4 Adjusted Rebound Data

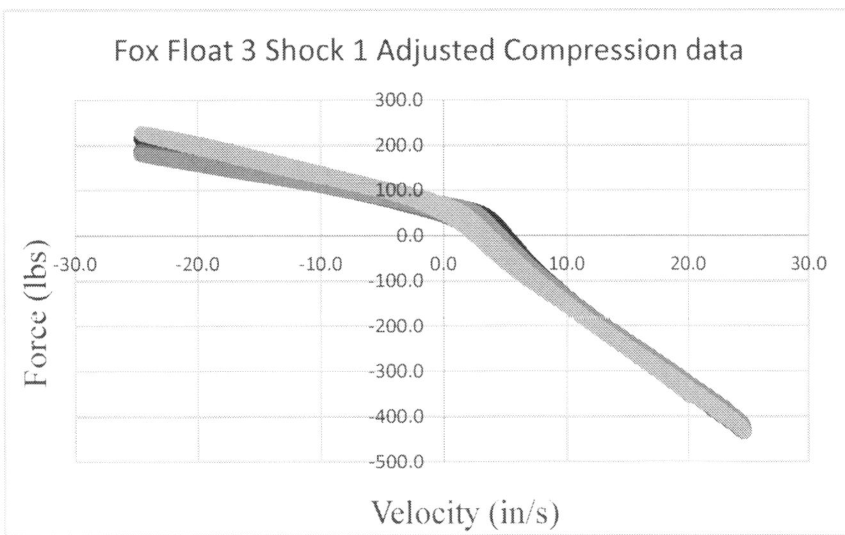

Figure 7: Fox Float 3 Shock 1 Adjusted Compression Data

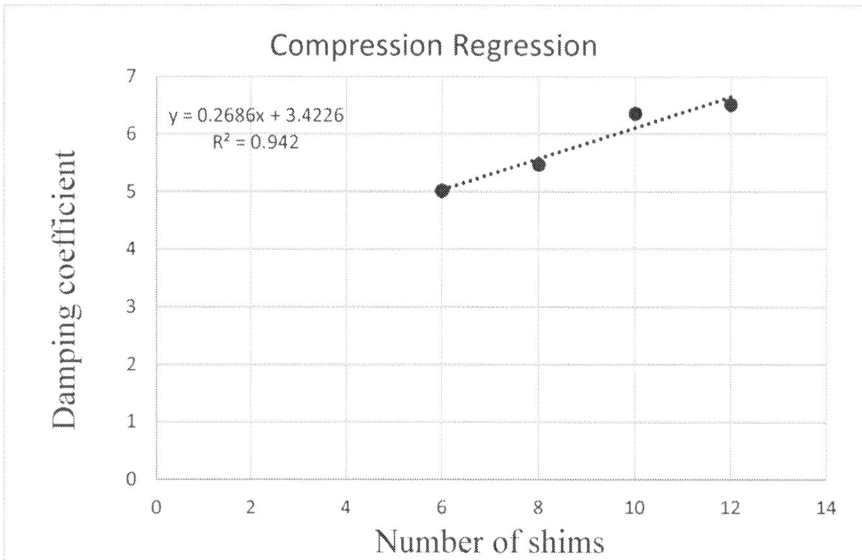

Figure 8: Compression Regression Analysis

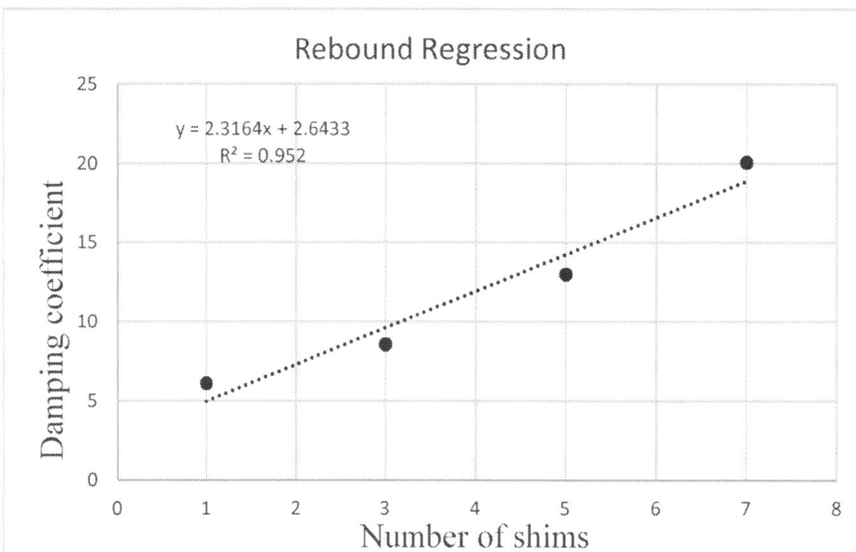

Figure 9: Rebound Regression Analysis

DISCUSSION:

To establish the baseline, three different shock absorbers were tested with the original company setting with twelve sims on the compression side and seven sims in the rebound side. Based on the regression analysis, both compression and rebound zones for all the shocks are quite linear with correlation coefficient close to 0.99. In case of compression, the average damping coefficient is 6.06 lb /(in/sec) with a standard deviation of 0.42 lb/(in/sec). In case of rebound, however, the average is 20.35 lb/(in/sec) with a standard deviation of 0.83 lb/(in/sec). Hence, the shocks show higher stiffness under rebound in factory condition.

Two particular shocks were selected for further investigation under different tune up settings which is accomplished by removing preplanned number of sims. One of them was used for carrying out different settings under compression while the other one was used for rebound. The experimental results again indicate that irrespective of the level of tuning, shocks have linear characteristics curves, the force is linearly proportional to the velocity. In all cases, the correlation coefficients are close to 0.99. Hence the damping coefficient can be viewed as constant for a given setting. This nature is quite in contrast with some of the references where the characteristic curve shows significant nonlinearity and in some cases power relationship. Four different tune up conditions were tried for each compression and rebound cycle including the initial factory setting. In each case the sims were removed in steps and the experiment was conducted. Although, the thickness of the individual sims varies a little with a mean of 0.1 inch, they can be treated identical all practical purposes with respect to the resistance they offer to the fluid flow from one chamber to the other (figure 6 & 7. The experiment reveals that the number of sims dictates the damping coefficient, and it is quite linear as well. In the compression side each sims adds 0.27 lb/(in/sec) to the overall damping coefficient. In the rebound side each sim adds 2.32 lb/(in/sec) to the overall damping coefficient. The data also indicates there are significant intercepts at zero, meaning even without any sims there are some resistances, nevertheless.

CONCLUSION:

The characteristic curve, force versus velocity, of shock absorber is an important aspect with respect to vehicle performance as well as human comfort angle. Identical model of three Fox Float shock absorbers were tested using the dyno available in the mechanical engineering laboratory at Georgia Southern University. They are also tested for different tune up setting by removing predetermined number of sims both in compression and rebound zone. As expected, the damping coefficient, the slope of the characteristic curve, is different for compression and rebound with rebound being stiffer. However, based on the data collected, it is quite apparent that the relationship between the force and the velocity is linear with high correlation coefficient. It has also been observed that the damping coefficient is proportional to the number of active sims, at least within the working range of the tune up settings.

AKNOWLEDGEMENTS:
I'd like to thank the Department of Mechanical Engineering for providing the funds for the shock dynamometer, and the shock absorbers used in this research.

REFERENCES:

[1] D. I. Narkhede and R. Sinha, "Behavior of Nonlinear Fluid Viscous Dampers for Control of Shock Vibrations," *J Sound Vib*, vol. 333, no. 1, pp. 80–98, Jan. 2014, doi: 10.1016/j.jsv.2013.08.041.

[2] S. I. Ao and International Association of Engineers., *World Congress on Engineering and Computer Science : WCECS 2010 : 20-22 October, 2010, San Francisco, USA*. Newswood Ltd., International Association of Engineers, 2010.

[3] D. Buczkowski and G. Nowak, "Increase in Tuning Ability of a Car Shock Absorber Valve using CFD," *Journal of Applied Fluid Mechanics*, vol. 12, no. 6, pp. 1847–1854, Feb. 2019, doi: 10.29252/JAFM.12.06.29581.

[4] P. Skačkauskas, V. Žuraulis, V. Vadluga, and S. Nagurnas, "Development and Verification of a Shock Absorber and its Shim Valve Model Based on the Force Method Principles," *Eksploatacja i Niezawodnosc*, vol. 19, no. 1, pp. 126–133, 2017, doi: 10.17531/ein.2017.1.18.

VIATC TRAINING CONFERENCE & EXPO
NIAGARA FALLS, NEW YORK JULY 26, 2023

COUPLED LATERAL-TORSIONAL VIBRATION ANALYSIS OF AN INTEGRALLY GEARED CENTRIFUGAL COMPRESSOR

Zhusan Luo
Lead Rotordynamics Engineer
Turbomachinery, Linde Inc.
zhusan.luo@linde.com

Zhusan Luo is a Lead Rotordynamics Engineer of Turbomachinery Department, Linde Inc. in Tonawanda, New York, USA. Dr. Luo is responsible for the rotordynamic design, analysis and testing of Linde turbomachinery. Prior to joining Linde (Praxair), he worked as a Lead Rotordynamics Engineer with Siemens Industrial Turbomachinery Ltd., Lincoln, UK. He received his B.E. and M.Sc. in Mechanical Engineering from Nanjing University of Aeronautics and Astronautics, Nanjing, China and Ph.D. in Applied Mechanics from the University of British Columbia, Vancouver, Canada. Dr. Luo is a member of the ASME, the Vibration Institute and the IMechE, and a Chartered Professional Engineer registered with UK Engineering Council.

ABSTRACT:
An integrally geared centrifugal compressor is driven by an electrical motor or an industrial turbine and operates at high speed and heavy load. The lateral vibration is usually analyzed for every shaft using an uncoupled rotordynamic model. The torsional vibration of the compressor drive train is analyzed separately from the lateral vibration. The gear meshing between shafts can modify the system dynamics, result in the variations in mode frequencies and shapes and generate additional modes that do not appear in uncoupled shafts. These coupling effects are not normally considered in the rotordynamic design and analysis. However, the coupling effects should be properly included to resolve some vibration issues.

This paper presents a study of the coupled torsional-lateral vibration characteristics of an integrally geared centrifugal compressor. The lateral vibration test data from a synchronous motor driven compressor is analyzed to demonstrate the coupling effects and adverse influence on the compressor reliability. Both torsional vibration of the drive train and the lateral vibrations of uncoupled shafts are firstly analyzed and taken as baseline. Then the coupling effects of the gear meshing on the torsional and lateral natural frequencies and mode shapes are examined in this paper. The dynamic phenomena observed in the compressor test and commissioning are simulated and the physics is in detail discussed in this paper.

KEY WORDS: Rotordynamics, Lateral Vibration, Torsional Vibration, Coupled Vibration, Integrally Geared Centrifugal Compressor

INTEGRALLY GEARED CENTRIFUGAL COMPRESSOR:

Integrally geared centrifugal compressors are widely found in the petro-chemical, industrial gas and process industries. An integrally geared compressor is usually made by fitting compressor impellers at the ends of pinion shafts. The compressor impeller size and speed together with inlet guided vanes and diffusers can be designed to achieve higher efficiency at given processing conditions. The pinions (usually 1-4) are driven by a central bullgear. As shown in Figure 1, the bullgear drives two pinions: low speed pinion (Pinion 1) and high speed pinion (Pinion 2). Two impellers are attached to each pinion shaft. The 4-stage compressor is driven by a synchronous motor through a flexible element coupling.

In the rotordynamic design, the lateral rotordynamics is usually analyzed for a single shaft, i.e., motor, bullgear and pinions, using a simplified model, which is also recommended in API Standards ([1] and [2]). The flexible element coupling is modelled as two overhung masses attached to the motor shaft and the bullgear shaft. A pair of gears is disengaged at the mesh point. The coupling effect of a gear mesh is usually ignored in the lateral vibration analysis. The simplified shaft model, or an uncoupled model, can normally give reliable results in comparison with test data, and so is widely accepted in the turbomachinery industry.

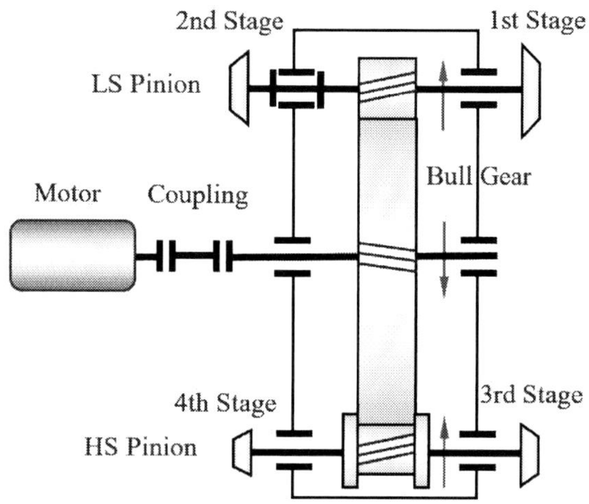

Figure 1 An integrally geared compressor

The torsional vibration of the whole compressor drive train is analyzed separately from the lateral vibration ([1] and [2]). The gear mesh stiffness is considered as a torsional stiffness between two shafts. The influences of the dynamic mesh force during a torsional vibration on the lateral vibrations are not included in the uncoupled lateral vibration.

Coupled lateral-torsional vibration physically exists in geared shaft systems. In most cases, the coupling effect is not so significant that the lateral and torsional vibrations can be analyzed separately. In other cases, however, the coupling effect is so strong that it should be considered to understand the dynamic phenomenon. This paper presents a coupled lateral-torsional vibration analysis of an integrally geared compressor. Three coupled lateral-torsional vibration phenomena observed in the compressor test and operation will be described. Typical damped modes due to the coupling effect will be depicted. A comparison of damped modes on coupled model with uncoupled model will be made. The mechanisms of the three dynamic phenomena will be discussed by means of the coupled vibration analysis.

COUPLED VIBRATION PHENOMENA:

A 9,000 HP synchronous motor drives this 4-stage integrally geared centrifugal compressor. The 6-pole motor runs at a synchronous speed of 1,200 rpm. The low speed and the high speed pinions run at 8,576 rpm and 9,767 rpm, respectively. During the test and commissioning, three coupled lateral-torsional vibration phenomena were observed on the two pinion shafts.

Startup Transient Vibration

A synchronous motor inherently produces a pulsating torque at 2x slip frequency during start. The excitation frequency decreases with increasing speed, i.e., 120 Hz at zero speed and 0 Hz at the synchronous speed. When it passes through the first torsional natural frequency, this torsional mode is certainly excited. The oscillating torque in the low speed drive train is typically 1±4 times the rotor rated torque.

Due to the excessive torsional vibration, the lateral vibrations of all shafts can also be detected as shown in Figure 2. The low speed pinion (Pinion 1) is subjected to high vibration with primary components at the first torsional natural frequency of 675 cpm (11.3 Hz). Since such a high vibration amplitude is close to the journal bearing clearance, the bearing life and the compressor reliability are likely compromised. Adachi and Baba (2022 [3]) also reported a component failure case due to the torsional resonance and then predicted the lateral vibration response of a pinion on a coupled vibration model. To understand the mechanism and mitigate the coupled vibration are of significance for the synchronous motor driven unit.

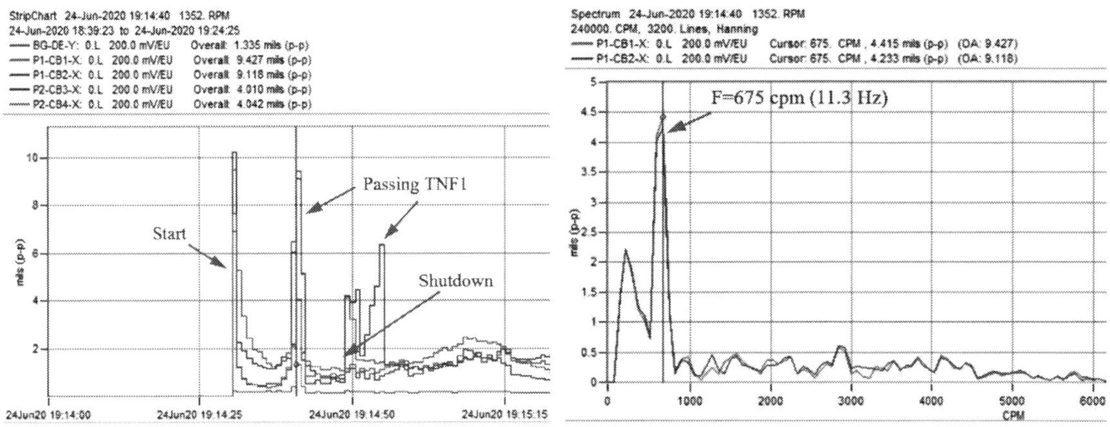

Figure 2 Startup lateral vibrations (left) and Pinion 1 frequency spectrum (right)

Additional Critical Speed

During a factory acceptance test, the compressor was driven by a variable frequency drive (VFD) motor. An additional critical speed of the high speed pinion (Pinion 2) was observed at 8,796 rpm during startup and shutdown, as shown in Figure 3. The abnormally narrow and sharp peak indicates a high amplification factor (AF) and is just below the operating speed of 9,767 rpm. The running speed does not have an adequate separation margin to this critical speed [1] and so the design is unacceptable. However, this critical speed was not predicted in the conventional lateral rotordynamic analysis. This issue should be resolved before completing the test. A close look at the torsional vibration analysis found that a torsional mode was near this unexpected critical speed. But how this torsional mode was excited and how the torsional vibration was transmitted to the lateral vibration are two questions to be answered.

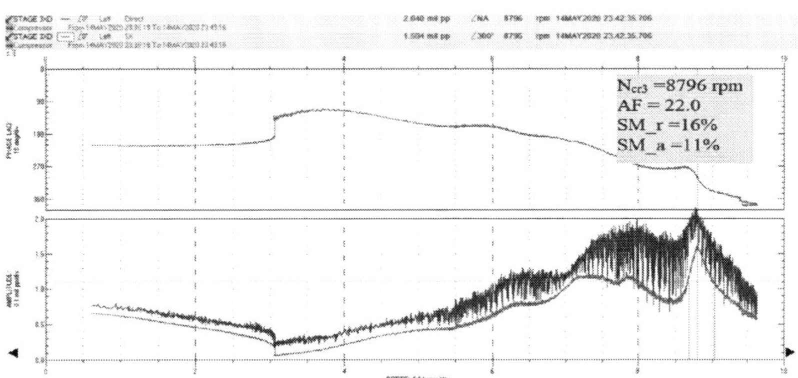

Figure 3 Unexpected critical speed of the high speed pinion

Interactive Vibrations of Two Pinions

The lateral vibration of one shaft can be detected by a probe on another shaft. This interactive vibration, or casually cross-talking, is a common phenomenon in gearboxes. The vibration of one shaft can be transmitted to a different shaft through the gear case or the gear mesh. If the interactive vibration is not high, it can be ignored.

The cross-talking phenomenon was also observed on the two pinions of this compressor. Figure 4 shows the frequency spectra for the low speed pinion (Pinion 1) and the high speed pinion (Pinion 2). The interactive vibrations are detected at low levels. When one pinion passes through a critical speed, higher vibration of this pinion can be observed on the other pinion.

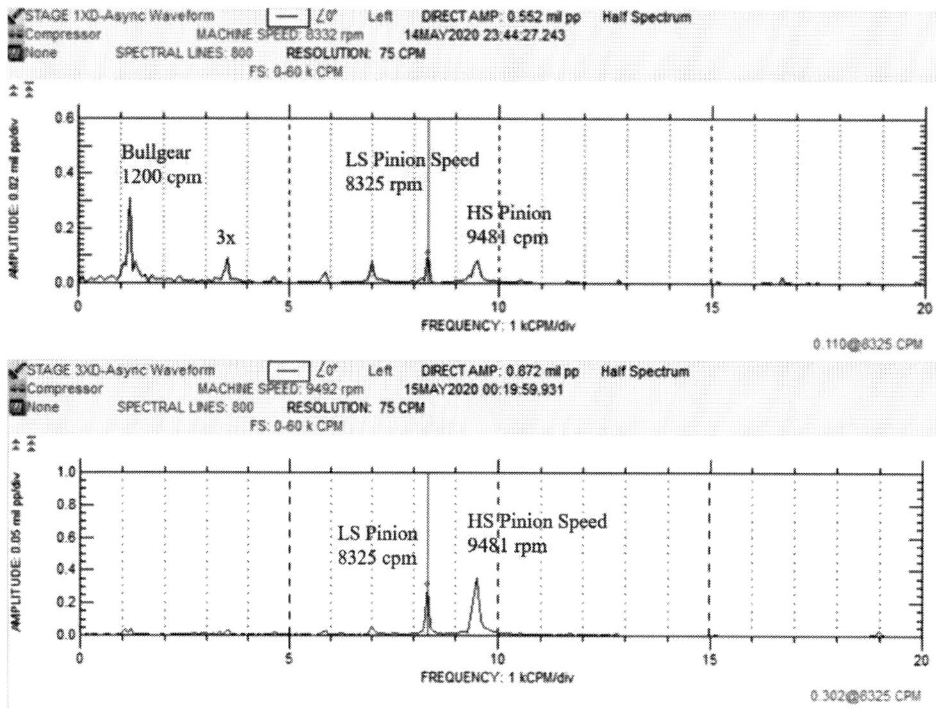

Figure 4 Frequency spectra of Pinion 1 (upper) and Pinion 2 (lower)

UNCOUPLED VIBRATION ANALYSIS:

The coupled vibration phenomena described above are closely associated with the lateral vibrations of the two pinions and the torsional vibration of the drive train. The rotordynamic performance of a rotor-bearing system can be described by the undamped critical speed map, damped modes presented in the Campbell diagram and a stability plot, and unbalance responses. Among them, the damped modes are basic and key performance parameters (Luo and Schwarz, 2019 [4]). This section presents the uncoupled models and the damped lateral modes of the two pinions together with uncoupled torsional model and undamped torsional mode. All the modes based on uncoupled models are listed for a comparison with the modes based on a coupled model in APPENDIX A.

An uncoupled lateral rotordynamic model and damped modes of Pinion 1 are shown in Figure 5. The first four pairs of backward and forward damped modes at 8,576 rpm are conical (Con), cylindrical (Cyl), the first bending (Bend1) and the second bending (Bend2) modes.

Figure 6 shows the uncoupled model and damped modes at 9,767 rpm of Pinion 2.

A torsional vibration analysis is performed to calculate the torsional natural frequencies and the startup transient responses. The torsional model and undamped modes are given in Figure 7. The first torsional natural frequency (TNF1) is 659 cpm (11.0 Hz), which is excited during start and was observed in commission (Figure 2). This mode is dominated by the coupling deflection shown in Figure 8.

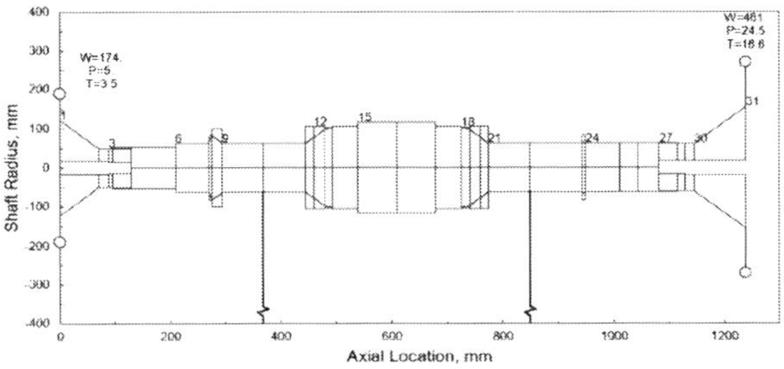

	Backward		Forward	
Mode	Freq	Log Dec	Freq	Log Dec
1 (Con)	1.256	0.271	2.647	0.608
2 (Cyl)	3.302	0.266	4.737	0.747
3 (Bend1)	6.533	1.514	14.422	2.203
4 (Bend2)	10.949	1.529	19.696	0.726

Figure 5 Uncoupled model and damped modes at 8,576 rpm of the low speed pinion

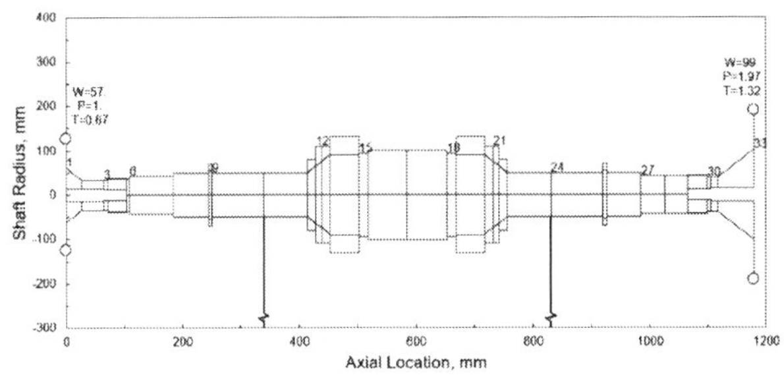

	Backward		Forward	
Mode	Freq	Log Dec	Freq	Log Dec
1 (Con)	3.478	0.376	4.921	0.717
2 (Cyl)	5.014	0.216	7.012	0.475
3 (Bend1)	12.181	4.318	13.874	4.701
4 (Bend2)	15.429	1.522	30.100	1.098

Figure 6 Uncoupled model and damped modes at 9,767 rpm of the high speed pinion

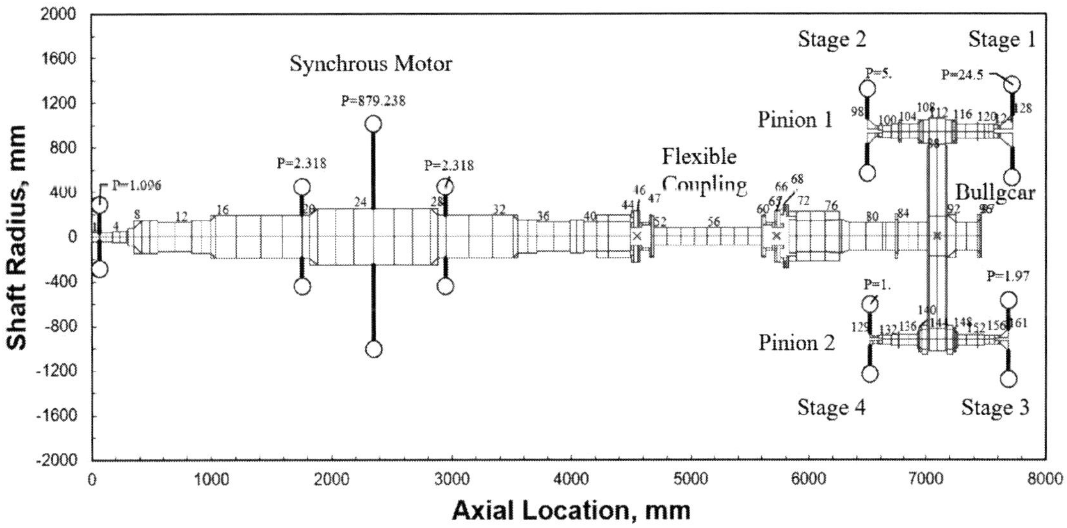

Mode	1	2	3	4	5	6
F (Hz)	11.0	84.8	127.3	147.2	168.9	199.5
F (cpm)	659	5,086	7,640	8,831	10,132	11,968
Shape	Coupling	P1+P2+BG	P1-S2 & P2-S3	P1-Stg2 & P2-Stg3	P2-Stg4	Exciter

Figure 7 Uncoupled torsional model and undamped modes of the compressor drive train

In addition, a torsional mode of Pinion 2 at 8,831cpm (147.2 Hz) is a mixed mode with primary motion of the third and fourth stage impellers. This was thought to be excited and related to the unexpected additional critical speed of the high speed pinion during testing.

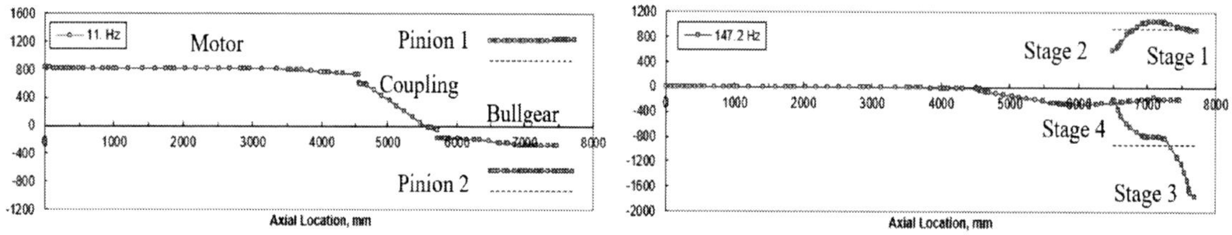

Figure 8 Two torsional modes: Mode 1 (left) and Mode 4 (right)

COUPLED LATERAL-TORSONAL VIBRATION ANALYSIS:

The lateral and torsional vibrations of a geared rotor-bearing system are inherently coupled to each other. The lateral vibration of a pair of spur gears is coupled with the torsional vibration. A pair of helical gears couples the vibrations in the lateral, torsional and axial directions. In the finite element formulation, each node has six degrees-of-freedom and a finite element model of a whole drive train is complicated. To deal with such a model more efficiently, previous researchers studied a substructure synthesis technique (Luo, Sun and Fawcett 1996 [5]) and modal reduction methods (Stringer, Sheth and Allaire 2010 [6], Kaplan et al 2013 [7]). This paper will focus on the coupled lateral-torsional vibrations for the helical gears. The purpose of this work is to understand the coupled vibration phenomena observed in the compressor test and commissioning using a rotordynamics software, XLROTOR (Murphy 2018 [8]), based on a theory developed by Rao, Shiau and Chang (1998 [9]).

Coupled Rotordynamic Model

A coupled lateral-torsional rotordynamic model is shown in Figure 9. This model looks like the uncoupled torsional model shown in Figure 7. In this coupled model, each station (node in FEA) has five degrees-of-freedom: displacements x and y in two orthogonal transverse directions, rotations θ_x and θ_y about axes X and Y , and angular displacement ϕ about axial axis Z. Each shaft is supported on two journal bearings. Two flexible elements of the diaphragm coupling are properly modelled by the radial, bending and torsional stiffness coefficients. A gear mesh is modelled by a linear spring that tangentially connects the bullgear base circle to a pinion base circle as shown in Figure 10.

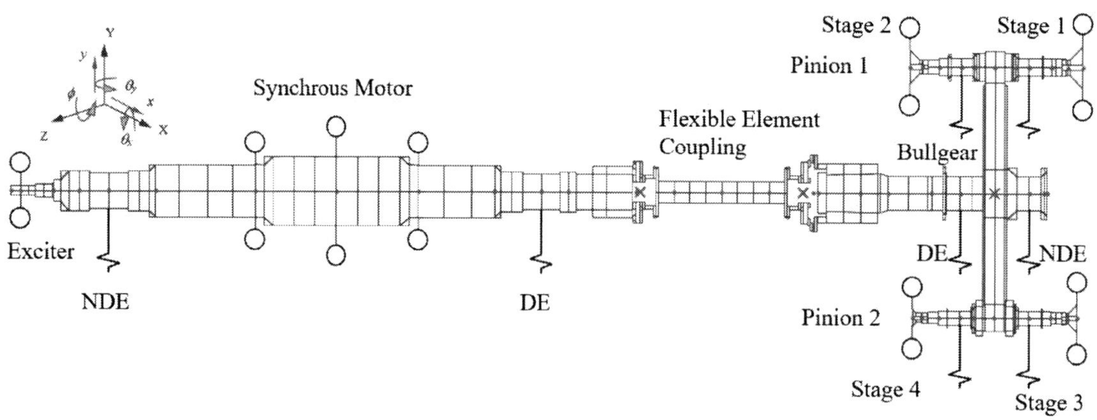

Figure 9 Coupled lateral-torsional vibration analysis model

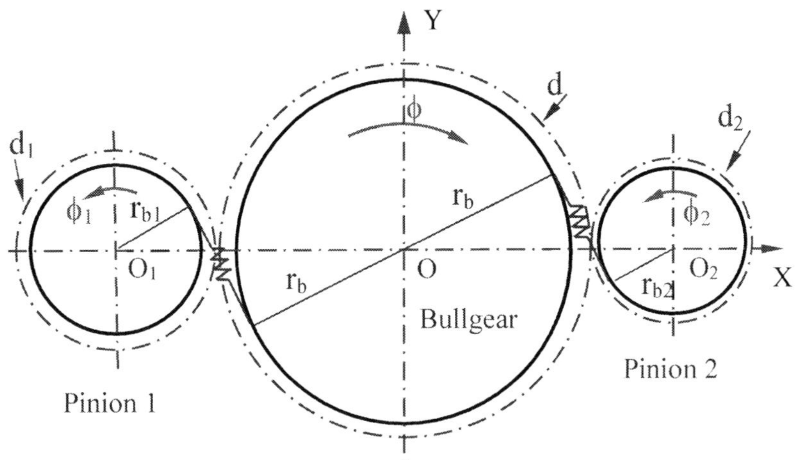

Figure 10 Gear mesh model

Mathematically, the equations of lateral motion of the system and the equations of torsional motion can be combined by properly arranging relevant elements in those matrices. The coupling relationship between the bullgear center O and the pinion center O_1 (or O_2) is determined by formulating the mesh force in terms of the spring constant and all ten displacements of a pair of gear centers. Although the gear mesh stiffness varies with time, an average stiffness coefficient is normally used and the damping is excluded in the analysis. Considering the bullgear rotational direction, the low speed pinion (Pinion 1) is upward meshed while the high speed pinion (Pinion 2) is downward meshed. The mesh direction makes certain differences in the bearing loads and rotordynamic performance, especially for a heavy rotor partly loaded during start or shutdown.

Damped Modes

A coupled lateral-torsional modal analysis can generate dozens of modes relevant to different shafts. Table 1 gives typical modes calculated on coupled and uncoupled models. All the modes are listed in APPENDIX.

To identify a torsional or lateral or mixed mode, the kinetic energy in the two directions can be calculated. Mode 3-T is the first torsional mode (Figure 8), a pure torsional mode with 0% lateral kinetic energy (L-KE). Mode 6-L is a pure lateral mode with E_L=100%, or the first backward (B1) and conical (Con) mode of Pinion 1 (P1) at Stage 1 (S1) impeller.

Table 1 Selected Damped Modes Based on Coupled and Uncoupled Models

Mode No.	Coupled Mode						Uncoupled		Diff. (%)	
	F (kcpm)	Log Dec	L-KE (%)	Comp	L-Shape	T-Shape	F (kcpm)	Log Dec	Freq	Log Dec
3-T	0.656	0.00	0.0	Train		Cplg	0.659	0.00	-0.4	
6-L	1.253	0.27	100	P1-S1	B1-Con		1.256	0.27	-0.2	-0.5
11-Extra	3.307	0.84	34.7	P1-P2	S2	S1	NA			
17-Extra	5.074	0.64	88.5	P1-P2	P1-P2	P1-P2	NA			
19-Mixed	6.515	0.36	25.2	P1-P2	P2-P1	P1-P2	5.086	0.00	28.1	
27-Mixed	9.361	0.08	2.3	P2-P1	P2	S2-S3	8.831	0.00	6.0	
30-Cplg	12.201	0.03	99.9	Cplg to P2	BG-P1-P2	P1-P2	12.104	0.00	0.8	

Modes 11 and 17 are two extra modes that do not appear in uncoupled models. They are coupled lateral-torsional modes. Their mode shapes are illustrated in Figure 11 and Figure 12.

Figure 13 shows that Mode 27 is another mixed mode dominated with the torsional motion of Pinion 2 and coupled with some lateral motion of Pinion 2 and small lateral motion of Pinion 1.

Figure 14 depicts Mode 30 at 12,201 cpm. It is primarily a new lateral mode with the bending of the coupling, bullgear and two pinions. The significant coupling deflection is involved in this mode. If the coupling runs near this mode (not for this case), high synchronous vibration may be expected on the adjacent rotors and may be difficultly trimmed on the rotors themselves.

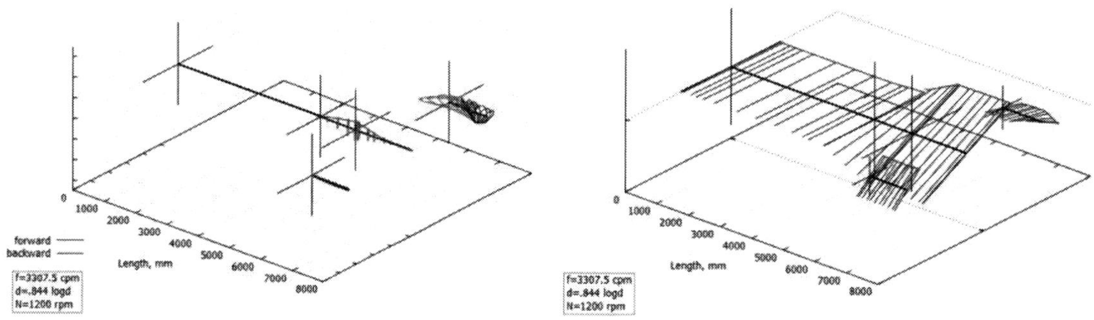

Figure 11 Mode 11, F=3307 cpm, E_L=34.7%, a torsional mode (right) with a lateral mode (left)

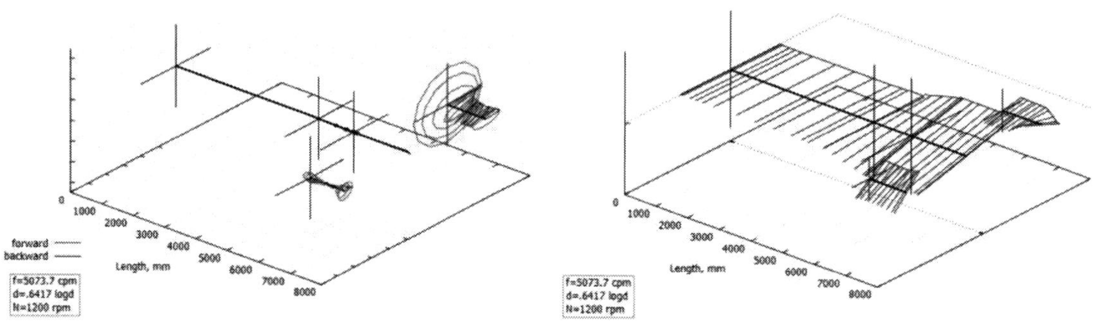

Figure 12 Mode 17, F=5074 cpm, E_L=88.5%, a lateral mode (left) with a torsional mode (right)

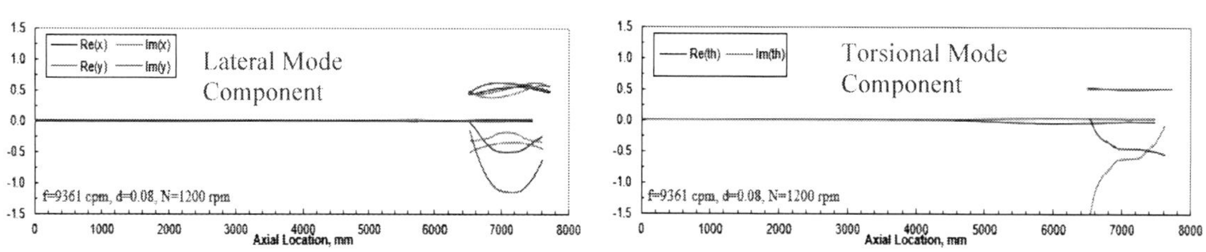

Figure 13 Mode 27, F=9361 cpm, E_L=2.3%, a torsional mode coupled with a lateral mode

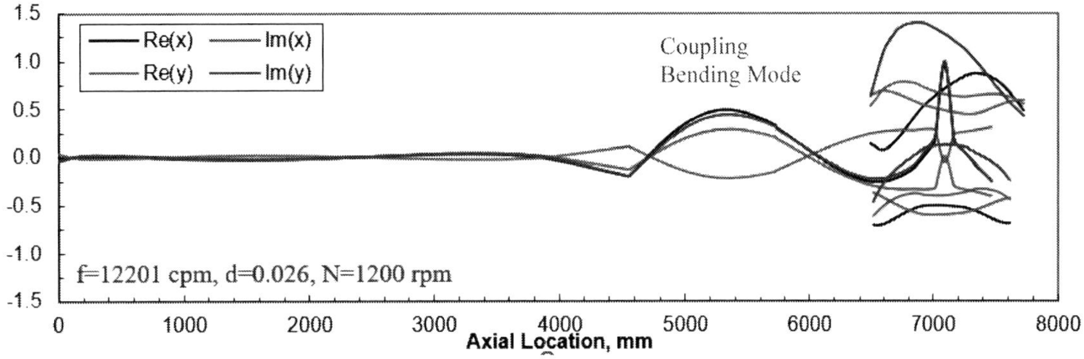

Figure 14 Mode 30, F=12,201 cpm, E_L=99.9%, a primary lateral mode

Coupled Lateral-Torsional Vibration Responses

The coupled lateral-torsional vibration responses will be presented in this section to simulate the vibration phenomena observed in the compressor test and commissioning.

Startup Transient Vibration

A startup transient vibration response analysis shows the compressor start time is about 12.5 sec (Figure 15). The first torsional mode is fully excited during the start. The torsional resonance generates excessive oscillating torque in the coupling. Because of the gear coupling effect, the high lateral vibration of Pinion 1 (P1) is predicted while the lateral vibration of Pinion 2 (P2) is pronounced. This trend explains the observance in the compressor commission. The high lateral vibration depends upon the torque response and the pinion dynamic behavior. A recent study on the mitigation of this vibration is reported in a separate paper (Luo and Schwarz, 2023 [10]).

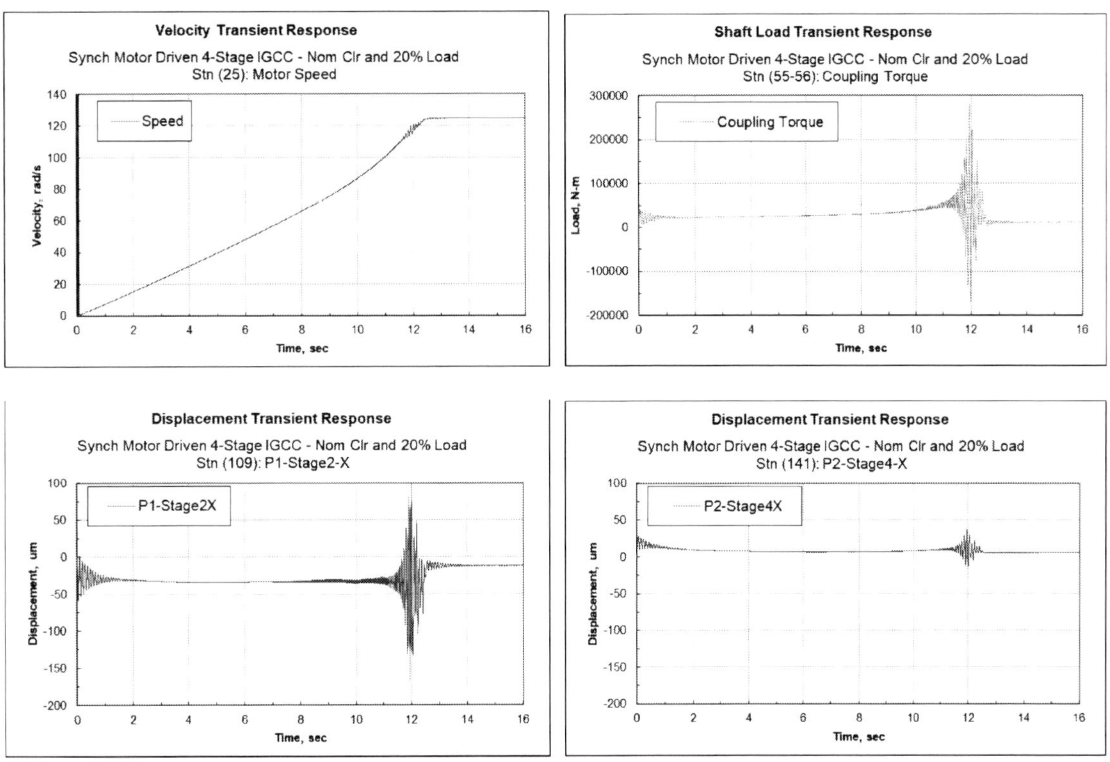

Figure 15 Startup transient responses

Additional Critical Speed of the High Speed Pinion

The unbalance responses of Pinion 2 to static unbalance shown in Figure 16 are calculated on both the uncoupled model (left) and the coupled model (right). The coupled model predicts a mixed lateral-torsional mode at 9,361 cpm shown in Figure 13. Figure 17 depicts the mesh force (left) and the angular displacement (right) of Pinion 2 varying with speed. They are peaked at both lateral and torsional critical speeds. This implies that the unbalances on the two impellers can not only excite the lateral vibration, but also resonate a torsional mode if this mode is a coupled mode. If this mode appears at a low frequency, this might be incorrectly taken as an instability induced subsynchronous vibration as studied by Rajagopalan and Vance (2008 [11]).

On the other hand, if a torsional excitation of 1% rated torque is applied at the fourth stage impeller (S4), the mesh force response is peaked at the torsional mode frequency and the lateral vibration is also peaked at this torsional mode (Figure 18).

The results show that the additional critical speed observed in testing may be excited by a VFD driver harmonic or unbalance. This critical speed was not observed in the compressor commissioning or operation.

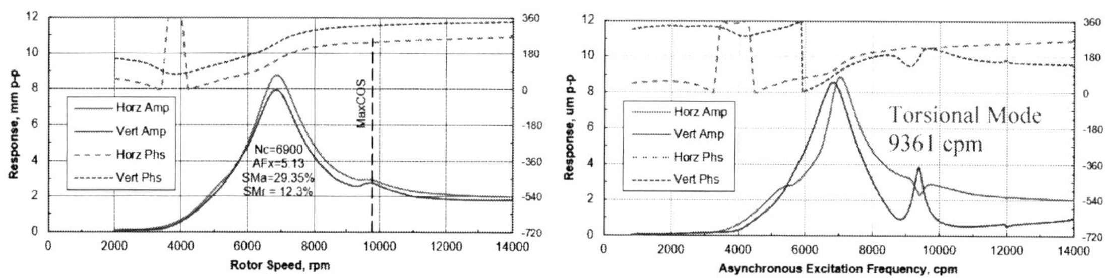

Figure 16 Lateral responses to static unbalance on P2-S4 probe calculated on uncoupled (left) and coupled (right) model

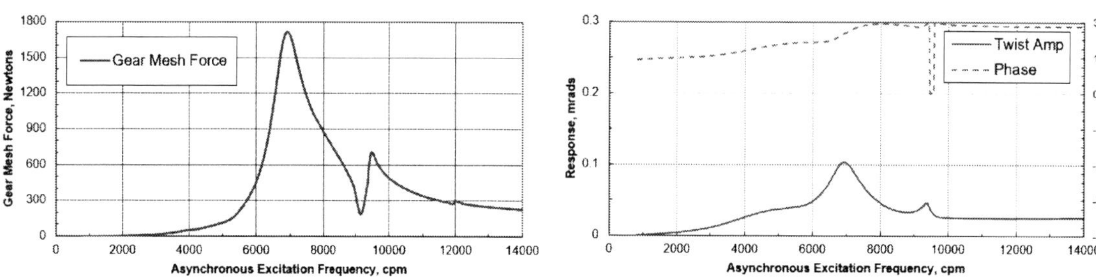

Figure 17 Mesh force (left) on Pinion 2 and twist (right) of Pinion 2

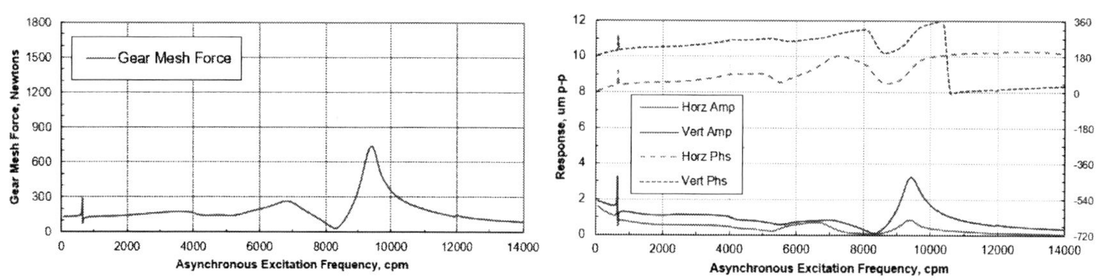

Figure 18 Responses to torsional excitation (1% nominal torque, z=1.75%) on P2-S4 probe: Mesh force of bull-P2 (left) and lateral response on P2-S4 probe (right)

Cross Talking between Two Pinions

When the same unbalances are applied to Pinion 2 (P2), the lateral vibration responses of Pinion 1 (P1) can be calculated (or vice versa). The mesh force applied to Pinion 1 is illustrated in Figure 19. The corresponding lateral vibration on the second stage (S2) probe is peaked at the critical speed and about 40% of the lateral response of Pinion 1. This indicates that the interactive vibrations, or cross-talking, can be transmitted from one pinion to the other, and simulated using the coupled vibration model. In addition, if a forward mode of one pinion is coincident with the operating speed of another pinion, this forward mode is likely excited in operation. This design should be avoided and improved whenever possible.

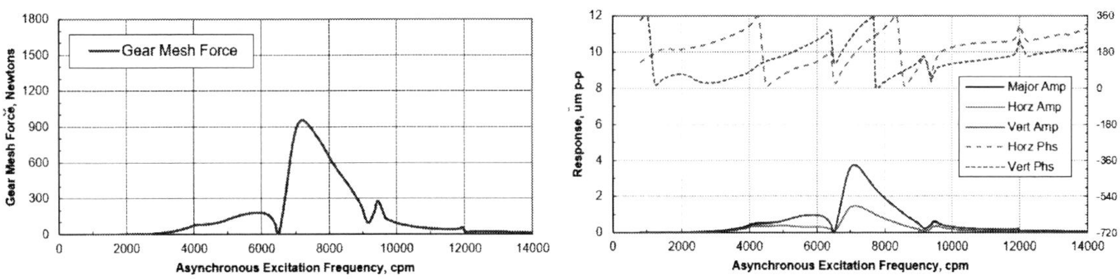

Figure 19 Mesh force on Pinion 1 (left) and lateral response on P1-S2 probe (right)

CONCLUSIONS:

The lateral rotordynamics and the torsional vibration of an integrally geared centrifugal compressor are usually analyzed separately. The lateral vibration of each shaft is analyzed on a decoupled rotor-bearing system and the torsional vibration on a purely torsional model. Three interesting dynamic phenomena were observed in the test and commissioning of a synchronous motor driven 4-stage integrally geared centrifugal compressor: (1) Excessive lateral vibration during startup, (2) Unexpected additional critical speed in the compressor test with a VFD driver, and (3) Interactive vibrations between the two pinions. A coupled lateral-torsional vibration analysis has been performed to simulate these phenomena and understand the physics behind them. Main findings are given below:

(1). All the lateral and torsional modes calculated on uncoupled models can be traced from the coupled model. A coupled modal analysis can search for additional coupled modes that do not appear in uncoupled models.

(2). Considering the gear mesh coupling effect, excessive torsional vibration can produce high lateral vibrations of relevant shafts. The lateral vibrations are dependent upon both the torque response and the dynamic behavior of the shafts.

(3). Additional lateral critical speed of the high speed pinion is the result from an excited torsional mode. The existence of torsional resonance is associated with the mode frequency (different drive trains for the test and operation) and excitation sources (motor torque harmonics and large rotor unbalance). Large unbalance on a rotor can also excite a torsional mode when the rotor speed passes through the torsional mode.

(4). Interactive (cross-talking) vibration responses of the two pinions are a common phenomenon in gearboxes due to the gear mesh coupling. A forward mode of one pinion is likely excited by an unbalance force on another shaft that runs at a speed coincident with the mode frequency. Such a design should be avoided and improved wherever possible. Furthermore, this coupled vibration analysis can examine the vibration transmission path, explain the observance in the compressor test, and explore remedy measures if it becomes an issue.

AKNOWLEDGEMENTS:

The author would like to thank Turbomachinery Department, Linde Engineering, Linde, Inc. for their strong support of this work and the permission for the publication.

REFERENCES:

[1] API Standard 617 (8th Edition), *"Axial and Centrifugal Compressors and Expander-Compressors"*, American Petroleum Institute, Part 3, Integrally Geared Centrifugal Compressors, September 2014.

[2] API Standard 672 (4th Edition), *"Packaged Geared Centrifugal Air Compressors for Petroleum, Chemical, and Gas Industry Services"*, American Petroleum Institute, March 2004.

[3] Adachi, A. and Baba, Y., 2022, *"Prediction of Lateral Vibration Behavior of Integrally Geared Centrifugal Compressor During Synchronous Motor Startup by Transient Torsional-Lateral Coupled Analysis"*, Proceedings of the 51st Turbomachinery Symposium, Texas A&M University, 1-19.

[4] Luo, Z. and Schwarz, C. L., 2019, *"Improvement of Rotordynamic Performance of Integrally Geared Centrifugal Compressors through a Systematic Design Modification Approach"*, Proceedings of ASME Turbo Expo 2019, Phoenix, Arizona, USA, June 17-21, 2019, GT2019-91995, 1-12.

[5] Luo, Z., Sun, X. and Fawcett, J. N., 1996, *"Coupled Torsional-Lateral-Axial Vibration Analysis of a Geared Shaft System using Substructure Synthesis"*, Mechanism and Machine Theory, Vol. 31, No. 3, 345-352, 1996.

[6] Stringer, D. B., Sheth, P. N. and Allaire, P. E., 2010, *"Modal reduction of a Geared Rotor Systems with General Damping and Gyroscopic Effects"*, Journal of Vibration and Control, 0(0), 1-13, 2010.

[7] Kaplan, J. A., Dousti, S., Allaire, P. E., Nichols, B. R., Dimond, T. W. and Untaroiu, A., 2013, *"Rotor Dynamic Modeling of Gears and Geared systems"*, Proceedings of ASME Turbo Expo 2013, San Antonio, Texas, USA, June 3-7, 2013, GT2013-94654, 1-11.

[8] Murphy, B., 2018, *"Coupled Lateral-Torsional Analysis with XLROTOR"*, XLRotor User's Manual, version 5.5.

[9] Rao, J. S., Shiau, T. N. and Chang, J. R., 1998, *"Theoretical Analysis of Lateral Response due to Torsional Excitation of Geared Rotors"*, Mechanism and Machine Theory, 33(6), 761-783, 1998.

[10] Luo, Z. and Schwarz, C. L., 2023, "Mitigation of Startup Coupled Lateral-Torsional Vibration of a Synchronous Motor Driven Integrally Geared Centrifugal Compressor", Proceedings of ASME Turbo Expo 2023, Bostom, Massachusetts, USA, June 26-30, 2023, GT2023-103162, 1-9.

[11] Rajagopalan, V. N. and Vance, J. M., 2008, *"Diagnosing Coupled Lateral-Torsional Vibrations in Turbomachinery"*, Proceedings of ASME Turbo Expo 2008, Berlin, Germany, June 9-13, 2008, GT2008-50125, 1-8

APPENDIX A: A COMPARISON OF COUPLED MODES WITH UNCOUPLED MODES

Mode No.	Coupled Mode						Uncoupled Mode		Difference (%)	
	F (kcpm)	Log Dec	L-KE (%)	Comp	L-Shape	T-Shape	F (kcpm)	Log Dec	Freq	Log Dec
1	0.557	23.94	95.9	BG	F1-Con		0.545	23.27	2.1	2.9
2	0.626	7.73	100	Motor	F1-Con		0.624	7.88	0.2	-2.0
3	0.656	0.00	0.0	Train		Cplg	0.659	0.00	-0.4	
4	0.680	8.03	100	Motor	F2-Cyl		0.668	8.05	1.7	-0.3
5	1.097	5.75	97.9	BG	B1-Con		1.088	5.70	0.8	1.0
6	1.253	0.27	100	P1-S1	B1-Con		1.256	0.27	-0.2	-0.5
7	1.956	2.19	97.4	BG	F2-Con		1.906	2.11	2.6	3.7
8	2.513	0.74	100	Motor	F3-Bend		2.507	0.74	0.2	-0.1
9	2.626	0.11	100	Motor	B3-Bend		2.616	0.11	0.4	-0.1
10	2.627	0.62	96.6	P1	F1-Con		2.647	0.61	-0.7	1.8
11	3.307	0.84	34.7	P1-P2	S2	S1	NA			
12	3.317	0.35	83.2	P1	B2-Cyl		3.302	0.27	0.5	32.0
13	3.475	0.37	99.9	P2	B1-Con		3.478	0.38	-0.1	-0.5
14	4.771	1.07	99.1	BG-Cplg	B2-Cyl		4.871	1.37	-2.0	-21.7
15	4.917	0.73	98.8	P2	F1-Con		4.921	0.72	-0.1	1.6
16	5.009	0.22	99.3	P2	B2-Cyl		5.014	0.22	-0.1	1.5
17	5.074	0.64	88.5	P1-P2	P1-P2	P1-P2	NA			
18	6.240	1.85	98.2	BG-Cplg	F3-Bend		6.394	2.15	-2.4	-14.0
19	6.515	0.36	25.2	P1-P2	P2-P1	P1-P2	5.086	0.00	28.1	
20	6.538	1.12	51.1	P1	B3-Bend		6.533	1.51	0.1	-26.0
21	7.138	0.08	100	Motor	B4-Exc		7.118	0.08	0.3	3.7
22	7.209	0.35	81.3	P2	F2-Cyl		7.012	0.48	2.8	-25.4
23	7.215	0.14	100	Motor	F4-Exc		7.179	0.15	0.5	-4.3
24	8.036	0.47	53.3	P1-P2		S2-S3	7.604	0.00	5.7	
25	8.273	0.60	100	Motor-Cplg	F5		8.276	0.65	0.0	-8.1
26	8.332	0.09	100	Motor-Cplg	B5		8.354	0.09	-0.3	3.5
27	9.361	0.08	2.3	P2-P1		S2-S3	8.831	0.00	6.0	
28	11.756	1.59	96.4	P1	B4-Bend	BG-P1	10.949	1.53	7.4	3.7
29	11.983	0.00	0.0	Motor		Exc	11.968	0.00	0.1	
30	12.201	0.03	99.9	Cplg to P2	BG-P1-P2	P1-P2	12.104	0.00	0.8	

BG = Bullgear
Cplg = Coupling
Exc = Exciter
P1, P2 = Pinion 1, Pinion 2
S*n* = the *n*-th stage impeller

Bend = Bending mode
B*n* = the *n*-th backward mode
F*n* = the *n*-th forward mode
Con = Conical mode
Cyl = Cylindrical mode

L-KE = Lateral kinetic energy
L-Shape = Lateral mode shape
T-Shape = Torsional mode shape